T0207494

# Communications in Computer and Information Science 1914

**Rationale**

The CCIS series is devoted to the publication of proceedings of computer science conferences. Its aim is to efficiently disseminate original research results in informatics in printed and electronic form. While the focus is on publication of peer-reviewed full papers presenting mature work, inclusion of reviewed short papers reporting on work in progress is welcome, too. Besides globally relevant meetings with internationally representative program committees guaranteeing a strict peer-reviewing and paper selection process, conferences run by societies or of high regional or national relevance are also considered for publication.

**Topics**

The topical scope of CCIS spans the entire spectrum of informatics ranging from foundational topics in the theory of computing to information and communications science and technology and a broad variety of interdisciplinary application fields.

**Information for Volume Editors and Authors**

Publication in CCIS is free of charge. No royalties are paid, however, we offer registered conference participants temporary free access to the online version of the conference proceedings on SpringerLink (http://link.springer.com) by means of an http referrer from the conference website and/or a number of complimentary printed copies, as specified in the official acceptance email of the event.

CCIS proceedings can be published in time for distribution at conferences or as post-proceedings, and delivered in the form of printed books and/or electronically as USBs and/or e-content licenses for accessing proceedings at SpringerLink. Furthermore, CCIS proceedings are included in the CCIS electronic book series hosted in the SpringerLink digital library at http://link.springer.com/bookseries/7899. Conferences publishing in CCIS are allowed to use Online Conference Service (OCS) for managing the whole proceedings lifecycle (from submission and reviewing to preparing for publication) free of charge.

**Publication process**

The language of publication is exclusively English. Authors publishing in CCIS have to sign the Springer CCIS copyright transfer form, however, they are free to use their material published in CCIS for substantially changed, more elaborate subsequent publications elsewhere. For the preparation of the camera-ready papers/files, authors have to strictly adhere to the Springer CCIS Authors' Instructions and are strongly encouraged to use the CCIS LaTeX style files or templates.

**Abstracting/Indexing**

CCIS is abstracted/indexed in DBLP, Google Scholar, EI-Compendex, Mathematical Reviews, SCImago, Scopus. CCIS volumes are also submitted for the inclusion in ISI Proceedings.

**How to start**

To start the evaluation of your proposal for inclusion in the CCIS series, please send an e-mail to ccis@springer.com.

Dmitry Balandin · Konstantin Barkalov ·
Iosif Meyerov
Editors

# Mathematical Modeling and Supercomputer Technologies

23rd International Conference, MMST 2023
Nizhny Novgorod, Russia, November 13–16, 2023
Revised Selected Papers

*Editors*
Dmitry Balandin
Lobachevsky State University of Nizhny Novgorod
Nizhny Novgorod, Russia

Konstantin Barkalov
Lobachevsky State University of Nizhny Novgorod
Nizhny Novgorod, Russia

Iosif Meyerov
Lobachevsky State University of Nizhny Novgorod
Nizhny Novgorod, Russia

ISSN 1865-0929 ISSN 1865-0937 (electronic)
Communications in Computer and Information Science
ISBN 978-3-031-52469-1 ISBN 978-3-031-52470-7 (eBook)
https://doi.org/10.1007/978-3-031-52470-7



This Springer imprint is published by the registered company Springer Nature Switzerland AG
The registered company address is: Gewerbestrasse 11, 6330 Cham, Switzerland

# Preface

The 23rd International Conference and School for Young Scientists "Mathematical modeling and supercomputer technologies" (MMST 2023) was held on November 13–16, 2023, in Nizhni Novgorod, Russia. The conference and school were organized by the Mathematical Center "Mathematics of Future Technologies" and the Research and Educational Center for Supercomputer Technologies of the Lobachevsky State University of Nizhni Novgorod. It was organized in a partnership with the International Congress "Russian Supercomputing Days".

The topics of the conference and school covered a wide range of problems in mathematical modeling of complex processes and numerical methods of research, as well as new methods of supercomputing aimed at solving topical problems in various fields of science, industry, business, and education.

This edition of the MMST conference was dedicated to the 60th anniversary of the founding of the Faculty of Computational Mathematics and Cybernetics at Lobachevsky University (UNN) and the 50th anniversary of its Software Department. The Faculty of Computational Mathematics and Cybernetics was established at UNN in 1963 on the initiative of Yuri Neimark and was the first such faculty in Russia. The primary goal of the UNN Faculty of Computational Mathematics and Cybernetics was to train professionals for the new field, who would combine thorough mathematical education with the ability to apply it to mathematical modeling and research (based on computational mathematics and new cybernetic approaches) of the most diverse processes and phenomena occurring in nature and human society.

Ten years later, in 1973, the Software Department was established at the Faculty of Computational Mathematics and Cybernetics under the leadership of Roman Strongin. The main research efforts of the department were focused on the development of software systems for decision-making support and processing of large information arrays. Over the last decade, the department's faculty members have obtained significant results in the fields of supercomputing in global optimization problems, high-performance computing for mathematical modeling, and computer graphics.

Over the period of its existence, the Faculty of Computational Mathematics and Cybernetics has trained more than 9000 programmers in "non-linear thinking". In 2015, the Institute of Information Technologies, Mathematics and Mechanics was established at Lobachevsky University as a result of the merger of the Faculty of Computational Mathematics and Cybernetics with the Faculty of Mechanics and Mathematics. By merging the faculties, it became possible to concentrate the efforts of the consolidated departments on priority areas of research and education in the fields of mathematical modeling, mechanics and supercomputer technologies.

The scientific program of the conference featured the following plenary lectures given by outstanding researchers:

- Valery Cherepennikov (Neimark IT Campus): Artificial intelligence in strategy games;
- Dmitry Kvasov (University of Calabria): Recent trends in Lipschitz global optimization;
- Nikolai Khokhlov (Moscow Institute of Physics and Technology): Modeling of wave perturbation propagation in heterogeneous media using the grid-characteristic method;
- Victor Zhukov (Keldysh Institute of Applied Mathematics of the Russian Academy of Sciences): Computational model of unsteady multicomponent gas-dynamic flows taking into account conjugate heat exchange processes;
- Dmitry Kasatkin (Institute of Applied Physics of the Russian Academy of Sciences): Synchronization processes in oscillatory networks with adaptive coupling.
- Denis Goldobin (Institute of Continuous Media Mechanics, Ural Branch of the Russian Academy of Sciences): The method of exponential time differencing for stiff systems with nondiagonal linear part.

These proceedings contain 18 full papers and 7 short papers carefully selected from the main track and special sessions of MMST 2023. The papers accepted for publication were reviewed by three referees from the members of the MMST 2023 Program Committee and independent reviewers. A total of 45 papers were submitted by conference participants for a single-blind peer review, and only 25 papers were chosen to be published in this book.

The proceedings editors would like to thank all the conference committees' members, especially the Organizing and Program Committee members as well as external reviewers for their contributions. We also thank Springer for producing these high-quality proceedings of MMST 2023.

November 2023

Dmitry Balandin
Konstantin Barkalov
Iosif Meyerov

# Organization

## Program Committee Chair

| | |
|---|---|
| Balandin, D. V. | Lobachevsky State University of Nizhni Novgorod, Russia |

## Program Committee

| | |
|---|---|
| Barkalov, K. A. | Lobachevsky State University of Nizhni Novgorod, Russia |
| Belykh, I. V. | Georgia State University, USA |
| Denisov, S. V. | Oslo Metropolitan University, Norway |
| Feygin, A. M. | Institute of Applied Physics RAS, Russia |
| Goldobin, D. S. | Institute of Continuous Media Mechanics UB RAS, Russia |
| Gonchenko, S. V. | Lobachevsky State University of Nizhni Novgorod, Russia |
| Hramov, A. E. | Innopolis University, Russia |
| Ivanchenko, M. V. | Lobachevsky State University of Nizhni Novgorod, Russia |
| Kazantsev, V. B. | Lobachevsky State University of Nizhni Novgorod, Russia |
| Koronovskii, A. A. | Saratov State University, Russia |
| Kvasov, D. E. | University of Calabria, Italy |
| Malyshkin V. E. | Institute of Computational Mathematics and Mathematical Geophysics SB RAS, Russia |
| Mareev, E. A. | Institute of Applied Physics RAS, Russia |
| Moshkov, M. Y. | King Abdullah University of Science and Technology, Saudi Arabia |
| Meyerov, I. B. | Lobachevsky State University of Nizhni Novgorod, Russia |
| Nekorkin, V. I. | Institute of Applied Physics RAS, Russia |
| Osipov, G. V. | Lobachevsky State University of Nizhni Novgorod, Russia |
| Sergeyev, Y. D. | University of Calabria, Italy |
| Turaev, D. V. | Imperial College London, UK |

| | |
|---|---|
| Wyrzykowski, R. | Czestochowa University of Technology, Poland |
| Yakobovskiy, M. V. | Institute of Applied Mathematics RAS, Russia |
| Zaikin, A. A. | University College London, UK |

## Organizing Committee

| | |
|---|---|
| Zolotykh, N. Yu. (Chair) | Lobachevsky State University of Nizhni Novgorod, Russia |
| Barkalov, K. A. | Lobachevsky State University of Nizhni Novgorod, Russia |
| Lebedev, I. G. | Lobachevsky State University of Nizhni Novgorod, Russia |
| Meyerov, I. B. | Lobachevsky State University of Nizhni Novgorod, Russia |
| Oleneva, I. V. | Lobachevsky State University of Nizhni Novgorod, Russia |
| Sysoyev, A. V. | Lobachevsky State University of Nizhni Novgorod, Russia |
| Vedernikov, A. C. | Lobachevsky State University of Nizhni Novgorod, Russia |
| Usova, M. A. | Lobachevsky State University of Nizhni Novgorod, Russia |

# Contents

# Computational Methods for Mathematical Models Analysis

# Dynamics of Honing of Deep Cylindrical Holes with Memory Effects in the Frictional Interaction

Aleksandra V. Grezina(✉), Leonid A. Igumnov, Vladimir S. Metrikin, and Adolf G. Panasenko

National Research Lobachevsky State University of Nizhny Novgorod, Nizhny Novgorod, Russia
aleksandra-grezina@yandex.ru

**Abstract.** In this paper a semi-analytical methodology is presented of the mathematical modelling of abrasive machining of deep cylindrical holes. The modelling accounts for the memory effects in the frictional interaction that governs the process. The mathematical model is presented in the form of a non-autonomous system of differential-difference equations of variable structure. The model is investigated using mapping of the Poincare surface, the boundaries of which vary in time in correspondence with the functional dependence of the coefficient of static friction. To this end an original semi-analytical approach is developed for the identification of the fixed points that correspond to periodic motions of arbitrary complexity including chaotic one. Obtained with the help of developed software package in a high-level language, the bifurcation diagrams enabled an in-depth analysis of the main regimes of abrasive machining in the presence of memory effects in the frictional interaction. These regimes were previously unknown when using frictional models without memory effects. In particular, the paper shows that imperfect smoothness of surfaces after honing is due to multiple temporary relative rests of the hone and the workpiece.

**Keywords:** Dynamical systems of variable structure · Mathematical model · Poincare maps · Stability · Chaos bifurcation diagram · Numerical modelling · Friction with memory · Abrasive machining · Honing

## 1 Introduction

A wide industrial application of new high-strength materials with special properties requires the development of efficient machining cutting methods. This issue is especially acute to date in high-precision processing of deep holes of small diameter in the fuel equipment of engines, parts of cooling systems as well as in

---

Supported by the Ministry of Science and Higher Education of the Russian Federation, project no. FSWR-2023-0034, and by the Research and Education Mathematical Center «Mathematics for Future Technologies».


D. Balandin et al. (Eds.): MMST 2023, CCIS 1914, pp. 3–14, 2024.
https://doi.org/10.1007/978-3-031-52470-7_1

the processing of holes in long cylindrical parts (artillery barrels, high-strength pipes for reactor batteries and high-precision products for oil production and oil refining industries, etc.). Due to insufficient tool rigidity, technological operations for machining deep holes are accompanied by tool or workpiece vibrations which significantly affects their processing accuracy. The main parts in the listed technical systems are long cylinders with accurate and clean deep holes. The most common operation that provides the required hole parameters is abrasive processing - honing [1–7], which makes it possible to consistently and economically obtain the highest quality indicators for parts in computer aided manufacturing. Honing is also widely used in various branches of engineering in the processing of liners, engine blocks, connecting rods, gears, hydraulic cylinders and shock absorbers [8–12], etc. Honing, compared with grinding, provides a much lower impact on the surface layer of manufactured part. This is due to the contact over the entire bar surface resulting in specific pressure and temperature decrease in the cutting zone as compared with grinding, when contact occurs only along the line, thus, the process is easier to control. To ensure the required surface quality, honing machines have been developed and used, on which the workpiece or honing head performs an additional oscillatory motion, due to which, with the right choice of parameters and processing modes, the honing process is of oscillatory character and cutting grains do not repeat the trajectories of the previous grains during their movement [13–16]. The tool used in honing is called a honing tool (see Fig. 1), or a honing head. Hone is a metal-cutting tool consisting of three or more abrasive bars (stones) located along the perimeter.

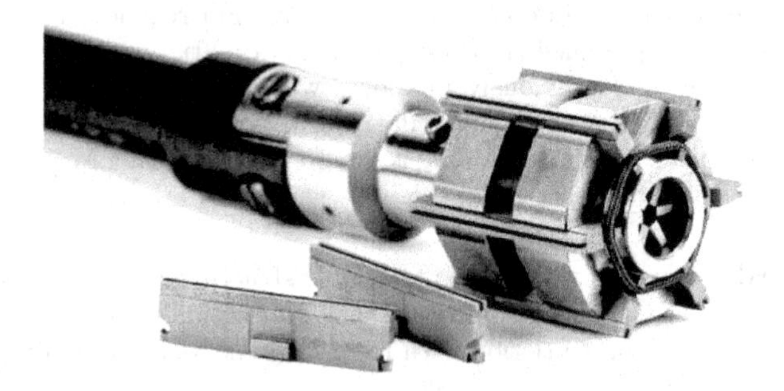

**Fig. 1.** Honing head.

In many hons the stones are removable. That is, when worn or broken, they can be easily replaced. As a result, their cutting properties are used to a greater extent, which leads, in particular, to better self-sharpening and an increase in the efficiency of the coolant [1]. Hole honing is used in cases where it is necessary to achieve minimal surface roughness and cylindricity, down to microns. That is why this type of processing received the greatest popularity in cylinders manufacture. The latter are obviously widely used in engineering, construction and car industry. To conduct a full-scale research into dynamics influence on process performance and machining accuracy, it is necessary to develop both adequate mathematical models that describe the dynamic behavior of the tool and workpiece and models of surface formation during abrasive machining. Well-known publications on the development and study of mathematical models ([4–7] and others) that describe the process of cylinder honing are mainly related to the study of dynamic characteristics of the process and forms inside the cylinder holes without taking into account friction with memory. The obtained theoretical results on dynamics of honing process using the developed mathematical models do not explain, as expected, the results of experimental data on non-purely cylindrical shapes of machined surfaces, obtained in experiments [13–15].

It is known that A.Yu. Ishlinsky and I.V. Kragelsky [17] showed that the coefficient of friction of relative rest (CFRR) during the motion of two rubbing bodies is not a constant value, but depends on time of their "previous" joint motion with the same speeds (friction with memory). This attracted the attention of scientists, albeit with a long time delay, involved in the study of systems with friction (see [18–21] and the literature cited in them). In these papers, a number of new results are presented on the example of the simplest nonlinear autonomous dynamical systems. It was shown that in comparison with the known studies of systems with constant CFRR, in systems with variable CFRR there are arbitrarily complex periodic and stochastic modes of motion, both with sliding and with relative rest. Dynamic vibration systems, the scheme of which coincides with the schemes studied in works [18–21], were also considered by foreign scientists (see [22–28] and the bibliography given in them), but the hypothesis of A.Yu. Ishlinsky and I.V. Kragelsky about the CFRR with memory was not applied. In these works, the classical model of static friction was used.

## 2 Mathematical Model

In industry, honing is performed on special machines of two types: horizontal and vertical. Horizontal honing machines are made with a horizontal spindle. Vertical machines have a vertical spindle arrangement respectively. The former are used both for long parts and outdoor operations. Vertical machines are employed for short parts and mainly for internal machining, i.e. holes. To build a mathematical model that describes the dynamics of the process of abrasive machining of deep cylindrical holes, we consider the fundamental physical model of the first type [14], shown in Fig. 2.

To study the role of the specific character of friction forces, let us consider a simple technique machining with a hone rigidly fixed on the shaft during a single-pass movement of the tool. For such a regime, a tool with a solid rod is suitable, the physical model of which is shown in [6] in Fig. 3.

Figure 3 shows: 1 – a shank that transmits torque to the head 2 with grinding pads attached to it. The shank is an elastic steel bar with the following characteristics: $r = d/2$ – radius, $l$ – length, $G$ – shear modulus, $\rho$ – density. The head with grinding pads attached to it is modeled by a short non-deformable cylinder of radius $R = D/2$ and moment of inertia $J_0$. The upper base of the shank rotates at a constant angular velocity $\omega_0$.

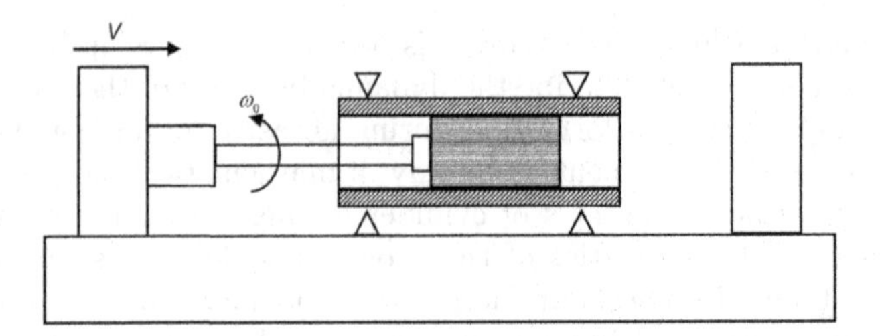


**Fig. 2.** Physical model of the machine.

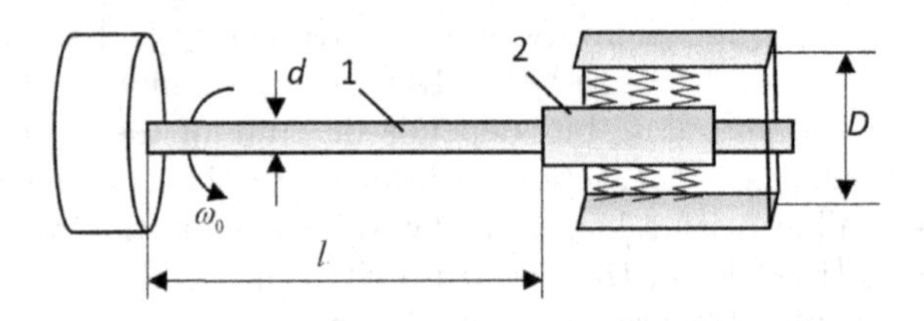


**Fig. 3.** Physical model of the machine.

It is convenient to describe the dynamics of the working tool in a system associated with the shank upper base. In this case, inertia forces can be ignored, since the centrifugal inertia force makes a constant contribution to the normal pressure force and the Coriolis forces are small due to the small tangential velocities of the head relative to the rotating system.

Thus, we arrive at a model in which a stationary working tool interacts with a rotating cylinder. The longitudinal movement of the tool with the speed V is not taken into account, since its speed is much less than the relative tangential speed of hone side surface and inner surface of the cylinder.

It is known that under the action $N$ of a torque on a solid bar, the angle of torsion of the lower base $\varphi_l$ relative to the fixed upper base is determined by the relation

$$\varphi_l = \frac{2l}{\pi r^4 G} N. \tag{1}$$

The angle of rotation in the section $z$ $(0 < z < l)$ in the static state is obviously equal to

$$\varphi(z) = \frac{2\,N}{\pi r^4 G} z = \frac{\varphi_l}{l} z. \tag{2}$$

Relation 2 will also be preserved in the case of variables $N$, if the period of change $N << l/c_{sound}$, where $c_{sound}$ is the sound speed in the bar.

The energy of elastic deformations during the bar twisting

$$U = \int_0^{\varphi} N(\varphi)\,\mathrm{d}\varphi = \frac{\pi r^4 G}{2l}\frac{\varphi_l^2}{2} = \chi\frac{\varphi_l^2}{2}. \tag{3}$$

The kinetic energy of a section with thickness $\mathrm{d}z$ is equal to

$$dK = J(\mathrm{d}z)\dot{\varphi}_l^2/2 = \pi r^4 \rho\,\mathrm{d}z\dot{\varphi}_l^2 z^2/(4l^2) \tag{4}$$

The kinetic energy of the elastic bar is therefore equal to $K_{bar} = \pi r^4 \rho l \dot{\varphi}_l^2/2$ and for working tool $K = (J_0 + \pi r^4 \rho l/6)\dot{\varphi}_l^2/2 = J_l\dot{\varphi}_l^2/2$.

Under the action of the moment of forces $N_{fr}$ for changing $\varphi_l$, we obtain $N_{fr}\,\mathrm{d}\varphi_l = \mathrm{d}K + \mathrm{d}U = \chi\varphi_l\,\mathrm{d}\varphi_l + J_l\dot{\varphi}_l\,\mathrm{d}\dot{\varphi}_l$, hence

$$J_l\ddot{\varphi} = -h(\dot{\varphi_l - \omega_0}) - \chi\varphi_l + N_{fr} \tag{5}$$

where $\varphi_l$ is the deflection angle of the tool head, $\varphi_l = 0$ corresponds to an unstressed bar.

Let us transform (5) into the form

$$\ddot{\varphi} + \tilde{\alpha}(\dot{\varphi} - \omega_0) + \omega^2\varphi = \frac{1}{J_l}N_{fr}, \tag{6}$$

where $\varphi_l \equiv \varphi$, $\omega^2 = \chi/J_l$, $\tilde{\alpha} = h/J_l$.

We accept that the friction force of the working tool on the cylinder obeys the Coulomb-Amanton law. In the case when the angular velocity of the working tool $\dot{\varphi} \neq \omega_0$, then the equations of dynamics of the system under consideration according to (6) will be written in the form

$$\ddot{\varphi} + \tilde{\alpha}(\dot{\varphi} - \omega_0) + \omega^2\varphi = J_l^{-1}RPf_*\,\mathrm{sign}(\omega_0 - \dot{\varphi}), \omega_0 \neq \dot{\varphi}, \tag{7}$$

where $f_*$ is the coefficient of sliding friction, $P$ is the pressure force of the working tool on the cylinder, and in the case when the angular velocity of hone $\dot{\varphi} = \omega_0$, then

$$|\chi\varphi| \leq Rf(t_k)P, \omega_0 = \dot{\varphi}, \tag{8}$$

where $f(t_k)$ is the coefficient of friction of relative rest (CFRR), the form of which, according to the hypothesis of A.Yu. Ishlinsky and I.V. Kragelsky is shown below in the Fig. 4.

Introducing dimensionless time $\tau = \omega t$, variable $\xi = \chi/(Rf_*P)$ and angular velocity $\theta = \omega_0\sqrt{\chi J_l}/(Rf_*P)$, Eqs. (7, 8) will take the form

$$\ddot{\xi} + \tilde{\alpha}(\dot{\xi} - \theta) + \xi + \mathrm{sign}(\dot{\xi} - \theta) = 0, \dot{\xi} \neq \theta \tag{9}$$

$$|\xi| \leq 1 + \varepsilon_k, \dot{\xi} = \theta, \varepsilon_k = \varepsilon(\tau_k), \varepsilon(\tau_k) = \frac{f(\tau) - f_*}{f_*}. \tag{10}$$

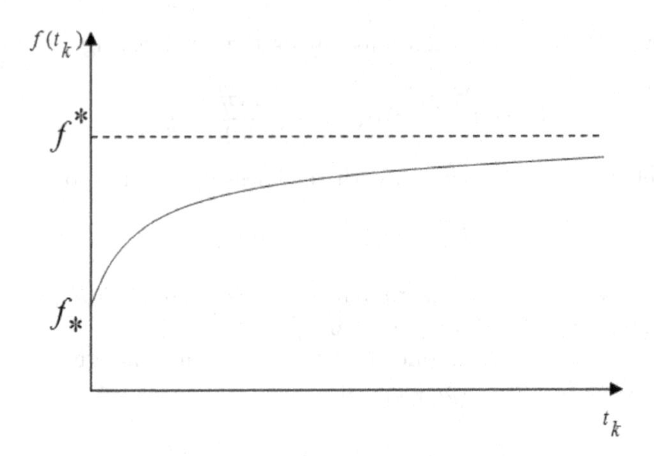


**Fig. 4.** CFRR as a function of time of relative rest $t_k$.

It is known that the specific form of the functional dependence of FCRR $f(t_k)$ on time $t_k$ depends on many factors [17], therefore, further in the work, the form of FCRR is taken as a piecewise linear nondecreasing continuous function

$$f(t_k) = \begin{cases} f_* + (f^* - f_*)t_k/t^*, & 0 < t_k < t^*, \\ f^*, & t_k \geq t^*. \end{cases} \tag{11}$$

Then, in the dimensionless form, the functional dependence of the FCRR looks like

$$\varepsilon(\tau_k) = \begin{cases} \tau_k, & 0 < t_k < \varepsilon_* \\ \varepsilon_*, & t_k \geq \varepsilon_* \end{cases}, \varepsilon_* = \frac{f^* - f_*}{f_*} > 0.$$

## 3 Constructing a Mapping Function

According (9) and (10), the phase portrait in the plane $(\xi, \dot{\xi})$ at $\tilde{\alpha} = 0$ is presented in Fig. 5.

Figure 5 shows that there is a segment $l(-1 \leq \dot{\xi} \leq +1, \dot{\xi} = \theta)$ of the junction of phase trajectories in the plane $(\xi, \dot{\xi})$ [28,31]. Once on the segment $l$, image point moves along it and, having reached its right end $(\xi = 1, \dot{\xi} = \theta)$, will always continue to move along a straight line $L(\dot{\xi} = \theta, \xi > 1)$ $(\varepsilon_k > 0)$ until the moment $\tau_{k+1}$, determined from the equation $\xi(\tau_{k+1}) = 1 + \varepsilon(\tau_{k+1})$. Phase trajectories in the half-plane $\Pi_+(\dot{\xi} > \theta)$ are circles $(\xi+1)^2 + \dot{\xi}^2 = R_1^2 = \dot{\xi}_0^2 + (\xi_0+1)^2$ and in the half-plane $\Pi_-(\dot{\xi} < \theta)$ – circles $(\xi - 1)^2 + \dot{\xi}^2 = R_2^2 = \dot{\xi}_0^2 + (\xi_0 - 1)^2$. Image point moves along the line $l$ and then until the moment $t_{k+1}$ it corresponds to the relative rest of the tool head and the upper base of the shank (hone). Let's call these movements as ones with long stops. If the image point falls on a straight line $\dot{\xi} = \theta$ in the region $|\xi| = 1$, then there is an instantaneous change in sign

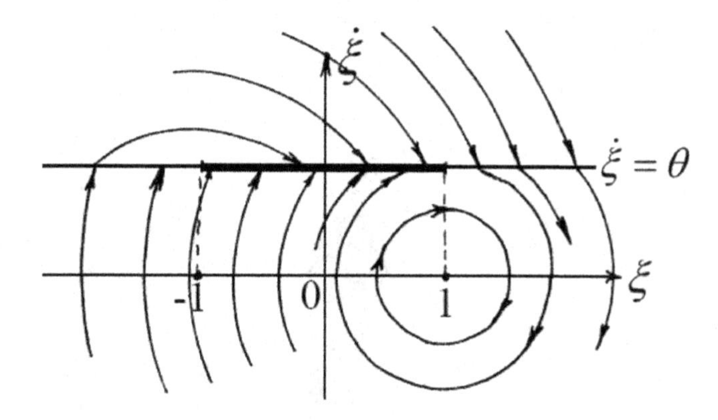


**Fig. 5.** The phase portrait in the plane $(\xi, \dot{\xi})$ at $\tilde{\alpha} = 0$.

of hone relative rotation velocity. Hone begins to either outpace the cylinder in speed (transition of the image point from half-plane $\Pi_-$ to half-plane $\Pi_+$) or lag behind it (transition of the representative point from half-plane $\Pi_+$ to half-plane $\Pi_-$). These moments of transitions will be called instantaneous stops. It should be noted that if the image point does not intersect the straight $\dot{\xi} = \theta$, while in space $\Pi_-$, then this corresponds to hone oscillations under the condition $\dot{\xi} < \theta$. The relation connecting the durations of two consecutive intervals of long-term contact $\tau_k$ and $\tau_{k+1}$ are found using the succession function [28,31] $\psi(\tau_{k+1}) = \Phi(\tau_k)$, where the functions $\Phi$ and $\phi$ have the form

$$\phi(\tau) = \theta\tau - \varepsilon(\tau), \Phi(\tau) = 1-(-1)^j(\varepsilon(\tau)-2j+1), 2(j-1) < \varepsilon(\tau) < 2j, j = 1, 2, 3, ...$$

## 4 Numerical Results

As the image point moving along the positive half-path from a starting point $M(\xi, \theta)$, $1 \leq \xi \leq \xi^*$, always hits (for selected $\tilde{\alpha}_1$ and $\varepsilon^*$) the segment of sliding motions. The investigation of the dynamics of the system considered can be reduced to studying the sequence properties $\tau_k$ $(k = 1, 2, 3, ...)$ of the duration of the stick phase. Since for any element of this sequence there must exist a successor, then it is necessary to define point mapping (11) in discontinuity points $\tau^{(i)}$, $i = \overline{1, k}$ (if any). We assume that $0 \leq \varphi(\tau^{(i)}) \leq \varphi(0)$ $(i = \overline{1, k})$. The bifurcation diagrams were plotted using so defined mapping function. Figures 6 and 7 show bifurcation diagrams plotted against shank velocity $\theta$ for $\varepsilon^* = 3$ and $\tilde{\alpha}_1 = 0.05$, $\tilde{\alpha}_1 = 0.11$ respectively. In these figures, the values of $\theta$ parameter ranged from 0.9 to 2.5 are plotted on the abscissa axis, whereas the values of the duration of the stick phase are plotted on the ordinate axis. It is clearly seen that due to an increase in the shank velocity or dependence on the parameter $\tilde{\alpha}_1$, either a period-doubling process occurs leading to a chaotic motion [9], (Fig. 6), or there exist only periodic motions of the system with one or several stick phases

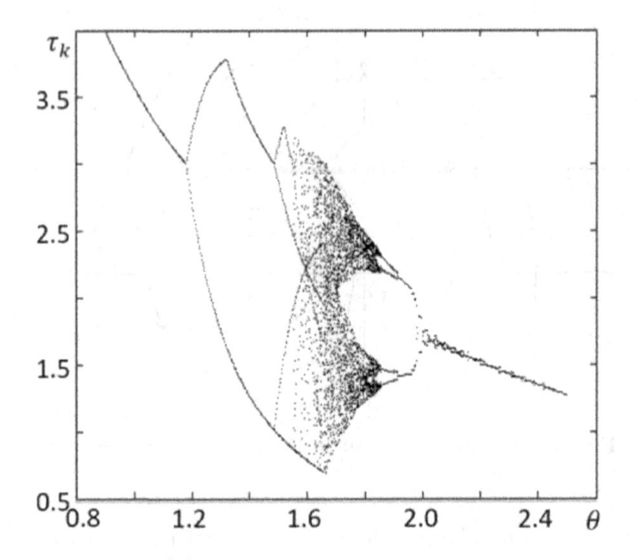


**Fig. 6.** Bifurcation diagram against $\theta$ for $\varepsilon^* = 3$, $\tilde{\alpha}_1 = 0.05$.

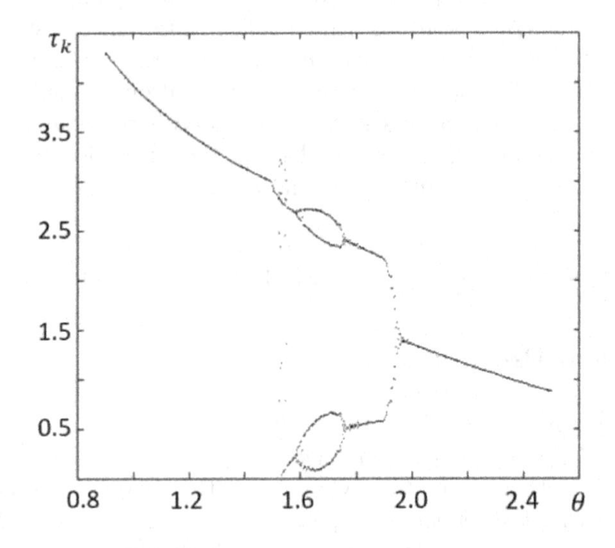


**Fig. 7.** Bifurcation diagram against $\theta$ for $\varepsilon^* = 3$, $\tilde{\alpha}_1 = 0.1$.

(Fig. 7). It follows from Figs. 6 and 7 that for periodic motions, an increase in shank velocity leads to a decrease in the duration of the stick phase. An increase in the value of viscous friction entails the disappearance of chaotic motion.

Figures 8 and 9 show bifurcation diagrams plotted against shank velocity $\theta$, $0.9 \leq \theta \leq 2.5$, for $\tilde{\alpha}_1 = 0.15$ and for various parameter values $\varepsilon^*$. Figure 8 corresponds to the value $\varepsilon^* = 2$, whereas Fig. 9 to the values $\varepsilon^* = 4$ and $\varepsilon^* = 5$ respectively. For $\varepsilon^* = 2$ depending on the parameter $\theta$, the system exhibits

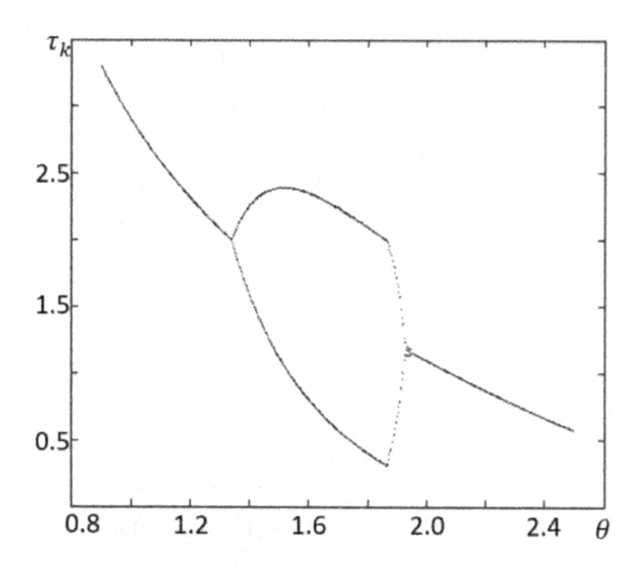


**Fig. 8.** Bifurcation diagram against $\theta$ for $\varepsilon^* = 2$, $\tilde{\alpha}_1 = 0.15$.

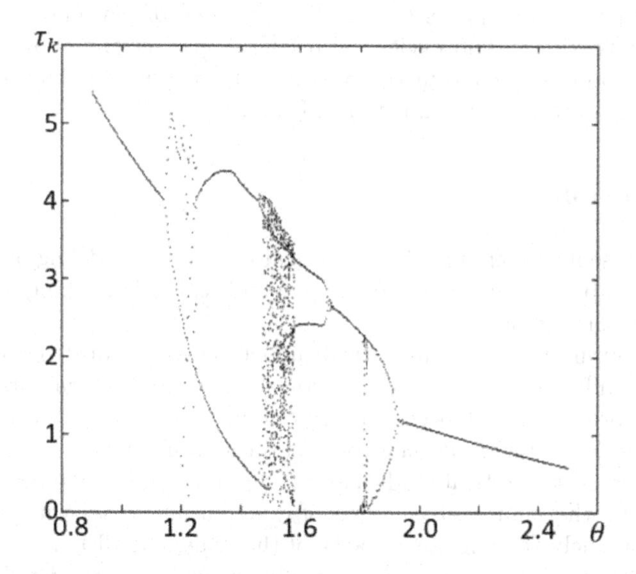


**Fig. 9.** Bifurcation diagram against $\theta$ for $\varepsilon^* = 4$, $\tilde{\alpha}_1 = 0.15$.

periodic motion regimes with one or two times of relative rest of the hone and the shank. With increasing $\varepsilon^*$ ($\varepsilon^* = 4$, $\varepsilon^* = 5$) stochastic motion regimes occur, which turn into periodic motion regimes with one prolonged stop due to further increase in shank velocity.

It follows from the figures that chaotic regimes during abrasive machining by honing, taking into account friction with memory occur at large $\varepsilon_* > 2$

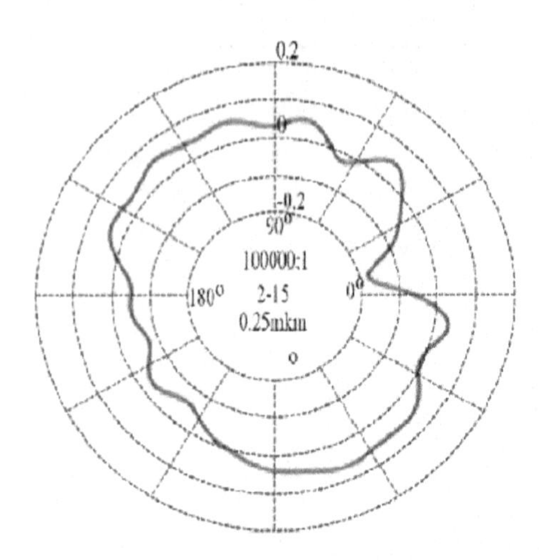


**Fig. 10.** Experimental profile of a cylindrical surface after honing [32,33]

and angular processing speeds $\theta > 1$. The presence of periodic movements of arbitrary complexity ($n$-fold cycles) with long stops (see the above bifurcation diagrams) theoretically confirm the presence of not purely cylindrical surfaces during honing, obtained experimentally (Fig. 10).

## 5 Conclusion

The paper presents a technique for numerical-analytical modeling of hone interaction during abrasive machining of deep cylindrical holes, taking into account the friction with memory.

A new computational mathematical model has been constructed including the system of differential-difference equations with a variable structure describing the physical processes that occur in the system under study during honing of deep cylindrical holes with account for friction with memory.

The presented theoretical studies and computer experiments made it possible to describe the main restructuring of dynamic processes with a change in parameters, namely the angular velocity of the hone and CFRR.

The results of numerical simulation are illustrated by bifurcation diagrams for the system parameters. By using mathematical model with account for friction with memory enabled us to explain the possibility of the existence of a non-purely cylindrical surface during honing of deep holes in full-scale experiments.

The developed mathematical model makes it possible to repeatedly carry out computational experiments with different input data without resorting to full-scale experiments.

## References


1. Naerman, M.S., Gorlov, V.V.: Active control during honing. Mach. Tools Tools **7**, 21–22 (1962)
2. Nedelin, Yu.L.: Study of the process of diamond honing of steels. Abstract of the dissertation for the degree of candidate of technical sciences, Kharkov, Kharkov Polytechnic Institute, 23 p. Speciality 05.17 (1971)
3. Orlov, P.N.: Technological Support for the Quality of Parts by Finishing Methods, 384 p. Mechanical Engineering, Moscow (1988)
4. Ostrovsky, V.I.: Theoretical Foundations of the Grinding Process, 144 p. Publishing House of Leningrad University, Leningrad (1981)
5. Sokolov, S.P.: Fine Grinding and Finishing, 88 p. M.-L. 6 Mashgiz (1961). (Library shlifovalshchika)
6. Voronov, S.A., Fatalchuk, A.V.: Simulation of dynamic processes of vibroabrasive processing of deep holes. Izv. RAN MTT, no. 6, pp. 167–173 (2000)
7. Voronov, S.A., Gouskov, A.M., Batzer, S.A.: Modeling vibratory drilling dynamics. Trans. ASME J. Vib. Acoust. **123**(4), 435–443 (2001)
8. Gouskov, A.M., Voronov, S.A., Butcher, E.A., et al.: Nonconservative Oscillations of a Tool for Deep Hole Honing. Commun. Nonlinear Sci. Numer. Simul. **11**(6), 685–708 (2006)
9. Voronov, S.A., Bobrenkov, O.A.: The dynamics of the honing process and its influence on the shape errors in the processing of holes. Izvestiya vysshikh obuchennykh zavedenii Mashinostroyeniye, no. 4, pp. 10–29 (2008)
10. Voronov, S.A.: The influence of dynamics on the process of hole formation during honing. In: Problems of Mechanical Engineering and Reliability of Machines, no. 3, pp. 75–83 (2008)
11. Voronov, S.A., Gouskov, A.M., Butcher, E.A., et al.: Influence of honing dynamics on surface formation. In: Proceedings of DETC 2003 ASME Design Engineering Technical Conference, Chicago, Illinois, pp. 1–7 (2003)
12. Voronov, S.A., Gouskov, A.M., Butcher, E.A., et al.: Simulation of machined surface formation while honing. In: Proceedings of IMECE 2004 The Influence of Process Dynamics on Traditionally Machined Surface, Anaheim, California, pp. 1–8 (2004)
13. Voronov, S.A., Guskov, A.M.: Investigation of non-linear spatial vibrations of a tool for deep drilling. In: Problems of Applied Mechanics, Dynamics and Strength of Machines, pp. 88–111. Izd. N.E. Bauman, Moscow (2005)
14. Voronov, S.A., Gouskov, A.M., Batzer, S.A.: Dynamic stability of rotating abrasive tool for deep hole honing. In: Proceedings of DETC 2001 ASME Design Engineering Technical Conference, Pittsburgh, pp. 1–8 (2001)
15. Voronov, S.A., Gouskov, A.M.: Dynamic models generalization of manufacturing systems with single-point cutting considering equations of new surface formation. In: Proceedings of 2nd Workshop on Nonlinear Dynamics and Control of Mechanical Processing, Budapest, Hungary, pp. 1–10 (2001)
16. Igumnov, L.A., Metrikin, V.S., Grezina, A.V., Panasenko, A.G.: The effect of dry friction forces on the process of dielectric wafer grinding. J. Vibroengineering Procedia **8**, 501–505 (2016)
17. Ishlinsky, A.Yu., Kragelsky, I.V.: On friction jumps. Zhurn. Tech. Phys. **14**(4/5), 276–282 (1944)
18. Kashchenevsky, L.Ya.: Stochastic self-oscillations in dry friction. Inzh.-Fiz. **47**(1), 143–147 (1984)

19. Vetyukov, M.M., Dobroslavsky, S.V., Nagaev, R.F.: Self-oscillations in a system with hereditary dry friction characteristic. Izv. Acad. Sci. USSR. MTT (1), 23–28 (1990)
20. Metrikin, V.S., Nagaev, R.F., Stepanova, V.V.: Periodic and stochastic self-oscillations in a system with hereditary dry friction. PMM **60**(5), 859–864 (1996)
21. Zaitsev, M.V., Metrikin, V.S.: On the theory of a non-autonomous dynamical system with hereditary type friction. Bull. Nizhny Novgorod Univ. N.I. Lobachevsky **4**(1), 470–475 (2014)
22. Leine, R.I., van Campen, D.H., De Kraker, A.: Stick-slip vibrations induced by alternate friction models. Nonlinear Dyn. **16**, 41–54 (1998). https://doi.org/10.1023/A:1008289604683
23. Leine, R.I., van Campen, D.H., De Kraker, A.: Approximate analysis of dry-friction-induced stick-slip vibrations by a smoothing procedure. Nonlinear Dyn. **19**, 157–169 (1999). https://doi.org/10.1023/A:1008306327781
24. Leine, R.I., van Campen, D.H.: Discontinuous fold bifurcations in mechanical systems. Arch. Appl. Mech. **72**, 138–146 (2002). https://doi.org/10.1007/s00419-001-0190-9
25. Liu, Y., Pavlovskaia, E., Wiercigroch, M., Peng, Z.: Forward and backward motion control of a vibro-impact capsule system. Int. J. Non-Linear Mech. **70**, 30–46 (2015)
26. Leine, R.I., van Campenb, D.H.: Eur. J. Mech. A Solids. Bifurcation phenomena in non-smooth dynamical systems **25**, 595–616 (2006)
27. Luo, G.W., Lv, X.H., Ma, L.: Periodic-impact motions and bifurcations in dynamics of a plastic impact oscillator with a frictional slider. Eur. J. Mech. A. Solids **27**, 1088–1107 (2008)
28. Feigin, M.I.: Forced Oscillations of Systems with Discontinuous Nonlinearities, 285 p. Nauka, Moscow (1994)
29. Shuster, G.: Deterministic Chaos: Introduction: Per. from English, 237 p. Mir, Moscow (1988)
30. Voronov, S.A.: Development of mathematical models and methods for analyzing the dynamics of the processes of abrasive machining of holes. Abstract of the dissertation for the degree of Doctor of Technical Sciences, Moscow, 35 p (2008)
31. Neimark, Yu.I.: Method of Point Mappings in the Theory of Nonlinear Oscillations, 472 p. Librokom, Moscow (2010)
32. Orlova, P.N. (ed.): Brief Guide Metalworker, 960 p. Mashinostroenie, Moscow (1986)
33. Novikov, N.V., Klimenko, S.A.: Honing. Tools made of Superhard Materials, 608 p. Mashinostroenie, Moscow (2014)

# Application of Elastoplastic Model to the Simulation of the Low-Speed Impact on an Ice Plate

Evgeniya K. Guseva[1,2(✉)], Vasily I. Golubev[1], Viktor P. Epifanov[2], and Igor B. Petrov[1]

[1] Moscow Institute of Physics and Technology (National Research University), Dolgoprudny, Russia
guseva.ek@phystech.edu

[2] Ishlinsky Institute for Problems in Mechanics of the Russian Academy of Sciences, Moscow, Russia

**Abstract.** Ice plays the central role in many processes in the Arctic region. However, the problem of the determination of the correct rheological model that can define its behavior is still unsolved. Thus, this work is a part of a series of works dedicated towards the creation of the method that allows to select ice model based on the laboratory and numerical experiments. The process of the low-speed impact on the artificially frozen ice plate by the ball striker is examined. The isotropic linear elasticity model serves as the governing system of equations. It is solved using grid-characteristic method on structured grids. Plasticity is introduced as a corrector of the elasticity with the von Mises yield criteria. Finally, ice is divided into two zones, the plasticity zone in the form of a hemisphere with the center in the first contacting point of the plate and the ball, and the elasticity zone in the rest part of the medium. The radius of the hemisphere is considered to be a function of the depth of the ball immersion into the ice. As a result of the simulation, wave and stress tensor patterns are obtained. The comparison with the laboratory experiment is conducted based on the analysis of the ball's coordinate and velocity. The influence of the model's parameters on the simulation results is analyzed.

**Keywords:** ice rheology · elastoplasticity · numerical experiment · grid-characteristic method

## 1 Introduction

Ice is a central part of the Arctic region and forms a lot of its features. The region is becoming more popular for investigation [1] as more oil and gas deposits are

The reported study was funded by the Russian Science Foundation, project no. 23-21-00384.


D. Balandin et al. (Eds.): MMST 2023, CCIS 1914, pp. 15–27, 2024.
https://doi.org/10.1007/978-3-031-52470-7_2

found in its shelf zone. Different types of ice, [2,3], have a great influence on many occurring natural and anthropogenous processes [4]. Moreover, because of its complex structure, there is no one universal model that is able to encompass its behavior in all cases. Thus, in each instance, different models are used. For simulations of acoustic wave propagation, the elasticity model [5] is enough [6]. In large-scale processes [7,8], there appears to be a need for models that describe plasticity and brittle failure. Furthermore, numerical modeling becomes more difficult because different factors, [9], significantly influence ice behavior. Temperature especially plays a great role due to the fact that ice naturally exists in the environment at temperatures close to its melting point, [10].

Therefore, this work is part of a series of works that focuses on the understudied process of low-speed impact with the ball indenter. The final goal is a survey of ice rheology based on laboratory and numerical experiments. The first step towards this objective is to study different models and formulate a method that allows for comparison of the obtained results. The aim of this work is the formulation and investigation of the elastoplastic model built with two zones: a plastic hemisphere in the impact zone with a changing radius and a center in the first contacting point between the target and the striker, and elasticity in the rest of the medium. Plasticity is realized as a modification of the Prandtl-Reuss model [11] with the von Mises yield flow. The considered model is applied to the simulation of the experiment conducted in a laboratory of the Ishlinsky Institute for Problems in Mechanics RAS. The comparison is carried out based on the ball's instantaneous velocity and coordinate. Several model parameter selection criteria are introduced and analyzed. The dependencies of two criteria on the parameters are approximated and used for predicting the parameters that reproduce the solutions close to the real experiment.

The computer modeling was conducted using a software package written in C++ by the Computational Physics Department and the Informatics and Computational Mathematics Department of the Moscow Institute of Physics and Technology, RECT (also used, for instance, in [12]). It is dedicated to the solution of hyperbolic equations by applying the grid-characteristic method, [13], on structured grids. The software supports several means of parallelization, such as OpenMP, MPI, and CUDA and does not use external libraries apart from stdlib. It was run on a personal laptop computer with an Intel Core i7-10510U, a 64-bit operating system, and 16 GB of operating memory.

## 2 Problem Formulation

The laboratory experiment was conducted in a cooling chamber with a temperature $-10\,^{\circ}$C. Granular polycrystalline ice disc was frozen and placed on a massive metal plate with the ability to slide freely on its surface. Above the ice, a hard steel ball was hanging on a string. The strike was direct. A piezoelectric accelerometer was set in the middle of the ball. Another sensor was fastened to the bottom surface of the ice in the middle of the plate. The initial distance between the disk and the striker defined the striking velocity, which equaled

$0.56\,\frac{\text{m}}{\text{s}}$. The results of the experiment are depicted in Fig. 1 on the right. In this work, the comparison is conducted based on the ball's acceleration.

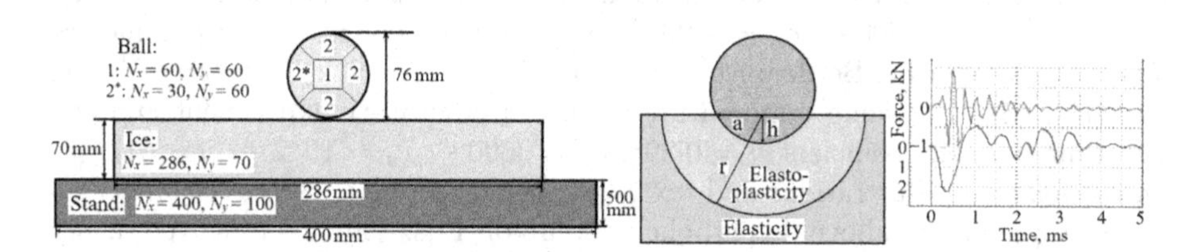


**Fig. 1.** On the left: computational domain, in the middle: elasticity model with elastoplastic inclusion, on the right: results of the laboratory experiment, for blue line transmitter is in the ball, for magenta – in the ice. (Color figure online)

For simulation of the experiment, the computational domain presented in Fig. 1 on the left is created. In our research, in order to save time and computational resources, the mechanical problem in a two-dimensional case is considered. Due to the fact that the software used works with structured grids, the ball was separated into five curvilinear grids. These grids were generated with a script written in Python. Domains 2 are formed as rotated domains 2*. The number of grid cells along the horizontal, $N_x$, and the vertical, $N_y$, axes is also depicted in Fig. 1. The time step was equal to $5 \cdot 10^{-8}$ s. The computation was conducted until the ball reflects off the ice.

Between the ball's parts, a full adhesion contact condition was applied. Between the contacting cells of the ball and the ice, and between the plate and the stand, a free slippage condition was set. At the ball surface and on the right, left, and top sides of the ice, a free boundary condition is used. On the other sides of the metal plate, no reflection boundary was placed. Grids were moved and rebuilt at each time step using the Lagrange corrector. When the contact grids did not coincide, an interpolation procedure was conducted. The striking velocity was set as an initial condition in the ball's grids. The grids were considered to be in contact if the distance between them was equal to or less than 0.05 mm. The same distance was initially set between the ball and the ice.

## 3 Rheological Models

At first, the isotropic linear elasticity model was used as a system of governing equations [5]. This model is commonly applied to describe granular polycrystalline ice, [14]:

$$\rho \dot{\mathbf{v}} = \nabla \cdot \sigma + \mathbf{f}, \tag{1}$$

$$\dot{\sigma} = \lambda(\nabla \cdot \mathbf{v})I + \mu(\nabla \otimes \mathbf{v} + (\nabla \otimes \mathbf{v})^T) + F. \tag{2}$$

Here, the velocity $\mathbf{v}$ and the stress tensor $\sigma$ are unknown. $\lambda$ and $\mu$ are the Lame parameters, $\rho$ is the medium density, $\mathbf{f}$ is the external force, and the additional term $F$ is equal to zero. This model reproduces primary and secondary waves with velocities: $c_p = \sqrt{\frac{\lambda+2\mu}{\rho}}$ and $c_s = \sqrt{\frac{\mu}{\rho}}$. As a result, $c_p$, $c_s$ and $\rho$ can be chosen as parameters that define elastic medium. In the ball and the metal stand, they were set to $c_p = 5700\,\frac{\text{m}}{\text{s}}$, $c_s = 3100\,\frac{\text{m}}{\text{s}}$, $\rho = 7800\,\frac{\text{kg}}{\text{m}^3}$. For ice, these parameters were calculated using Berdennicov's formula ($E = (87,6-0,21T-0,0017T^2)\cdot 10^8$ Pa, [15]) with the same temperature as in the laboratory experiment and the constant Poisson coefficient $\nu = 0.295$: $c_p = 3600\,\frac{\text{m}}{\text{s}}$, $c_s = 1942\,\frac{\text{m}}{\text{s}}$, $\rho = 917\,\frac{\text{kg}}{\text{m}^3}$. In the following illustrations and text, this model will be noted as "elasticity".

The presented above hyperbolic system of Eqs. (1)–(2) is solved using the grid-characteristic method [13]. Following the method, the coordinate-wise splitting was performed, and the system is reduced to a system of independent one-dimensional transport equations using the Riemann invariants as new variables. Each equation was solved using the third approximation order Rusanov scheme [16] monotonized by the grid-characteristic monotonicity criterion [17,18]. As the chosen scheme is explicit, geometrical parallelization can be easily performed. Thus, using the software package's support of OpenMP and three allocated CPU cores, computations were accelerated.

Next, in order to introduce nonlinear behavior, splitting on physical processes was performed. The following models were realized as correctors that modify elastic solutions after each calculation step. A simplified version of the Prandtl-Reuss flow rule [11] was applied accordingly in ice to simulate plasticity. It consists in the correction of the stress deviator, $s_{ij} = \sigma_{ij} - \frac{\sigma_{ll}}{\delta_{mm}}\delta_{ij}$, when the von Mises yield criterion is fulfilled $\frac{1}{2}s_{ij}s_{ij} - k^2 > 0$:

$$s_{ij} = s_{ij}^0 \frac{\sqrt{2}k}{\sqrt{s_{el}s_{el}}}, \tag{3}$$

Here a new parameter, the maximum shear stress $k$, is introduced. The value $\frac{1}{2}s_{ij}s_{ij}$ is called the von Mises stress. In the pictures and the text, this model will be named "elastoplasticity".

The last and central model for this work model is the combination of the models described above (see Fig. 1 in the middle). This model originates from several works that can be divided into two groups. In the first articles, [19,20], it was stated that during a low-speed impact, ice shows elastoplastic behavior. This notion is acknowledged in many works focused on simulations of practical tasks, such as ice-ship interactions [21]. Moreover, plastic behavior is typical of ice at low deformation rates until a certain limit when brittle failure starts to occur, [22]. In the other articles, [23,24], the target medium in the process of indentation is divided into zones in the form of hemispheres. This idea is also proved by experiments, [25,26], in which separation is demonstrated by the appearance of distinct areas of recrystallization and microfractures. Thus, in this work, elastoplasticity is used to define such complex behavior.

Thus, it is thought that the ice shows plastic behavior in a small area in the ball-ice contact zone. This area was simulated as a hemisphere (a semicircle in the two-dimensional case) with a radius $r$ and the center at the first contacting point between the striker and the plate. In the cells that fell inside the hemisphere, the above-described elastoplasticity model with the von Mises yield criterion was applied. For the rest of the medium, the elasticity model was used. Therefore, apart from the maximum shear stress $k$, a new variable, $r$ appears. It is calculated based on the size of the indentation, $a$. It can be enumerated during the simulation from the depth of the ball's immersion, $h$, using the standard geometrical formula. In [23,24] the relationship between $r$ and $a$ is considered to be linear. Thus, in this work, the constant $\frac{r}{a}$ becomes this model's parameter. The name of this elastoplastic model with two zones is omitted in the figures.

## 4 Simulation Results

As a result of the simulations, the velocity and the stress tensor patterns were obtained. In Fig. 2 the von Mises stress distribution in the elastoplastic area is depicted. The tooth-shaped structures are formed 0.6 ms from the collision's beginning. These patterns are congruent with the ring fractures [27]. The picture also demonstrates the formation of a dent, which is similar to the observations in the real experiment. Moreover, the red area is limited by the elastoplastic hemisphere. Thus, Fig. 2 displays the increase in the radius of the elastoplastic zone. As a result, the model shows the great potential for qualitative reproduction of the laboratory experiment.

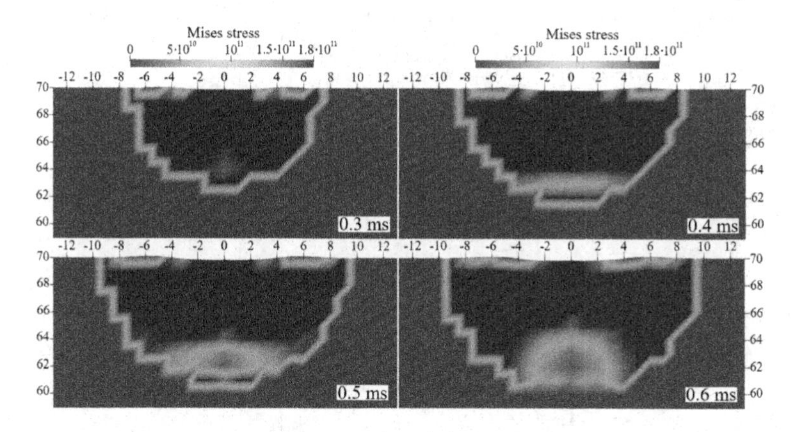


**Fig. 2.** The von Mises stress distribution, $r = 4 \cdot a$, $k = 3 \cdot 10^5$ Pa, dimensions are in millimeters. (Color figure online)

The following comparison with the experiment was conducted based on the ball's acceleration. In order to do this, the ball's velocity was estimated: $v(t) = 0.56 + \frac{1}{m}\int_0^t F(\tau)\,d\tau$, where $F(\tau)$ is the force in Fig. 1, $m = 1.76$ kg is the ball's mass. However, the velocity did not reach zero. This can be explained by the fact that the ice surface cannot be fully smooth; thus, tangential velocity appears. Therefore, the transformation is conducted: $v(t) = 0.56(1 - \frac{v-0{,}56}{\min(v-0{,}56)})$.

Next, the ball's coordinate was calculated: $x(t) = \int_0^t v(\tau) sign(v(\tau))\, d\tau$, with the added function *sign* equal to 1 before velocity reaches zero and $-1$ after this point. Integration was conducted using the Simpson rule. The results of the simulation compared to the estimated curves are presented in Fig. 3. The illustration shows the modulus of the vertical projection of the stress tensor, $\sigma_{yy}$, the ball's coordinate, $x$, and the modulus of its smoothened velocity, $v_y$, in the process of the collision. The computed velocity in case of the elastoplastic model with two zones had a lot of oscillations, which made it harder to analyze the graphs. Thus, the curves were smoothened by the Savitzky-Golay filter with the length of the filter window set to 51 and linear polynomials used to fit the samples. Results by elasticity and elastoplasticity models were added as a reference.

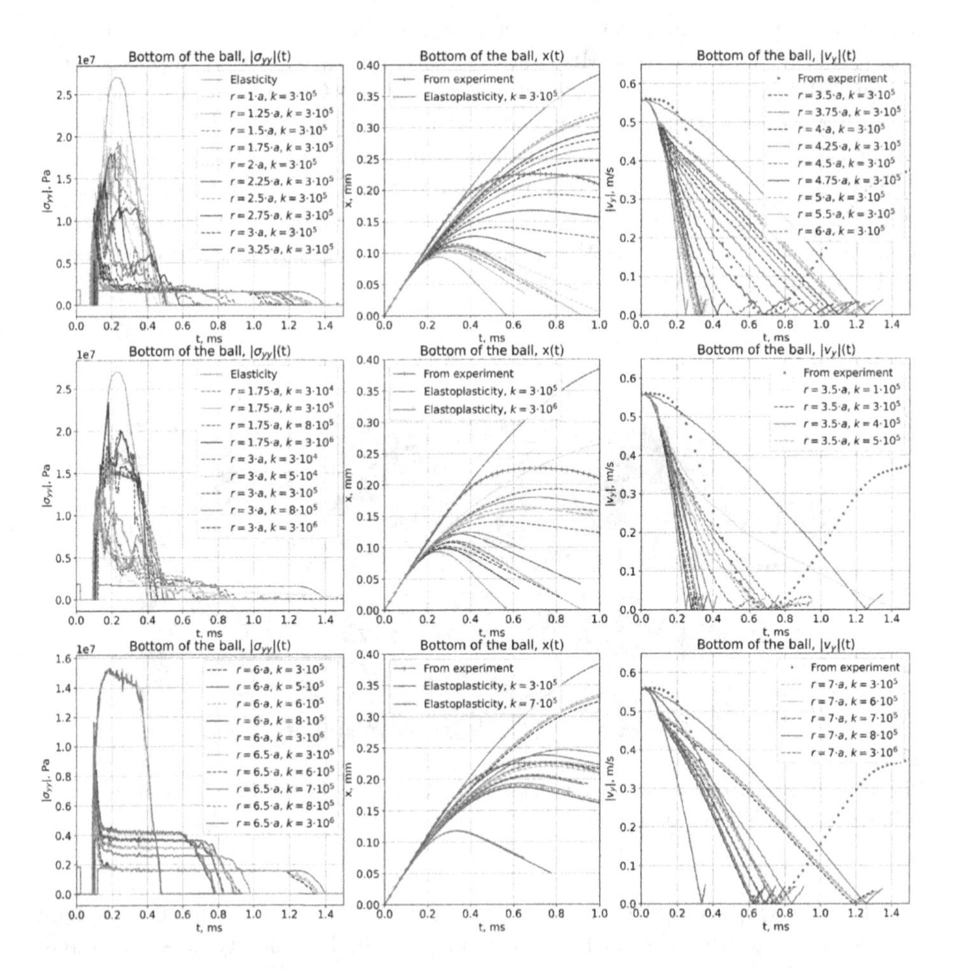


**Fig. 3.** Results of the simulation from the receiver in the bottom point of the ball. Left column – modulus of vertical projection of the stress tensor, middle column – the ball's coordinate, right column – modulus of the ball's velocity. Rows share the same legend. (Color figure online)

Figure 3 demonstrates the basic characteristics of the elastoplastic model with two zones. The curve $\sigma_{yy}$ contains a lot of oscillations. These effects can be explained by the conducted corrections of the deviator stress. On the other hand, similar phenomena can be seen in the stress-displacement graphs in other experiments [27]. Figure 3 also shows the influence of the main parameters, $r$, $k$, on the results. According to the first row of Fig. 3, the increase in the $r$ value generally leads to a lower amplitude of $\sigma_{yy}$, bigger indentation, and moves the velocity curve to the right. The graphs are visibly located between the elasticity and elastoplasticity models' results. However, the curves never reach this limit. This may be explained by the fact that the curves coincide until about 0.09–0.1 ms. At this point in time, the elastoplastic zone increases its volume from one cell to nearby areas.

The graphs can also be conditionally divided into three groups. For $r \leq 2.75 \cdot a$ the lines are closer to each other, and the form of the stress curves is similar to the elastic curve. As the left picture on the second row of Fig. 3 shows, $\sigma_{yy}$ amplitude is close to the amplitude in the case of the elastoplastic model with $k = 3 \cdot 10^6$ Pa. For $r \geq 4 \cdot a$ the results also get closer, and in the later stages of the collision, the stress almost completely coincides with the elastoplasticity curves with the same $k$. However, for $k < 3 \cdot 10^6$ Pa the elastoplastic model produces $x$ higher than the model with two zones and $v_y$ on the left from its velocity. At the beginning of the impact, the amplitude $\sigma_{yy}$ is large but almost immediately drops. Nonetheless, at this stage, it is still possible to see that curves with smaller $r$ have higher stress curves.

Finally, for $r \geq 3 \cdot a$ it is possible to select the model's parameters that can reproduce the amplitude of the coordinate, $x_{max}$, and the time when the modulus of velocity reaches zero, $t_{v=0}$, close to the laboratory experiment's results. With the constant $r$ the increase in $k$ leads to the rise of $\sigma_{yy}$ amplitude, smaller indentation, and a movement of the velocity curves to the left. Thus, for almost each $r$ the proper $k$ can be chosen. However, there are still several limitations, which are more prominent for smaller $r$. At first, when $k$ is too large, the von Mises criterion is never satisfied, and the results tend toward elasticity. A further increase in its value won't change the outcome. On the other hand, with too small $k$ the yield criterion begins to constantly work almost in every elastoplastic cell. Therefore, the decrease in $k$ won't affect the results either. The last point explains why the bound of $r \geq 3 \cdot a$ appears.

Figure 3 shows examples of the found parameters. The parameters that reconstruct $t_{v=0}$ close to the experimental value of 0.7 ms are: $r = 3.5 \cdot a$ and $k = 4 \cdot 10^5$ Pa (gray line in the second row of Fig. 3, $t_{v=0} = 0.7065$ ms), $r = 3 \cdot a$ and $k = 5 \cdot 10^4$ Pa (brown line in the second row, 0.6995 ms), $r = 6.5 \cdot a$ and $k = 7 \cdot 10^5$ Pa (teal line in the third row, 0.693 ms), $r = 7 \cdot a$ and $k = 7 \cdot 10^5$ Pa (magenta line in the third row, 0.694 ms). For $x_{max}$ close to 0.2253 mm from the experiment, they were found to be: $r = 6.5 \cdot a$ and $k = 6 \cdot 10^5$ Pa (dark violet line in the third row, 0.2258 mm), $r = 7 \cdot a$ and $k = 6 \cdot 10^5$ Pa (sky blue line in the third row, 0.2276 mm). A relatively close value of $x_{max} = 0.2203$ mm is obtained with $r = 6 \cdot a$ and $k = 6 \cdot 10^5$ Pa. The values $x_{max}$ and $t_{v=0}$ turned out to be pretty good representations of the computed $x$ and $v_y$ curves. Thus, they were chosen as selection criteria for the parameters of the elastoplastic model with two zones. In order to evaluate the parameters' impact on them, the trends were depicted in Fig. 4. The picture helps to define the form of dependencies of the criteria on $k$ and $\frac{r}{a}$. The shapes of $x_{max}$ and $t_{v=0}$ curves are visibly similar to each other.

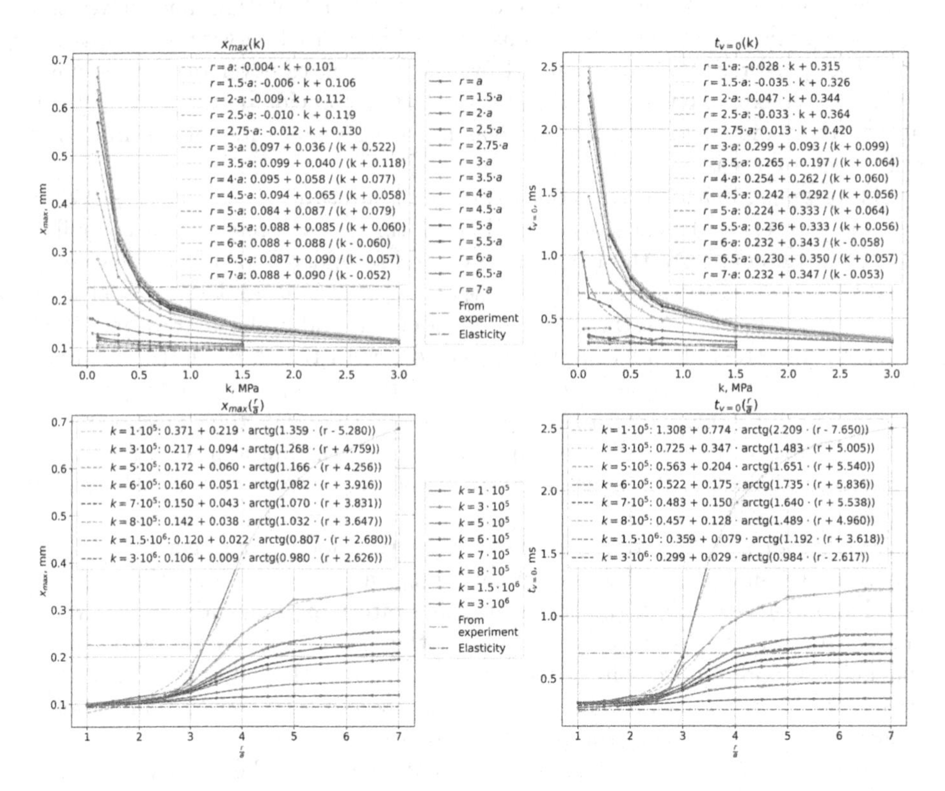


**Fig. 4.** Dependencies of maximum depth of the ball's immersion and the time when the ball's velocity changes its sign on the model's parameters. Solid lines in the rows share the same legend.(Color figure online)

The first row in Fig. 4 demonstrates the limitations on the parameters' selection due to the fact that the lines $x_{max}$ with $r \leq 3 \cdot a$ cannot reach the experimental value, similarly to $t_{v=0}$ with $r \leq 2.75 \cdot a$. The graphs with $r < 3 \cdot a$ can be approximated with the linear trend with almost constant dependence for small $\frac{r}{a}$. The less $\frac{r}{a}$ is, the generally lower the lines lie and the closer they get to the values of $x_{max}$ and $t_{v=0}$ in the case of the elasticity model. However, this lower limit is not reached. The graphs with $r > 3 \cdot a$ can be characterized by hyperbolic curves. For $r > 5 \cdot a$ curves are visibly constrained by a boundary. Therefore, the proper parameters can only be achieved for $k < 7 \cdot 10^5$ Pa. The second row in Fig. 4 shows that the dependence of $x_{max}$ and $t_{v=0}$ on $\frac{r}{a}$ can be described with an arctangent function. The quality of approximation appears to be worse for smaller $k$. Nevertheless, all built trends allow predicting the model's parameters that reproduce solutions close to the estimations from the laboratory experiment. The calculated values are presented in the Table 1.

**Table 1.** Predicted parameters that reconstruct maximum depth of the ball's immersion and the time when the ball's velocity changes its sign close to the estimations from the laboratory experiment. Calculation was conducted based on approximating curves presented in Fig. 4.

| Row № | Known parameter | Prediction from $x_{max}$ | Prediction from $t_{v=0}$ |
|---|---|---|---|
| 1 | $r = 3 \cdot a$ | × | $k = 1.32 \cdot 10^5$ Pa |
| 1 | $r = 3.5 \cdot a$ | $k = 2.01 \cdot 10^5$ Pa | $k = 3.89 \cdot 10^5$ Pa |
| 1 | $r = 4 \cdot a$ | $k = 3.65 \cdot 10^5$ Pa | $k = 5.29 \cdot 10^5$ Pa |
| 1 | $r = 4.5 \cdot a$ | $k = 4.4 \cdot 10^5$ Pa | $k = 5.83 \cdot 10^5$ Pa |
| 1 | $r = 5 \cdot a$ | $k = 5.35 \cdot 10^5$ Pa | $k = 6.38 \cdot 10^5$ Pa |
| 1 | $r = 5.5 \cdot a$ | $k = 5.571 \cdot 10^5$ Pa | $k = 6.62 \cdot 10^5$ Pa |
| 1 | $r = 6 \cdot a$ | $k = 5.574 \cdot 10^5$ Pa | $k = 6.76 \cdot 10^5$ Pa |
| 1 | $r = 6.5 \cdot a$ | $k = 5.98 \cdot 10^5$ Pa | $k = 6.88 \cdot 10^5$ Pa |
| 1 | $r = 7 \cdot a$ | $k = 6.07 \cdot 10^5$ Pa | $k = 6.89 \cdot 10^5$ Pa |
| 2 | $k = 1 \cdot 10^5$ Pa | $r = 3.31 \cdot a$ | $r = 3.01 \cdot a$ |
| 2 | $k = 3 \cdot 10^5$ Pa | $r = 3.82 \cdot a$ | $r = 3.33 \cdot a$ |
| 2 | $k = 5 \cdot 10^5$ Pa | $r = 4.69 \cdot a$ | $r = 3.84 \cdot a$ |
| 2 | $k = 6 \cdot 10^5$ Pa | $r = 6.66 \cdot a$ | $r = 4.3 \cdot a$ |
| 2 | $k = 7 \cdot 10^5$ Pa | × | $r = 8.26 \cdot a$ |

Unfortunately, overall, the difference between the predicted parameters from $x_{max}$ and $t_{v=0}$ proves the inability to select the parameters that will fulfill both criteria at the same time. This problem generates an important direction for further work, which consists in mesh refinement of the graphs from the laboratory experiment and in the computational domain of the numerical calculations. Nonetheless, in order to deal with this issue, a new set of criteria was formulated in this work. These criteria consist in the $L_1$, $L_2$, $L_\infty$ norms of the difference between the simulated and evaluated from the laboratory experiment ball's coordinate $x$ and velocity $v_y$. However, the main issue lies in the fact that the curves have different timelines.

In order to deal with this problem, at first, in each case, the values were taken from the time grid with the biggest step, or a new grid with the least common denominator as a time step was introduced. The next question is the point of time until the comparison should be made. The computations stop reconstructing the process after the impact within a short time after the ball stops contacting the ice. A similar problem arises in the laboratory experiment. Thus, the norms were calculated up until 0.7 ms for $v_y$ and up until 1.1 ms for

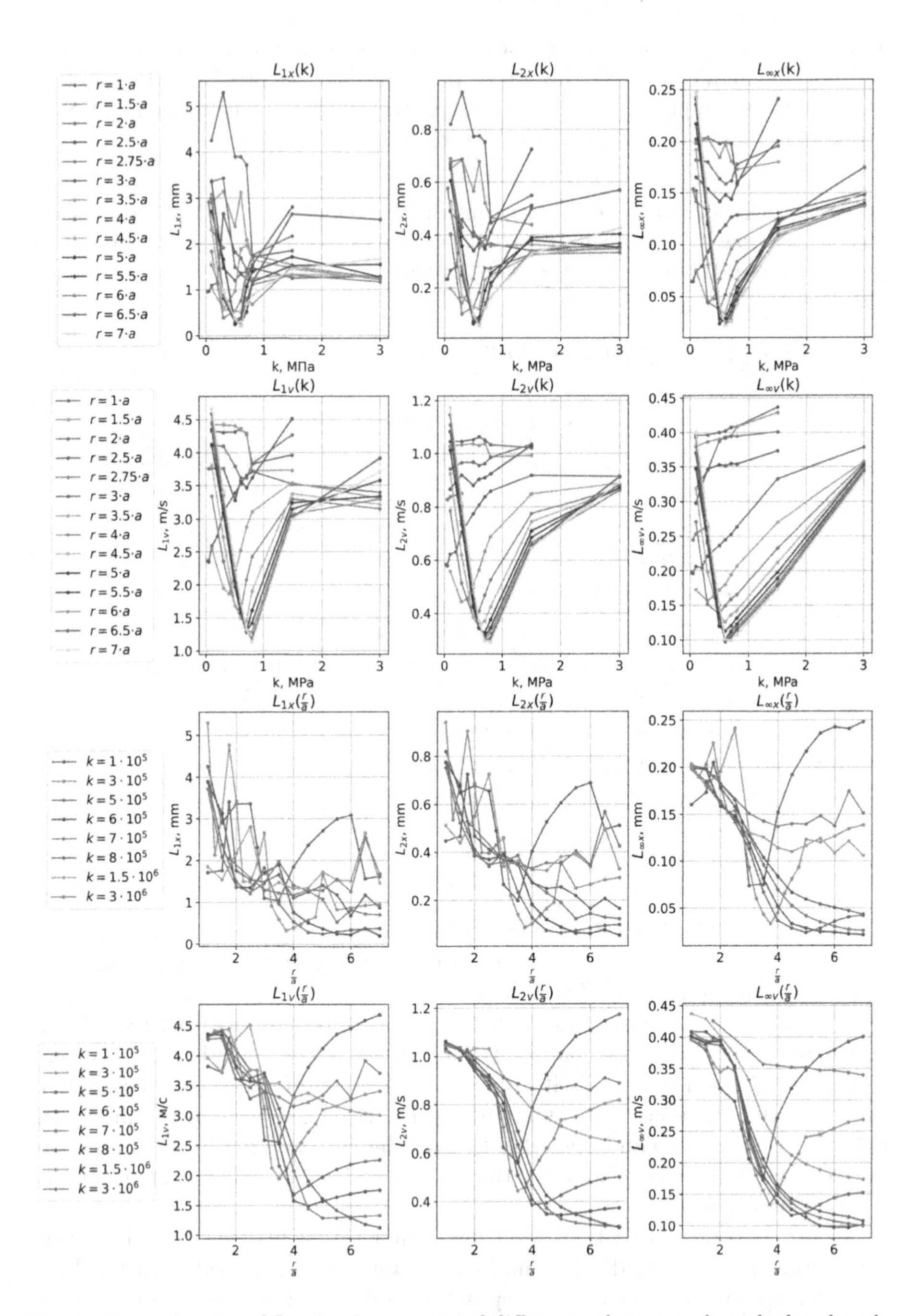


**Fig. 5.** Dependencies of $L_1$, $L_2$, $L_\infty$ norms of differences between the calculated and estimated from the laboratory experiment ball's coordinate and velocity. The rows share the same legend. (Color figure online)

$x$ or until the timeline of the simulation reaches its final point. In the case of velocity norms, the computational results, $|v_y|$, were smoothened in the same way as presented in Fig. 3. It was checked that the norms with and without this correction changes insignificantly. Similarly, the reduction of the timeline in the calculated curves until they reach their minimum value also does not drastically influence the results. Finally, the computed norms are presented in Fig. 5.

These newly formulated criteria allow us to estimate which parameters, overall, reconstruct $x$, $v_y$ closer to the evaluated curves from the laboratory experiment. According to Fig. 5, most of the depicted curves have minimums. Moreover, by using the least minimum value for all curves, there is also the possibility of choosing the best universal parameters. It may also be noted that the lines' minimum generally coincides with the closest parameters to the experimental lime lines in Fig. 4. However, the additional information helps to improve the analysis.

In accordance with the first two rows in Fig. 5, the least norms' values can be achieved with $r \geq 5 \cdot a$ (dark green line). The yellow line $r = 7 \cdot a$ generally shows the best behavior, with a minimum in $k = 6 \cdot 10^5$ Pa in the first row and $k = 8 \cdot 10^5$ Pa in the second one. These results are closely followed by the turquoise line $r = 6 \cdot a$ in the top row and the magenta line $r = 6.5 \cdot a$ lower. Most of the curves in the specified limit reach their minimum at $k = [5, 6] \cdot 10^5$ Pa and $k = [7, 8] \cdot 10^5$ Pa in the first and second rows, respectively. Overall, the graphs have a form of a sharp pit, which drastically distinguishes them from the lines in the third and fourth rows, where the dependence seems to be a lot smoother. However, similarly, there is a boundary of $k < 1.5 \cdot 10^5$ Pa. Within this limit, there is a clearly visible global minimum in each line. The minimums lie in the right part of the graph, starting from the point $r = 3 \cdot a$. The universal minimums coincide with the results obtained from the first two rows. However, the possibility of reaching the minimum for $k \geq 8 \cdot 10^5$ Pa appears with $r \geq 7 \cdot a$. This point can be noted as one of the directions for further work.

## 5 Conclusion

In this work, ice rheology in the process of low-speed impact was investigated. The elastoplastic model, which consists in identification of the plasticity zone in the elastic medium as a hemisphere with changing radius and center at the first contacting point between a striker and a target, is formulated. This model was applied for the simulation of ice behavior in the laboratory experiment. The output was compared based on the ball's instantaneous velocity and coordinate during the collision. The model managed to qualitatively reproduce several phenomena that occur in the real experiment. In order to evaluate the model parameters' influence on the results, several criteria were introduced. Approximating curves of two selection criteria's dependence on the parameters were constructed and used for calculation of the set that can reconstruct the solutions closest to the experimental curves. Unfortunately, the predictions based on different criteria proved to not coincide. Nonetheless, the established framework shows good potential to work in an improved setting of the current experiment, including simulations in a three-dimensional case. Another direction for further work is the consideration of more complex rheology models that take into account ice strengthening and brittle failure.

**Acknowledgements.** The reported study was funded by the Russian Science Foundation, project no. 23-21-00384.

## References

1. Petrov, I.B.: Problems of simulation of natural and anthropogenous processes in the Arctic zone of the Russian Federation. Matem. Mod. **30**(7), 103–136 (2018)
2. Staroszczyk, R.: Formation and types of natural ice masses. In: Staroszczyk, R. (ed.) Ice Mechanics for Geophysical and Civil Engineering Applications. GEPS, pp. 7–19. Springer, Cham (2019). https://doi.org/10.1007/978-3-030-03038-4_2
3. Michel, B., Ramseier, R.O.: Classification of river and lake ice. Can. Geotech. J. **8**(1), 36–45 (1971)
4. Tarovik, O., Yakimov, V., Dobrodeev, A., Li, F.: Influence of seasonal and regional variation of ice properties on ship performance in the Arctic. Ocean Eng. **257**, 111563 (2022)
5. Novatskii, V.: Theory of Elasticity. Mir, Moscow (1975). (in Russian)
6. Favorskaya, A.V., Petrov, I.B., Petrov, D.I., Khokhlov, N.I.: Numerical modeling of wave processes in layered media in the Arctic region. Math. Models Comput. Simul. **8**(4), 348–357 (2016). https://doi.org/10.1134/S2070048216040074
7. Obisesan, A., Sriramula, S.: Efficient response modelling for performance characterisation and risk assessment of ship-iceberg collisions. Appl. Ocean Res. **74**, 127–141 (2018)
8. Sain, T., Narasimhan, R.: Constitutive modeling of ice in the high strain rate regime. Int. J. Solid Struct. **48**(5), 817–827 (2011)
9. Timco, G.W., Weeks, W.F.: A review of the engineering properties of sea ice. Cold Reg. Sci. Technol. **60**(2), 107–129 (2010)
10. Neumeier, J.J.: Elastic constants, bulk modulus, and compressibility of $H_2O$ ice I*h* for the temperature range 50 K–273 K. J. Phys. Chem. Ref. Data **47**, 033101 (2018)
11. Bruhns, O.T.: The Prandtl-Reuss equations revisited. Z. Angew. Math. Mech. **94**(3), 187–202 (2014)
12. Petrov, I.B., Khokhlov, N.I.: Modeling 3D seismic problems using high-performance computing systems. Math. Models Comput. Simul. **6**, 342–350 (2014). https://doi.org/10.1134/S2070048214040061
13. Petrov, I.B.: Grid-characteristic methods. 55 years of developing and solving complex dynamic problems. Comput. Math. Inf. Technol. **6**(1), 6–21 (2023)
14. Sinha, N.K.: Elasticity of natural types of polycrystalline ice. Cold Reg. Sci. Technol. **17**(2), 127–135 (1989)
15. Berdennikov, V.P.: Izuchenie modulya uprugosti lda. Trudi GGI **7**(61), 13–23 (1948). (in Russian)
16. Rusanov, V.: The calculation of the interaction of non-stationary shock waves with barriers. J. Comput. Math. Phys. USSR **1**, 267–279 (1961)
17. Kholodov, A.S., Kholodov, Ya.A.: Monotonicity criteria for difference schemes designed for hyperbolic equations. Comput. Math. Math. Phys. **46**(9), 1560–1588 (2006)

18. Guseva, E.K., Golubev, V.I., Petrov, I.B.: Linear, quasi-monotonic and hybrid grid-characteristic schemes for hyperbolic equations. Lobachevskii J. Math. **44**, 296–312 (2023). https://doi.org/10.1134/S1995080223010146
19. Epifanov, V.P.: Contact fracture behavior of ice. Ice Snow **60**(2), 274–284 (2020). (in Russian)
20. Guseva, E.K., Beklemysheva, K.A., Golubev, V.I., Epifanov, V.P., Petrov, I.B.: Investigation of ice rheology based on computer simulation of low-speed impact. In: Balandin, D., Barkalov, K., Meyerov, I. (eds.) MMST 2022. CCIS, vol. 1750, pp. 176–184. Springer, Cham (2022). https://doi.org/10.1007/978-3-031-24145-1_15
21. Zhang, N., Zheng, X., Ma, Q., Hu, Z.: A numerical study on ice failure process and ice-ship interactions by Smoothed Particle Hydrodynamics. Int. J. Naval Archit. Ocean Eng. **11**(2), 796–808 (2019)
22. Ince, S.T., Kumar, A., Paik, J.K.: A new constitutive equation on ice materials. Ships Offshore Struct. **12**(5), 610–623 (2017)
23. Johnson, K.L.: The correlation of indentation experiments. J. Mech. Phys. Solids **18**, 115–126 (1970)
24. Studman, J., Moore, M.A., Jones, S.E.: On the correlation of indentation experiments. J. Phys. D Appl. Phys. **10**, 949–956 (1977)
25. Barrette, P., Pond, J., Jordaan, I.: Ice damage and layer formation in small-scale indentation experiments. In: Ice in the Environment: Proceedings of the 16th International Association for Hydraulic Engineering and Research (IAHR) International Symposium on Ice, 246–253 (2002)
26. Browne, T., Taylor, R., Jordaan, I., Gürtner, A.: Small-scale ice indentation tests with variable structural compliance. Cold Reg. Sci. Technol. **88**, 2–9 (2013)
27. Epifanov, V.P.: Physical mechanisms of ice contact fracture. Dokl. Phys. **52**(1), 19–23 (2007). https://doi.org/10.1134/S1028335807010053

# On the Period Length Modulo $D$ of Sequences of Numerators and Denominators of Convergents for the Square Root of a Non-square $D$

Sergey V. Sidorov and Pavel A. Shcherbakov(✉)

Lobachevsky State University of Nizhny Novgorod, Nizhny Novgorod, Russia
sergey.sidorov@itmm.unn.ru, shcherbakov.pavel.a@gmail.com

**Abstract.** In this paper we investigate the properties of sequences of numerators and denominators of convergents for $\sqrt{D}$, where $D$ is not a perfect square. In our previous work we prove that $L = l, 2l$ or $4l$, where $L$ is the period length modulo $p$ of the numerators of convergents for $\sqrt{p}$ and $l$ is the period length of the continued fraction for $\sqrt{p}$, where $p$ is a prime. Namely, if $p = 2$ or $p \equiv 7 \pmod 8$, then $L = l$; if $p \equiv 3 \pmod 8$, then $L = 2l$; if $p \equiv 1 \pmod 4$, then $L = 4l$. Here we generalize this result and prove that $L = l, 2l$ or $4l$ for convergents for $\sqrt{D}$, where $D$ is not a perfect square. Moreover, we prove that the shortest period length $L_B$ of the sequence of denominators of convergents for $\sqrt{D}$ modulo $D$ is equal to $\frac{D}{\gcd(D, B_{L-1})} \cdot L$. In addition, we prove under the assumption of Ankeny-Artin-Chowla and Mordell conjectures that $L_B = pL$ for a prime $p$. We also discuss a possible application of our results to image processing and encoding using a discrete version of the Arnold's cat map.

**Keywords:** Arnold's cat map · Continued fraction · Convergent · Denominator · Prime number · Numerator · Pell's equation · Square root

## 1 Introduction

An extensive literature (see, for example, [5,7,8,14,15,17]) is devoted to the topic of continued fractions. It is a classical object in number theory, which is used since ancient times for approximation of real numbers with rational numbers (so-called diophantine approximation, see [9]). The structure of solutions to Pell's equation $x^2 - Dy^2 = 1$ and its generalized version $x^2 - Dy^2 = N$ ($D > 1$ and is not a perfect square, $N \in \mathbb{Z}$) is also related to continued fractions (see [11]). Pell's equation can be used to create a public key cryptosystem (see [19]).

Continued fractions are also used in computer science. Note that in the standard floating point arithmetic we can only control relative round-off error during calculations, which in some cases (such as series summation) may be fatal for accuracy. When calculations crucially depend on round-off errors (as an example, in determining the stability of solutions to differential equations), we are


D. Balandin et al. (Eds.): MMST 2023, CCIS 1914, pp. 28–43, 2024.
https://doi.org/10.1007/978-3-031-52470-7_3

in need of methods which allow multiple or even arbitrary precision. Continued fractions are such a method. Some examples of the usage of continued fractions for arbitrary precision arithmetic can be found in [10,20,25].

An application of the generalized Pell's equaiton arises in quantum information theory. It can be used to study so-called SIC-POVMs (symmetric, informationally complete, positive operator-valued measures). See [2,3] for more details on the connection between algebraic number theory and SICs (symmetric, informationally complete measurements).

## 2 Preliminaries

It is known that any real number $x$ can be represented as a continued fraction

$$x = [q_0, q_1, q_2, \ldots, q_n, \ldots] = q_0 + \cfrac{1}{q_1 + \cfrac{1}{q_2 + \ldots}},$$

where $q_0 \in \mathbb{Z}$, and $q_i \in \mathbb{N}$ for $i = 1, 2, \ldots$. A continued fraction is finite if and only if $x \in \mathbb{Q}$. The integers $q_0, q_1, q_2, \ldots$ are called the coefficients, the terms or *partial quotients* of the continued fraction. The continued fraction $\alpha_i = [q_i, q_{i+1}, q_{i+2}, \ldots]$ is the $i$th *complete quotient* of $x$ and the continued fraction $[q_0, q_1, \ldots, q_i]$ is the $i$th *convergent* to $x$. A convergent is defined as a finite continued fraction. It is always a rational number, so $[q_0, q_1, \ldots, q_i] = \frac{A_i}{B_i}$, where $A_i, B_i$ are positive integers (except possibly $A_0 = q_0$ that can be negative).

There are the following recursive formulas for the numerator and the denominator of the $i$th convergent $\frac{A_i}{B_i}$:

$$\begin{cases} A_{-1} = 1, \\ A_0 = q_0, \\ A_{i+1} = q_{i+1} A_i + A_{i-1}, i \geq 0, \end{cases} \qquad \begin{cases} B_{-1} = 0, \\ B_0 = 1, \\ B_{i+1} = q_{i+1} B_i + B_{i-1}, i \geq 0. \end{cases}$$

**Theorem 1** ([15]). *The following relation holds for $i \geq 0$ :*

$$B_i A_{i-1} - B_{i-1} A_i = (-1)^i.$$

The following theorem gives the relationship between $x = \alpha_0$ and the $j$th complete quotient $\alpha_j$.

**Theorem 2** ([14,15,17]). *The following relation holds for $j \geq 1$ :*

$$\alpha_0 = \frac{\alpha_j A_{j-1} + A_{j-2}}{\alpha_j B_{j-1} + B_{j-2}}.$$

A continued fraction is called a periodic continued fraction if its terms eventually repeat from some point onwards. The minimal number of repeating terms is called the period length of the continued fraction. In general, a periodic continued fraction has the form $[q_0, q_1, \ldots, q_m, \overline{q_{m+1}, \ldots, q_{m+l}}]$, where the bar indicates the periodic part, which is repeated indefinitely, $l$ is the period length. Lagrange proved in 1770, that a continued fraction is periodic if and only if $x$ is a quadratic irrationality. That is, $x$ arises as the solution of quadratic equation with integral coefficients.

If $D > 0$ is an integer which is not a perfect square, then the continued fraction for $\sqrt{D}$ is not only periodic, but also has remarkable properties. Namely (see [5,15]),

$$\sqrt{D} = [q_0, \overline{q_1, q_2, \ldots, q_{l-2}, q_{l-1}, q_l}],$$

where $l$ is the period length, $q_l = 2q_0$ and $q_i = q_{l-i}$ for $i = 1, \ldots, l-1$, i.e. the sequence $q_1, q_2, \ldots, q_{l-2}, q_{l-1}$ is a palindrome.

There is a special algorithm that allows you to calculate the terms $q_i$ of the continued fraction for $\sqrt{D}$. This algorithm is given below and it is sometimes called the $PQ$-algorithm (see [13,15,18]).

Initial settings:

$$P_0 = 0, \quad Q_0 = 1, \quad \alpha_0 = \frac{P_0 + \sqrt{D}}{Q_0} = \sqrt{D}, \quad q_0 = \left\lfloor \frac{P_0 + \sqrt{D}}{Q_0} \right\rfloor.$$

For $i \geq 0$ set

$$P_{i+1} = q_i Q_i - P_i, \quad Q_{i+1} = \frac{D - P_{i+1}^2}{Q_i},$$

$$\alpha_{i+1} = \frac{P_{i+1} + \sqrt{D}}{Q_{i+1}}, \quad q_{i+1} = \left\lfloor \frac{P_{i+1} + \sqrt{D}}{Q_{i+1}} \right\rfloor.$$

It is easy to verify that this algorithm reproduces the continued fraction expansion of $\sqrt{D}$.

To find the period length of the continued fraction for $\sqrt{D}$, we need to finish the calculation at $q_i = 2q_0$. Under this condition, we get the period length $l = i$.

The following properties of sequences $\{A_i\}_{i=-1}^{\infty}$, $\{B_i\}_{i=-1}^{\infty}$, $\{P_i\}_{i=1}^{\infty}$, $\{Q_i\}_{i=0}^{\infty}$ are well known.

**Theorem 3** ([15,17]). *If $l$ is the period length of the continued fraction for $\sqrt{D}$, where $D$ is not a perfect square, then sequences $\{P_i\}_{i=1}^{\infty}$, $\{Q_i\}_{i=0}^{\infty}$ are purely periodic with the period length $l$, that is*

$$P_{l+i+1} = P_{i+1}, \quad Q_{l+i} = Q_i \quad \textit{for all} \quad i \geq 0,$$

*and*

$$P_{i+1} = P_{l-i}, \quad 0 \leq i \leq l-1,$$
$$Q_i = Q_{l-i}, \quad 0 \leq i \leq l.$$

*In addition,*

$$1 \leq Q_i \leq 2q_0 = 2\lfloor\sqrt{D}\rfloor \quad \textit{for all} \quad i \geq 0,$$

*and*

$$Q_i = 1 \quad \textit{if and only if} \quad l \mid i.$$

**Theorem 4 ([15], p. 92).** *If $D$ is not a perfect square, then for all $i \geq -1$ the following equality holds*

$$A_i^2 - DB_i^2 = (-1)^{i+1}Q_{i+1}.$$

**Theorem 5 ([5], p. 108).** *Let $l$ be the length of the period of the continued fraction for $\sqrt{p}$, where $p$ is an odd prime number. Then*

1. *$l$ is odd if and only if $p \equiv 1 \pmod 4$.*
2. *$l$ is even if and only if $p \equiv 3 \pmod 4$.*

The following theorem, proved in [22], reveals the structure of the periodic part $[A_{-1}, A_0, \ldots, A_{L-2}]$ of the sequence $\{A_i \bmod p\}_{i=-1}^{\infty}$ of numerators of convergents for $\sqrt{p}$.

**Theorem 6 ([22]).** *Suppose $p$ is a prime number, $l$ is the length of the period of the continued fraction for $\sqrt{p}$, and $L$ is the length of the period of the numerators modulo $p$ of the corresponding convergents.*

1. *If $p = 2$, then $L = l = 1$.*
2. *If $p \equiv 7 \pmod 8$, then $L = l$, $l = 4t$ for some $t \in \mathbb{N}$, and the periodic part is*

$$\underbrace{1, A_0, A_1, \ldots, A_{2t-3}, A_{2t-2}, A_{2t-1}, -A_{2t-2}, A_{2t-3}, \ldots, A_1, -A_0}_{l}.$$

3. *If $p \equiv 3 \pmod 8$, then $L = 2l$, $l = 4t + 2$ for some $t \in \mathbb{N} \cup \{0\}$, and the periodic part is*

$$1, A_0, A_1, \ldots, A_{2t-1}, A_{2t}, -A_{2t-1}, \ldots, -A_1, A_0,$$

$$-1, -A_0, -A_1, \ldots, -A_{2t-1}, -A_{2t}, A_{2t-1}, \ldots, A_1, -A_0.$$

4. *If $p \equiv 1 \pmod 4$, then $L = 4l$, $l = 2t + 1$ for some $t \in \mathbb{N} \cup \{0\}$, and the periodic part is*

$$1, A_0, A_1, \ldots, A_{t-2}, A_{t-1}, (-1)^t rA_{t-1}, (-1)^{t-1} rA_{t-2}, \ldots, rA_1, -rA_0,$$

$$r, rA_0, rA_1, \ldots, rA_{t-2}, rA_{t-1}, (-1)^{t-1}A_{t-1}, (-1)^{t-2}A_{t-2} \ldots, -A_1, A_0,$$

$$-1, -A_0, -A_1, \ldots, -A_{t-2}, -A_{t-1}, (-1)^{t-1}rA_{t-1}, (-1)^{t-2}rA_{t-2}, \ldots, -rA_1, rA_0,$$

$$-r, -rA_0, -rA_1, \ldots, -rA_{t-2}, rA_{t-1}, (-1)^t A_{t-1}, (-1)^{t-1}A_{t-2}, \ldots, A_1, -A_0,$$

*where $r$ satisfies the congruence $r^2 \equiv -1 \pmod p$.*

To the end of this paper, we will be interested in the properties of the sequences $\{A_i\}_{i=-1}^{\infty}$ and $\{B_i\}_{i=-1}^{\infty}$ for $\sqrt{D}$, where $D$ is not a perfect square.

# 3 The Main Result

## 3.1 Properties of Sequences $\{A_i\}_{i=-1}^{\infty}$, $\{B_i\}_{i=-1}^{\infty}$ for Any Non-square $D$

**Lemma 1.** *Let $l$ be the length of the period of the continued fraction for $\sqrt{D}$, where $D$ is not a perfect square. Then the following equalities hold for $j \geq -1$*

$$A_{l+j} = A_{l-1}A_j + DB_{l-1}B_j,$$
$$B_{l+j} = A_{l-1}B_j + B_{l-1}A_j.$$

*Proof.* We prove both equalities by induction on $j$. For $j = -1$ the statement is obvious since $B_{-1} = 0$, $A_{-1} = 1$. For $j = 0$ we have $B_0 = 1$, $A_0 = q_0$. Since by Theorem 2 we have $\alpha_0 = \frac{\alpha_l A_{l-1}+A_{l-2}}{\alpha_l B_{l-1}+B_{l-2}}$, $\alpha_l = q_0 + \alpha_0$ and $\alpha_0^2 = D$, then $\alpha_0 = \frac{(q_0+\alpha_0)A_{l-1}+A_{l-2}}{(q_0+\alpha_0)B_{l-1}+B_{l-2}}$, so $DB_{l-1} + \alpha_0(q_0 B_{l-1} + B_{l-2}) = \alpha_0 A_{l-1} + (q_0 A_{l-1} + A_{l-2})$. It follows from the last equality that $q_0 B_{l-1} + B_{l-2} = A_{l-1}$ and $DB_{l-1} = q_0 A_{l-1} + A_{l-2}$. Then we get

$$A_l = q_l A_{l-1} + A_{l-2} = 2q_0 A_{l-1} + A_{l-2} = 2q_0 A_{l-1} + (DB_{l-1} - q_0 A_{l-1}) = q_0 A_{l-1} + DB_{l-1} = A_{l-1}A_0 + DB_{l-1}B_0,$$
$$B_l = 2q_0 B_{l-1} + B_{l-2} = 2q_0 B_{l-1} + (A_{l-1} - q_0 B_{l-1}) = A_{l-1} + q_0 B_{l-1} = A_{l-1}B_0 + B_{l-1}A_0.$$

The induction base is proved.

Since

$$A_{l+j} = q_{l+j}A_{l+j-1} + A_{l+j-2} = q_j A_{l+j-1} + A_{l+j-2},$$
$$B_{l+j} = q_{l+j}B_{l+j-1} + B_{l+j-2} = q_j B_{l+j-1} + B_{l+j-2},$$

then applying the induction hypothesis we obtain

$$A_{l+j} = q_j(A_{l-1}A_{j-1} + DB_{l-1}B_{j-1}) + (A_{l-1}A_{j-2} + DB_{l-1}B_{j-2}) = A_{l-1}(q_j A_{j-1} + A_{j-2}) + DB_{l-1}(q_j B_{j-1} + B_{j-2}) = A_{l-1}A_j + DB_{l-1}B_j,$$
$$B_{l+j} = q_j(A_{l-1}B_{j-1} + B_{l-1}A_{j-1}) + (A_{l-1}B_{j-2} + B_{l-1}A_{j-2}) = A_{l-1}(q_j B_{j-1} + B_{j-2}) + B_{l-1}(q_j A_{j-1} + A_{j-2}) = A_{l-1}B_j + B_{l-1}A_j.$$

□

**Corollary 1.** *Let $l$ be the length of the period of the continued fraction for $\sqrt{D}$, where $D$ is not a perfect square. Then the following congruences modulo $D$ hold:*

$$A_{l+j} \equiv A_{l-1}A_j \pmod{D},$$
$$B_{l+j} \equiv A_{l-1}B_j + B_{l-1}A_j \pmod{D}$$

*for $j \geq -1$.*

*Proof.* The corollary obviously follows from Lemma 1. □

**Corollary 2.** *If $D$ is not a perfect square, then*

$$A_{kl+j} \equiv A_{l-1}^{k} A_j \pmod{D},$$
$$B_{kl+j} \equiv A_{l-1}^{k-1}(A_{l-1}B_j + kB_{l-1}A_j) \pmod{D}$$

*for $j \geq -1$, $k \geq 1$.*

*Proof.* Let's prove the corollary by induction on $k$. For $k = 1$ it's true by Corollary 1. By the induction hypothesis, we have

$$A_{(k-1)l+j} \equiv A_{l-1}^{k-1} A_j \pmod{p}, \ B_{(k-1)l+j} \equiv A_{l-1}^{k-2}(A_{l-1}B_j + (k-1)B_{l-1}A_j) \pmod{p}.$$

Therefore, by Corollary 1,

$$A_{kl+j} \equiv A_{l-1}A_{(k-1)l+j} \equiv A_{l-1}A_{l-1}^{k-1}A_j = A_{l-1}^{k}A_j \pmod{p},$$
$$B_{kl+j} \equiv A_{l-1}B_{(k-1)l+j} + B_{l-1}A_{(k-1)l+j} \equiv$$
$$A_{l-1}A_{l-1}^{k-2}(A_{l-1}B_j + (k-1)B_{l-1}A_j)) + B_{l-1}A_{l-1}^{k-1}A_j \equiv$$
$$A_{l-1}^{k-1}(A_{l-1}B_j + (k-1)B_{l-1}A_j + B_{l-1}A_j) \equiv A_{l-1}^{k-1}(A_{l-1}B_j + kB_{l-1}A_j) \pmod{p}.$$

□

**Corollary 3.** *If $D$ is not a perfect square, then*

$$A_{kl-1} \equiv A_{l-1}^{k} \pmod{D},$$
$$B_{kl-1} \equiv kA_{l-1}^{k-1}B_{l-1} \pmod{D}, \quad k \geq 1.$$

*Proof.* It follows from Corollary 2 with $j = -1$ ($A_{-1} = 1$, $B_{-1} = 0$). □

**Corollary 4.** *If $D$ is not a perfect square, then*

$$A_{kl+j} \equiv A_{kl-1}A_j \pmod{D},$$
$$B_{kl+j} \equiv A_{kl-1}B_j + B_{kl-1}A_j \pmod{D}$$

*for $j \geq -1$, $k \geq 1$.*

*Proof.* It follows from Corollary 2 and Corollary 3. □

**Theorem 7.** *If $D$ is not a perfect square, $l$ is the length of the period of the continued fraction for $\sqrt{D}$, and $L$ is the length of the shortest period of the sequence $\{A_i \bmod D\}_{i=-1}^{\infty}$, then this sequence is purely periodic and*

1. *If $l$ is even and $A_{l-1} \equiv 1 \pmod{D}$, then $A_{l+j} \equiv A_j \pmod{D}$, $j \geq -1$, and $L = l$ is even.*
2. *If $l$ is even and $A_{l-1} \not\equiv 1 \pmod{D}$, then $A_{2l+j} \equiv A_j \pmod{D}$, $j \geq -1$, and $L = 2l$ is even.*
3. *If $l$ is odd, then $A_{4l+j} \equiv A_j \pmod{D}$, $j \geq -1$, and $L = 4l$ is even.*

*Proof.* According to Theorem 4 and Theorem 3, we have $A_{l-1}^2 \equiv (-1)^l \pmod{D}$, so $A_{l-1}^4 \equiv 1 \pmod{D}$. Hence by Corollary 3 and Corollary 4 we obtain $A_{2l+j} \equiv A_{l-1}^2 A_j \equiv A_j \pmod{D}$ for even $l$ and $A_{4l+j} \equiv A_{l-1}^4 A_j \equiv A_j \pmod{D}$ for odd $l$, so the sequence $\{A_i \bmod D\}_{i=-1}^{\infty} = \{1, q_0, \ldots\}$ is purely periodic. Each periodic part (not just the shortest one) starts with 1. If $[1, A_0, \ldots, A_{L-2}]$ is the shortest periodic part, then $A_{L-1} \equiv 1 \pmod{D}$. Since by Theorem 4 $A_{L-1}^2 \equiv (-1)^L Q_L \pmod{D}$, then $(-1)^L Q_L \equiv 1 \pmod{D}$. Hence $Q_L \equiv (-1)^L \pmod{D}$, so $Q_L = 1$ or $Q_L = D-1$. Note that by Theorem 3 for $l \nmid L$ the inequality $2 \le Q_L \le 2\lfloor\sqrt{D}\rfloor$ holds. But if $D \ge 6$, then $2\lfloor\sqrt{D}\rfloor \le D-2$, so $2 \le Q_L \le D-2$ and the congruence $Q_L \equiv (-1)^L \pmod{D}$ cannot hold. Thus, we conclude that if $D \ge 6$ and $A_{L-1} \equiv 1 \pmod{D}$, then $l \mid L$. For $D = 2, 3, 5$ the theorem is true. It's easy to verify by direct calculations or refer to Theorem 6. Let's consider the cases of even and odd $l$.

If $l$ is even, then $A_{l-1}^2 \equiv 1 \pmod{D}$.

*Case 1.* If $A_{l-1} \equiv 1 \pmod{D}$, then by Corollary 2 we have $A_{l+j} \equiv A_{l-1}A_j \equiv A_j \pmod{D}$. Let's prove that $L = l$. Indeed, $A_{L-1} \equiv 1 \pmod{D}$ and $1 \le L \le l$ imply $l \mid L$, so $L = l$.

*Case 2.* If $A_{l-1}^2 \equiv 1 \pmod{D}$, $A_{l-1} \not\equiv 1 \pmod{D}$, then by Corollary 2 we have $A_{2l+j} \equiv A_{l-1}^2 A_j \equiv A_j \pmod{D}$. Let's prove that $L = 2l$. Indeed, $A_{L-1} \equiv 1 \pmod{D}$ and $1 \le L \le 2l$ imply $l \mid L$, so $L = l$ or $L = 2l$. But $A_{l-1} \not\equiv 1 \pmod{D}$. Therefore, $L = 2l$.

If $l$ is odd, then $A_{l-1}^2 \equiv -1 \pmod{D}$ and by Corollary 2 we have $A_{4l+j} \equiv A_{l-1}^4 A_j \equiv A_j \pmod{D}$. Let's prove that $L = 4l$. Indeed, $A_{L-1} \equiv 1 \pmod{D}$ and $1 \le L \le 4l$ imply $l \mid L$, so $L = l$, $L = 2l$, $L = 3l$ or $L = 4l$. But $A_{l-1} \not\equiv 1 \pmod{D}$, $A_{2l-1} \equiv A_{l-1}^2 \equiv -1 \pmod{D}$, $A_{3l-1} \equiv A_{l-1}^3 \equiv -A_{l-1} \not\equiv 1 \pmod{D}$. Therefore, $L = 4l$. □

**Lemma 2.** *Let $L$ be the length of the period of numerators modulo $D$ of the continued fraction's convergents for $\sqrt{D}$, where $D$ is not a perfect square. Then the following congruence modulo $D$ holds for $j \ge -1$*

$$B_{L+j} \equiv B_j + B_{L-1}A_j \pmod{D}$$

*and*

$$B_{L-1} = \begin{cases} B_{l-1} \pmod{D}, & \text{if } l \text{ is even and } A_{l-1} \equiv 1 \pmod{D}, \\ B_{2l-1} \equiv -2B_{l-1} \pmod{D}, & \text{if } l \text{ is even and } A_{l-1} \not\equiv 1 \pmod{D}, \\ B_{4l-1} \equiv -4A_{l-1}B_{l-1} \pmod{D}, & \text{if } l \text{ is odd.} \end{cases}$$

*Proof.* By Corollary 4 we have $B_{kl+j} \equiv A_{kl-1}B_j + B_{kl-1}A_j \pmod{D}$. By Theorem 7 we have $L \in \{l, 2l, 4l\}$, so

$$B_{L+j} \equiv A_{kl-1}B_j + B_{L-1}A_j \equiv A_{l-1}^k B_j + B_{L-1}A_j \pmod{D},$$

where $k \in \{1, 2, 4\}$. By Corollary 3 we have $B_{kl-1} \equiv kA_{l-1}^{k-1}B_{l-1} \pmod{D}$.

If $L = l$, then $A_{l-1} \equiv 1 \pmod{D}$ and $B_{L+j} \equiv B_j + B_{L-1}A_j \pmod{D}$.

If $L = 2l$, then $A_{l-1}^2 \equiv 1 \pmod{D}$ and $B_{L+j} \equiv A_{l-1}^2 B_j + B_{L-1}A_j \equiv B_j + B_{l-1}A_j \pmod{D}$, $B_{L-1} = B_{2l-1} \equiv 2A_{l-1}B_{l-1} \equiv -2B_{l-1} \pmod{D}$.

If $L = 4l$, then $A_{l-1}^2 \equiv -1 \pmod{D}$ and $B_{L+j} \equiv A_{l-1}^4 B_j + B_{L-1}A_j \equiv B_j + B_{L-1}A_j \pmod{D}$, $B_{L-1} = B_{4l-1} \equiv 4A_{l-1}^3 B_{l-1} \equiv -4A_{l-1}B_{l-1} \pmod{D}$.

□

**Lemma 3.** *Let $L$ be the length of the period of numerators modulo $D$ of the continued fraction's convergents for $\sqrt{D}$, where $D$ is not a perfect square. Then the following congruence modulo $D$ holds for $k \geq 0$, $j \geq -1$*

$$B_{(k+1)L+j} \equiv B_{kL+j} + B_{L-1}A_j \pmod{D}.$$

*Proof.* Let us prove this Lemma by induction on $k$. For $k = 0$ it's Lemma 2. We have

$$B_{(k+1)L+j} = q_{(k+1)L+j}B_{(k+1)L+j-1} + B_{(k+1)L+j-2}.$$

Since $L$ is divisible by $l$ by Theorem 7, then $q_{(k+1)L+j} = q_{kL+j} = q_j$, $A_{L+j} \equiv A_j \pmod{D}$. Hence

$$B_{(k+1)L+j} \equiv q_{kL+j}(B_{kL+j-1} + B_{L-1}A_{j-1}) + (B_{kL+j-2} + B_{L-1}A_{j-2}) \equiv \underbrace{(q_{kL+j}B_{kL+j-1} + B_{kL+j-2})}_{B_{kL+j}} + B_{L-1}\underbrace{(q_jA_{j-1} + A_{j-2})}_{A_j} \equiv B_{kL+j} + B_{L-1}A_j \pmod{D}.$$

□

**Corollary 5.** *The following congruence holds for $k \geq 0$, $j \geq -1$*

$$B_{kL+j} \equiv B_j + kB_{L-1}A_j \pmod{D}.$$

*In particular,*

$$B_{kL-1} \equiv kB_{L-1} \pmod{D}.$$

*Proof.* Induction on $k$. For $k = 0$ it's obvious. By Lemma 3 and by the induction hypothesis

$$B_{kL+j} \equiv B_{(k-1)L+j} + B_{L-1}A_j \equiv (B_j + (k-1)B_{L-1}A_j) + B_{L-1}A_j \equiv B_j + kB_{L-1}A_j \pmod{D}.$$

□

**Corollary 6.** *The sequence $\{B_i \bmod D\}_{i=-1}^{\infty}$ is purely periodic with the period length of $sL$, where $s = \frac{D}{\gcd(D, B_{L-1})}$, and the following congruence holds for $i \geq 0$, $j \geq -1$*

$$B_{isL+j} \equiv B_j \pmod{D}.$$

*In particular,*

$$B_{isL-1} \equiv 0 \pmod{D}, \quad B_{isL} \equiv 1 \pmod{D}.$$

*Moreover, the periodic part* $[B_{-1}, B_0, \ldots, B_{sL-2}]$ *of the sequence* $\{B_i \bmod D\}_{i=-1}^{\infty}$ *can be written as*

$$\begin{pmatrix} 0 & B_0 & \ldots & B_{L-2} \\ B_{L-1} & B_L & \ldots & B_{2L-2} \\ B_{2L-1} & B_{2L} & \ldots & B_{3L-2} \\ \vdots & \vdots & \ldots & \vdots \\ B_{(s-1)L-1} & B_{(s-1)L} & \ldots & B_{sL-2} \end{pmatrix} \equiv \begin{pmatrix} 0 & 1 \\ B_{L-1} & 1 \\ 2B_{L-1} & 1 \\ \vdots & \vdots \\ (s-1)B_{L-1} & 1 \end{pmatrix} \cdot \begin{pmatrix} 1 & A_0 & A_1 & \ldots & A_{L-2} \\ 0 & B_0 & B_1 & \ldots & B_{L-2} \end{pmatrix} \pmod D.$$

*Proof.* Since $isB_{L-1} = i \cdot \frac{B_{L-1}}{\gcd(D, B_{L-1})} \cdot D$, then $isB_{L-1} \equiv 0 \pmod D$ and by Corollary 5 we have $B_{isL+j} \equiv B_j \pmod D$. For $j = -1$ and $j = 0$, it follows, respectively, that $B_{isL-1} \equiv B_{-1} = 0 \pmod D$, $B_{isL} \equiv B_0 = 1 \pmod D$.

Since $B_{sL+j} \equiv B_j \pmod D$, $j \geq -1$, then $[B_{-1}, B_0, \ldots, B_{sL-2}]$ is the periodic part and it has the length of $sL$. The matrix congruence follows from Corollary 5.

□

**Theorem 8.** *For any non-square* $D$ *the shortest period of the sequence* $\{B_i \bmod D\}_{i=-1}^{\infty} = \{0, 1, B_1, \ldots\}$ *is equal to*

$$L_B = \frac{D}{\gcd(D, B_{L-1})} \cdot L.$$

*Proof.* For simplicity let's denote $s = \frac{D}{\gcd(D, B_{L-1})}$.

1. If $B_{kL+j-1} \equiv B_{-1} \equiv 0 \pmod D$ and $B_{kL+j} \equiv B_0 \equiv 1 \pmod D$ for some $0 \leq k \leq s-1$, $0 \leq j \leq L-3$, then by Corollary 5 we have

$$\begin{cases} B_{j-1} + kB_{L-1}A_{j-1} \equiv 0 \pmod D, \\ B_j + kB_{L-1}A_j \equiv 1 \pmod D, \end{cases} \begin{pmatrix} B_{j-1} & A_{j-1} \\ B_j & A_j \end{pmatrix} \begin{pmatrix} 1 \\ kB_{L-1} \end{pmatrix} \equiv \begin{pmatrix} 0 \\ 1 \end{pmatrix} \pmod D.$$

Since by Theorem 1 $B_j A_{j-1} - B_{j-1} A_j = (-1)^j$, then $B_{j-1} A_j - B_j A_{j-1} = (-1)^{j+1}$ and $\begin{pmatrix} B_{j-1} & A_{j-1} \\ B_j & A_j \end{pmatrix}^{-1} = (-1)^{j+1} \begin{pmatrix} A_j & -A_{j-1} \\ -B_j & B_{j-1} \end{pmatrix}$, so

$$\begin{pmatrix} 1 \\ kB_{L-1} \end{pmatrix} \equiv (-1)^{j+1} \begin{pmatrix} A_j & -A_{j-1} \\ -B_j & B_{j-1} \end{pmatrix} \begin{pmatrix} 0 \\ 1 \end{pmatrix} \equiv \begin{pmatrix} (-1)^j A_{j-1} \\ (-1)^{j+1} B_{j-1} \end{pmatrix} \pmod D.$$

Hence

$$\begin{cases} A_{j-1} \equiv (-1)^j \pmod D, \\ B_{j-1} \equiv (-1)^{j+1} kB_{L-1} \pmod D. \end{cases}$$

Since by Theorem 4 the congruence $A_{j-1}^2 \equiv (-1)^j Q_j \pmod D$, $j \geq 0$ holds, then under the condition $A_{j-1} \equiv (-1)^j \pmod D$ we obtain $Q_j \equiv (-1)^j \pmod D$. But for $D \geq 6$ the inequality $1 \leq Q_j \leq 2\lfloor\sqrt{D}\rfloor \leq D-2$ holds, so $Q_j = 1$ and $j$ is an even integer. By Theorem 3 we have $Q_j = 1$ if and only if $l \mid j$.

So we get that

$$\begin{cases} A_{j-1} \equiv 1 \pmod{D}, \\ B_{j-1} \equiv -kB_{L-1} \pmod{D}, \\ l \mid j, \\ 0 \leq j \leq L-3, \\ 0 \leq k \leq s-1. \end{cases}$$

Let's consider three possible cases (see Theorem 7).

Case 1. $L = l$. Since $j$ is divisible by $l$ and $0 \leq j \leq L-3 = l-3$, then $j = 0$, so $0 = B_{-1} \equiv -kB_{L-1} \pmod{D}$. Since $s = \frac{D}{\gcd(D, B_{L-1})}$, then the congruence $kB_{L-1} \equiv 0 \pmod{D}$ is equivalent to $s \mid k$. But $0 \leq k \leq s-1$, so $k = 0$.

Case 2. $L = 2l$. Since $j$ is divisible by $l$ and $0 \leq j \leq L-3 = 2l-3$, then $j = 0$, $k = 0$ or $j = l$. If $j = l$, then $A_{l-1} \equiv 1 \pmod{D}$. It's a contradiction with Theorem 7 since $A_{l-1} \not\equiv 1 \pmod{D}$.

Case 3. $L = 4l$. Since $j$ is divisible by $l$ and $0 \leq j \leq L-3 = 4l-3$, then $j = 0$, $k = 0$ or $j = l$, $j = 2l$, $j = 3l$. If $j = lk$, $k = 1, 2, 3$ then $A_{j-1} = A_{lk-1} \equiv 1 \pmod{D}$. It's a contradiction with Theorem 7 since $A_{l-1} \equiv r \pmod{D}$, $A_{2l-1} \equiv -1 \pmod{D}$, $A_{3l-1} \equiv -r \pmod{D}$, where $r^2 \equiv -1 \pmod{D}$.

Therefore, all cases imply $j = 0$, $k = 0$.

For $D = 2, 3, 5$ we need to directly check that $L_B = \frac{D}{\gcd(D, B_{L-1})} \cdot L$ (see examples below).

**2**. If $B_{kL-2} \equiv B_{-1} \equiv 0 \pmod{D}$ and $B_{kL-1} \equiv B_0 \equiv 1 \pmod{D}$ for some $1 \leq k \leq s-1$, then by Corollary 5 we have

$$B_{L-2} + (k-1)B_{L-1}A_{L-2} \equiv 0 \pmod{D},$$
$$kB_{L-1} \equiv 1 \pmod{D},$$

so

$$\begin{cases} B_{L-2} - B_{L-1}A_{L-2} \equiv -A_{L-2} \pmod{D}, \\ kB_{L-1} \equiv 1 \pmod{D}. \end{cases}$$

By Theorem 7 we have $A_{L-1} \equiv 1 \pmod{D}$. Since $A_L = q_L A_{L-1} + A_{L-2}$ and $q_L = 2q_0$, $A_L \equiv A_0 = q_0 \pmod{D}$, then $A_{L-2} \equiv -q_0 \pmod{D}$, so

$$1 = (-1)^L = A_{L-1}B_{L-2} - B_{L-1}A_{L-2} \equiv -A_{L-2} \equiv q_0 \pmod{D}.$$

Hence

$$q_0 \equiv 1 \pmod{D}.$$

Since $q_0 = \lfloor \sqrt{D} \rfloor$, then the congruence $q_0 \equiv 1 \pmod{D}$ implies $q_0 = 1$, so $D = 2$ or $D = 3$. For both cases, the theorem has been verified. □

### 3.2 Properties of Sequences $\{A_i\}_{i=-1}^{\infty}$, $\{B_i\}_{i=-1}^{\infty}$ for Any Prime $p$

**Corollary 7.** *If $p$ is a prime and $L$ is the length of the period of numerators modulo $p$ of the continued fraction's convergents for $\sqrt{p}$, then*

$$B_{L-1} = \begin{cases} B_{l-1} \pmod{p}, & \text{if } p = 2 \text{ or } p \equiv 7 \pmod{8}, \\ B_{2l-1} \equiv -2B_{l-1} \pmod{p}, & \text{if } p \equiv 3 \pmod{8}, \\ B_{4l-1} \equiv -4A_{l-1}B_{l-1} \pmod{p}, & \text{if } p \equiv 1 \pmod{4}. \end{cases}$$

*Proof.* It's follows from Theorem 6 and Lemma 2. □

**Corollary 8.** *If $p$ is a prime and $B_{l-1} \not\equiv 0 \pmod{p}$, then $B_{L-1} \not\equiv 0 \pmod{p}$.*

*Proof.* The corollary immediately follows from Corollary 7 and the congruence $A_{l-1}^2 \equiv (-1)^l \pmod{p}$. □

**Lemma 4.** *If $p$ is a prime, then $p \nmid A_i$ for all $i \geq 0$.*

*Proof.* Indeed, if $D = p$, then by Theorem 4 we have $A_i^2 \equiv (-1)^{i+1}Q_{i+1} \pmod{p}$, so $p \mid A_i$ if and only if $p \mid Q_{i+1}$. Since by Theorem 3 for an odd prime $p$ we have $1 \leq Q_{i+1} \leq 2\lfloor\sqrt{p}\rfloor < p$ for all $i \geq 0$, so $1 \leq Q_{i+1} < p$. Thus, $p \nmid Q_{i+1}$, so $p \nmid A_i$. When $p = 2$, we have $Q_i = 1$ for $i \geq 0$, so $A_i^2 \equiv 1 \pmod{2}$ and $2 \nmid A_i$, $i \geq 0$. □

**Corollary 9.** *Let $p$ be a prime. Then the sequence $\{B_i \bmod p\}_{i=-1}^{\infty}$ is purely periodic with the period length of $pL$, and the following congruence holds for $i \geq 0$, $j \geq -1$*

$$B_{ipL+j} \equiv B_j \pmod{p}.$$

*In particular,*

$$B_{ipL-1} \equiv 0 \pmod{p}, \quad B_{ipL} \equiv 1 \pmod{p}.$$

*Moreover, the periodic part $[B_{-1}, B_0, \ldots, B_{pL-2}]$ of the sequence $\{B_i \bmod p\}_{i=-1}^{\infty}$ can be written as*

$$\begin{pmatrix} 0 & B_0 & \ldots & B_{L-2} \\ B_{L-1} & B_L & \ldots & B_{2L-2} \\ B_{2L-1} & B_{2L} & \ldots & B_{3L-2} \\ \vdots & \vdots & \ldots & \vdots \\ B_{(p-1)L-1} & B_{(p-1)L} & \cdots & B_{pL-2} \end{pmatrix} \equiv \begin{pmatrix} 0 & 1 \\ B_{L-1} & 1 \\ 2B_{L-1} & 1 \\ \vdots & \vdots \\ (p-1)B_{L-1} & 1 \end{pmatrix} \cdot \begin{pmatrix} 1 & A_0 & A_1 & \ldots & A_{L-2} \\ 0 & B_0 & B_1 & \ldots & B_{L-2} \end{pmatrix} \pmod{p}.$$

*If $B_{L-1} \not\equiv 0 \pmod{p}$, then each column of the left matrix is a permutation of elements $\{0, 1, \ldots, p-1\}$.*

*Proof.* These congruences follow from Corollary 5 for $k = ip$, $j = -1$, $j = 0$ ($B_{-1} = 0$, $B_0 = 1$). Since $B_{pL+j} \equiv B_j \pmod{p}$, $j \geq -1$, then $[B_{-1}, B_0, \ldots, B_{pL-2}]$ is the periodic part and it has the length of $pL$.

If two matrix elements $B_{iL+j}$ and $B_{kL+j}$ ($0 \leq i \neq k \leq p-1$) in some column are the same modulo $p$, then by Corollary 5

$$B_j + iB_{L-1}A_j \equiv B_j + kB_{L-1}A_j \pmod{p}.$$

Hence $(i-k)B_{L-1}A_j \equiv 0 \pmod{p}$, so by Lemma 4 we obtain $i - k \equiv 0 \pmod{p}$. It's a contradiction. □

For now, the question remains whether a period length $pL$ of the sequence $\{B_i \bmod p\}_{i=-1}^{\infty}$ is the shortest period. It may turn out that the periodic part $[B_{-1}, B_0, \ldots, B_{pL-2}]$ is divided into smaller periods. From Theorem 8 and Corollary 7 it follows that $L_B = p \cdot L$ if and only if $\gcd(p, B_{L-1}) = \gcd(p, B_{l-1}) = 1$.

Whether $B_{l-1}$ is divisible by $p$ is the key to finding $L_B$. It is well known (e.g., see [5,14,15,17]) that $(A_{l-1}, B_{l-1})$ is the fundamental solution of Pell's equation $x^2 - py^2 = \pm 1$. If $l$ is odd, then $A_{l-1}^2 - pB_{l-1}^2 = -1$. If $l$ is even, then $A_{l-1}^2 - pB_{l-1}^2 = 1$. Note that if $l$ is odd, i.e. $p$ has the form $p \equiv 1 \pmod 4$, the condition $B_{l-1} \not\equiv 0 \pmod p$ is known as the Ankeny-Artin-Chowla (AAC) conjecture (see [1]). If $l$ is even, i.e. $p$ has the form $p \equiv 3 \pmod 4$, the condition $B_{l-1} \not\equiv 0 \pmod p$ is known as the Mordell conjecture (see [12]). Many interesting algebraic results follow from these conjectures. One of these results is the description of canonical matrices of similarity classes over integers of $2 \times 2$ matrices with zero trace and a determinant equal to an odd power of a prime number (see [21]).

**Theorem 9.** *Assuming the Ankeny-Artin-Chowla conjecture and the Mordell conjecture, the shortest period of the sequence $\{B_i \bmod p\}_{i=-1}^{\infty}$ ($p$ is a prime) is equal to*

$$L_B = p \cdot L = \begin{cases} 2, & \text{if } p = 2, \\ pl, & \text{if } p \equiv 7 \pmod 8, \\ 2pl, & \text{if } p \equiv 3 \pmod 8, \\ 4pl, & \text{if } p \equiv 1 \pmod 4. \end{cases}$$

*Proof.* Assuming the hypotheses of Ankeny-Artin-Chowla and Mordell, we have $B_{l-1} \not\equiv 0 \pmod p$, so $B_{L-1} \not\equiv 0 \pmod p$ by Corollary 8. Hence $\gcd(p, B_{L-1}) = 1$ and by Theorem 8 we obtain $L_B = pL$. It remains only to apply Theorem 6. □

We attach some examples of the periodic parts of sequences $\{B_i \bmod p\}_{i=-1}^{\infty}$ for some prime numbers $p$. The periods are marked in bold in the tables.

1. $\sqrt{2} = [1, \overline{2}]$. Here $l = 1$, $L = l = 1$, $B_{L-1} = B_0 = 1$, $\gcd(2, B_{L-1}) = 1$, $L_B = 2L = 2$.

| $n$ | $-1$ | 0 | 1 | 2 |
|---|---|---|---|---|
| $q_n$ | | 1 | **2** | 2 |
| $A_n$ mod 2 | **1** | 1 | 1 | 1 |
| $B_n$ mod 2 | **0** | **1** | 0 | 1 |

$$\begin{pmatrix} 0 \\ 1 \end{pmatrix} \equiv \begin{pmatrix} 0 & 1 \\ 1 & 1 \end{pmatrix} \cdot \begin{pmatrix} 1 \\ 0 \end{pmatrix} \pmod 2.$$

2. $\sqrt{3} = [1, \overline{1, 2}]$. Here $l = 2$, $L = 2l = 4$, $B_{L-1} = B_3 \equiv 1 \pmod 3$, $\gcd(3, B_{L-1}) = 1$, $L_B = 3L = 12$.

| $n$ | $-1$ | 0 | 1 | 2 | 3 | 4 | 5 | 6 | 7 | 8 | 9 | 10 | 11 | 12 |
|---|---|---|---|---|---|---|---|---|---|---|---|---|---|---|
| $q_n$ | | 1 | **1** | **2** | 1 | 2 | 1 | 2 | 1 | 2 | 1 | 2 | 1 | 2 |
| $A_n$ mod 3 | **1** | **1** | **2** | **2** | 1 | 1 | 2 | 2 | 1 | 1 | 2 | 2 | 1 | 1 |
| $B_n$ mod 3 | **0** | **1** | **1** | **0** | **1** | **2** | **0** | **2** | **2** | **0** | **2** | **1** | 0 | 1 |

$$\begin{pmatrix} 0 & 1 & 1 & 0 \\ 1 & 2 & 0 & 2 \\ 2 & 0 & 2 & 1 \end{pmatrix} \equiv \begin{pmatrix} 0 & 1 \\ 1 & 1 \\ 2 & 1 \end{pmatrix} \cdot \begin{pmatrix} 1 & 1 & 2 & 2 \\ 0 & 1 & 1 & 0 \end{pmatrix} \pmod 3.$$

3. $\sqrt{5} = [2, \overline{4}]$. Here $l = 1$, $L = 4l = 4$, $B_{L-1} = B_3 \equiv 2 \pmod 5$, $\gcd(5, B_{L-1}) = 1$, $L_B = 5L = 20$.

| $n$ | −1 | 0 | 1 | 2 | 3 | 4 | 5 | 6 | 7 | 8 | 9 | 10 | 11 | 12 | 13 | 14 | 15 | 16 | 17 | 18 | 19 | 20 |
|---|---|---|---|---|---|---|---|---|---|---|---|---|---|---|---|---|---|---|---|---|---|---|
| $q_n$ | | 2 | 4 | 4 | 4 | 4 | 4 | 4 | 4 | 4 | 4 | 4 | 4 | 4 | 4 | 4 | 4 | 4 | 4 | 4 | 4 | 4 |
| $A_n$ mod 5 | 1 | 2 | 4 | 3 | 1 | 2 | 4 | 3 | 1 | 2 | 4 | 3 | 1 | 2 | 4 | 3 | 1 | 2 | 4 | 3 | 1 | 2 |
| $B_n$ mod 5 | 0 | 1 | 4 | 2 | 2 | 0 | 2 | 3 | 4 | 4 | 0 | 4 | 1 | 3 | 3 | 0 | 3 | 2 | 1 | 1 | 0 | 1 |

$$\begin{pmatrix} 0\,1\,4\,2 \\ 2\,0\,2\,3 \\ 4\,4\,0\,4 \\ 1\,3\,3\,0 \\ 3\,2\,1\,1 \end{pmatrix} \equiv \begin{pmatrix} 0\,1 \\ 2\,1 \\ 4\,1 \\ 1\,1 \\ 3\,1 \end{pmatrix} \cdot \begin{pmatrix} 1\,2\,4\,3 \\ 0\,1\,4\,2 \end{pmatrix} \pmod 5.$$

4. $\sqrt{7} = [2, \overline{1, 1, 1, 2}]$. Here $L = l = 4$, $B_{L-1} = B_3 \equiv 3 \pmod 7$, $\gcd(7, B_{L-1}) = 1$, $L_B = 7L = 28$.

| $n$ | −1 | 0 | 1 | 2 | 3 | 4 | 5 | 6 | 7 | 8 | 9 | 10 | 11 | 12 | 13 | 14 | 15 | 16 | 17 | 18 | 19 | 20 | 21 | 22 | 23 | 24 | 25 | 26 |
|---|---|---|---|---|---|---|---|---|---|---|---|---|---|---|---|---|---|---|---|---|---|---|---|---|---|---|---|---|
| $q_n$ | | 2 | 1 | 1 | 1 | 2 | 1 | 1 | 1 | 2 | 1 | 1 | 1 | 2 | 1 | 1 | 1 | 2 | 1 | 1 | 1 | 2 | 1 | 1 | 1 | 2 | 1 | 1 |
| $A_n$ mod 7 | 1 | 2 | 3 | 5 | 1 | 2 | 3 | 5 | 1 | 2 | 3 | 5 | 1 | 2 | 3 | 5 | 1 | 2 | 3 | 5 | 1 | 2 | 3 | 5 | 1 | 2 | 3 | 5 |
| $B_n$ mod 7 | 0 | 1 | 1 | 2 | 3 | 0 | 3 | 3 | 6 | 6 | 5 | 4 | 2 | 5 | 0 | 5 | 5 | 4 | 2 | 6 | 1 | 3 | 4 | 0 | 4 | 2 | 6 | 1 |

$$\begin{pmatrix} 0\,1\,1\,2 \\ 3\,0\,3\,3 \\ 6\,6\,5\,4 \\ 2\,5\,0\,5 \\ 5\,4\,2\,6 \\ 1\,3\,4\,0 \\ 4\,2\,6\,1 \end{pmatrix} \equiv \begin{pmatrix} 0\,1 \\ 3\,1 \\ 6\,1 \\ 2\,1 \\ 5\,1 \\ 1\,1 \\ 4\,1 \end{pmatrix} \cdot \begin{pmatrix} 1\,2\,3\,5 \\ 0\,1\,1\,2 \end{pmatrix} \pmod 7.$$

# 4 Related Works and Further Research

## 4.1 Usage in Image Processing

One possible use of our results is image processing. Consider a discrete image $p-1$ by $p-1$ pixels, where $p$ is prime. Let the set $S = \{0, 1, \dots, p-1\} \times \{0, 1, \dots, p-1\}$ denote the coordinates of the pixels, where $\times$ is the cartesian product. Thus, a vector $v = (x, y) \in S$ represents the position of the pixel. Let $M = \begin{pmatrix} a\,b \\ c\,d \end{pmatrix} \in \mathrm{GL}(2, \mathbb{Z})$ be a unimodular matrix transformation. Repetitively apply this transformation by the rule:

$$v_{\text{new}} = M \cdot v_{\text{old}} \pmod p = \begin{pmatrix} a\,b \\ c\,d \end{pmatrix} \cdot \begin{pmatrix} x \\ y \end{pmatrix} \pmod p = \begin{pmatrix} (ax+by) \mod p \\ (cx+dy) \mod p \end{pmatrix}.$$

Now it follows from our theorems (under the assumption of hypotheses) that there exists $M$ for which the exact minimal number of repetitions until the image comes back is equal to $p, 2p$ or $4p$ depending on the type of $p$. According to Lemma 1, Corollaries 2, 3, 4 we have for $j \geq -1$, $k \geq 0$:

$$\begin{pmatrix} A_{j+l} \\ B_{j+l} \end{pmatrix} \equiv \begin{pmatrix} A_{l-1} & 0 \\ B_{l-1} & A_{l-1} \end{pmatrix} \cdot \begin{pmatrix} A_j \\ B_j \end{pmatrix} \pmod p,$$

$$\begin{pmatrix} A_{j+kl} \\ B_{j+kl} \end{pmatrix} \equiv \begin{pmatrix} A_{l-1} & 0 \\ B_{l-1} & A_{l-1} \end{pmatrix}^k \cdot \begin{pmatrix} A_j \\ B_j \end{pmatrix} \equiv \begin{pmatrix} A_{l-1}^k & 0 \\ kA_{l-1}^{k-1} B_{l-1} & A_{l-1}^k \end{pmatrix} \cdot \begin{pmatrix} A_j \\ B_j \end{pmatrix} \pmod p,$$

$$\begin{pmatrix} A_{kl-1} \\ B_{kl-1} \end{pmatrix} \equiv \begin{pmatrix} A_{l-1} & 0 \\ B_{l-1} & A_{l-1} \end{pmatrix}^k \cdot \begin{pmatrix} 1 \\ 0 \end{pmatrix} \equiv \begin{pmatrix} A_{l-1}^k \\ kA_{l-1}^{k-1} B_{l-1} \end{pmatrix} \pmod p.$$

In Theorem 9 we have proved that the minimal $k \geq 1$, such that:

$$\begin{pmatrix} A_{kl-1} \mod p \\ B_{kl-1} \mod p \end{pmatrix} = \begin{pmatrix} 1 \\ 0 \end{pmatrix}$$

is equal to $p$, $2p$ or $4p$ depending on the type of $p$. Therefore, the order of the matrix $M = \begin{pmatrix} A_{l-1} & 0 \\ B_{l-1} & A_{l-1} \end{pmatrix}$ as an element of the matrix group $\mathrm{GL}(2, \mathbb{Z}/p\mathbb{Z})$ is equal to $p$, $2p$ or $4p$ depending on the type of $p$. Namely, $M^p \equiv E \pmod p$ for $p = 2$ or $p \equiv 7 \pmod 8$; $M^{2p} \equiv E \pmod p$ for $p \equiv 3 \pmod 8$; $M^{4p} \equiv E \pmod p$ for $p \equiv 1 \pmod 4$.

In general, the map described above is chaotic and is a discrete version of so-called Arnold's cat map (see [6] for an introduction). Chaotic properties of such maps find their usage in image encryption (see [4,26]).

### 4.2 Further Research

Theorem 9 implies that if $\frac{L_B}{L} \neq D$, then $D$ can not be prime (of course, if the Ankeny-Artin-Chowla conjecture and the Mordell conjecture are true). We attach a table of the first 26 not prime and not square $D$ with their ratios $\frac{L_B}{L}$:

| $D$ | 6 | 8 | 10 | 12 | 14 | 15 | 18 | 20 | 21 | 22 | 24 | 26 | 27 | 28 | 30 | 32 | 33 | 34 | 35 | 38 | 39 | 40 | 42 | 44 | 45 | 46 |
|---|---|---|---|---|---|---|---|---|---|---|---|---|---|---|---|---|---|---|---|---|---|---|---|---|---|---|
| $L/l$ | 2 | 2 | 4 | 2 | 1 | 2 | 2 | 2 | 2 | 2 | 2 | 4 | 2 | 2 | 2 | 2 | 2 | 1 | 2 | 2 | 2 | 2 | 2 | 2 | 2 | 1 |
| $L_B/l$ | 6 | 8 | 20 | 6 | 7 | 30 | 18 | 10 | 14 | 22 | 24 | 52 | 54 | 14 | 30 | 32 | 66 | 17 | 70 | 38 | 78 | 40 | 42 | 22 | 30 | 1 |
| $L_B/L$ | 3 | 4 | 5 | 3 | 7 | 15 | 9 | 5 | 7 | 11 | 12 | 13 | 27 | 7 | 15 | 16 | 33 | 17 | 35 | 19 | 39 | 20 | 21 | 11 | 15 | 1 |

Numbers for which $L_B = L$, such as 46 in the table above, are special in the sense that for them Mordell's and Ankeny-Artin-Chowla's conjectures don't work (i.e. $B_{l-1} \equiv 0 \pmod D$). The sequence of numbers $D$ for which the condition $B_{l-1} \equiv 0 \pmod D$ is satisfied is $46, 430, 1817, 58254, 209991, 1752299, 3124318, \ldots$ (see [23]). They arise in a search for three consecutive powerful numbers (a powerful number is the product of a square and a cube). For more details see [16,24].

## References

1. Ankeny, N., Artin, E., Chowla, S.: The class-number of real quadratic number fields. Ann. Math. **56**(3), 479–493 (1952)
2. Appleby, M., Flammia, S., McConnell, G., Yard, J.: SICs and algebraic number theory. Found. Phys. **47**(8), 1042–1059 (2017)
3. Bengtsson, I.: The number behind the simplest SIC-POVM. Found. Phys. **47**(8), 1031–1041 (2017)
4. Chen, G., Mao, Y., Chui, C.K.: A symmetric image encryption scheme based on 3d chaotic cat maps. Chaos, Solitons Fractals **21**(3), 749–761 (2004)
5. Davenport, H.: The Higher Arithmetic: An Introduction to the Theory of Numbers. Cambridge University Press, Cambridge (1999)
6. Hariyanto, E., Rahim, R.: Arnold's cat map algorithm in digital image encryption. Int. J. Sci. Res. (IJSR) **5**(10), 1363–1365 (2016)
7. Khinchin, A.Y., Teichmann, T.: Continued fractions. American Institute of Physics (1964)
8. Khovansky, A.: The Application of Continued Fractions and Their Generalizations to Problems in Approximation Theory. P. Noordhoff, Library of applied analysis and computational mathematics (1963)
9. Lang, S.: Introduction to Diophantine Approximations. Springer, Cham (1995)
10. Litvinov, G., Rodionov, A., Chourkin, A.: Approximate rational arithmetics and arbitrary precision computations. arXiv: preprint math/0101152 (2001)
11. Matthews, K.: The diophantine equation $x^2 - dy^2 = n, d > 0$. Expo. Math. **18**(4), 323–332 (2000)
12. Mordell, L.: On a Pellian equation conjecture (II). J. Lond. Math. Soc. **1**(1), 282–288 (1961)
13. Niven, I., Zuckerman, H.S., Montgomery, H.L.: An Introduction to the Theory of Numbers. Wiley & Sons, Hoboken (1991)
14. Olds, C.: Continued Fractions. Random House (1963)
15. Perron, O.: Die Lehre von den Kettenbrüchen: Band I: Elementare Kettenbrüche. Springer-Verlag, Cham (2013)
16. Reinhart, A.: On orders in quadratic number fields whose set of distances is peculiar. arXiv preprint arXiv:2305.09267 (2023)
17. Rockett, A., Szusz, P.: Continued fractions. World Scientific (1992)
18. Rosen, K.H.: Elementary Number Theory. Pearson Education, London (2011)
19. Sarma, K., Avadhani, P.: Public key cryptosystem based on Pell's equation using the GNU Mp library. Int. J. Comput. Sci. Eng. **3**(2), 739–743 (2011)
20. Seidensticker, R.B.: Continued fractions for high-speed and high-accuracy computer arithmetic. In: 1983 IEEE 6th Symposium on Computer Arithmetic (ARITH), pp. 184–193. IEEE (1983)
21. Sidorov, S.V.: On similarity classes of second order matrices with zero trace over the ring of integers. Russ. Math. **60**(4), 65–71 (2016)
22. Sidorov, S.V., Shcherbakov, P.A.: On the period length modulo $p$ of the numerators of convergents for the square root of a prime number $p$. In: Balandin, D., Barkalov, K., Meyerov, I. (eds.) MMST 2022, vol. 1750, pp. 136–147. Springer, Cham (2022). https://doi.org/10.1007/978-3-031-24145-1_11
23. Sloane, N.: The On-Line Encyclopedia of Integer Sequences, A135735 (2023). http://oeis.org/A135735
24. Stephens, A.J., Williams, H.C.: Some computational results on a problem concerning powerful numbers. Math. Comput. **50**(182), 619–632 (1988)

25. Vuillemin, J.: Exact real computer arithmetic with continued fractions. In: Proceedings of the 1988 ACM Conference on LISP and Functional Programming, pp. 14–27 (1988)
26. Ye, G., Wong, K.W.: An efficient chaotic image encryption algorithm based on a generalized Arnold map. Nonlinear Dyn. **69**(4), 2079–2087 (2012)

# Linearizing Algorithm for Solving a Nonlinear Initial Boundary Value Problem of Radiation Transfer in Spatially Multidimensional Domains

Aleksey Busalov[1], Aleksey Kalinin[1,2,3(✉)], and Alla Tyukhtina[1,2]

[1] National Research Lobachevsky State University, 23 Gagarin Avenue, 603022 Nizny Novgorod, Russian Federation
avk@mm.unn.ru

[2] Institute of Applied Physics, Russian Academy of Sciences, 46 Ulyanova Street, 603950 Nizhny Novgorod, Russian Federation

[3] National Research University Higher School of Economics, Moscow, Russian Federation

**Abstract.** The present work is devoted to the study of nonlinear integro-differential equations of the theory of radiation transfer and statistical equilibrium. Previously developed methods for solving the corresponding one-dimensional boundary value and initial boundary value problems are applied to problems of higher dimension. The issues of the existence and uniqueness of a non-negative solution of the corresponding nonlinear problems in bounded domains in spatially two-dimensional and three-dimensional cases are considered. The paper proposes and substantiates a linearizing iterative algorithm for solving nonlinear problems. The results of numerical implementation of this algorithm in the form of software modules within the framework of the finite element method and the finite difference method are presented. An integro-interpolation method is used for finite-difference approximation of the kinetic transfer equation, isoparametric elements are used in the finite element implementation. The efficiency and effectiveness of the proposed algorithm is numerically illustrated on model problems for specific media under various assumptions about the optical density of matter. The relations for obtaining Einstein coefficients in the media under consideration are discussed in detail.

**Keywords:** System of radiation transfer equations · Nonlinear integro-differential equations · Iterative method · Finite element method · Integro-interpolation method

## 1 Introduction

Nonlinear problems of interaction of radiation with matter in the absence of local thermodynamic equilibrium are one of the important problems of modern

The work was supported by the Russian Science Foundation (project 23-71-30008).


D. Balandin et al. (Eds.): MMST 2023, CCIS 1914, pp. 44–58, 2024.
https://doi.org/10.1007/978-3-031-52470-7_4

mathematical physics, which has numerous applications. Mathematical models of these phenomena are described by nonlinear systems of integro-differential equations, including the kinetic equation of radiation transfer and equations characterizing the behavior of the medium [1–5]. Mathematical issues and numerous applications of linear transport theory are reflected in [6–14]. Nonlinear problems of the theory of radiation transfer are quite diverse and are determined by the specifics of the simulated physical processes [2,3,5,15–17]. Among the main methods for numerical solving the transport equation, one can highlight the finite-difference method, the finite element method, the method of spherical harmonics, the method of characteristics, and Monte Carlo methods [5,9,11,18–23]. The complexity of numerical modeling of the processes of transfer and interaction of radiation with matter is associated with the high dimension of discrete equations obtained by approximating the original problem.

In the case of stationary media, a nonlinear system of integro-differential equations of radiation transfer and statistical equilibrium is used to describe the interaction of radiation with matter, an important special case of which is the approximation of a two-level atom [3]. The two-level atom model reflects important substantive problems, and can also be considered as an integral part of the study of more general problems for a multi-level atom. Among the main formulations for integro-differential equations of radiation transfer within the framework of a two-level approximation, it is possible to distinguish stationary, non-stationary and quasi-stationary problems, as well as problems related to the specifics of dimension. In particular, the features of mathematical research and numerical implementation in a plane-parallel layer are discussed in [8].

The numerical solution of the nonlinear problems under consideration involves the construction of iterative linearizing algorithms. In [24,25] the iterative linearizing algorithm was proposed and justified for a nonstationary nonlinear system of integro-differential equations of the theory of radiative transfer and statistical equilibrium for the general model of a multi-level atom in three-dimensional space. In [26,27] the well-posedness of initial-boundary nonstationary problems for a two-level atom is proved, the properties of solutions to these problems are studied. For stationary problems, the linearization iterative algorithm was investigated in [28,29]. The issues of correctness and stabilization of the solution for the quasi-stationary model of a two-level atom were considered in [30].

This paper provides the justification of the convergence of a linearizing iterative algorithm for a nonlinear system of radiative transfer equations in a quasi-stationary approximation. Theoretical research essentially uses methods of the theory of ordered spaces. The proposed algorithm is numerically implemented as a software module, including the finite element method with isoparametric elements [23] and the finite difference method [19]. For finite-difference approximation of the kinetic transport equation, the integro-interpolation method is used.

## 2 Linearizing Algorithm. Theoretical Analysis

The nonlinear quasi - stationary system of integro-differential equations of radiation transfer and statistical equilibrium [30]

$$\frac{1}{c}\frac{\partial\varphi(x,\nu,\omega,t)}{\partial t}+(\omega,\nabla)\varphi(x,\nu,\omega,t)$$

$$+h\nu_{12}\frac{\kappa(\nu)}{4\pi}\left(B_{12}C_1(x,t)-B_{21}C_2(x,t)\right)\varphi(x,\nu,\omega,t)=h\nu_{12}\frac{\kappa(\nu)}{4\pi}A_{21}C_2(x,t), \tag{1}$$

$$\left(C_{12}n_e(x)+B_{12}\int_I\int_\Omega\frac{\kappa(\nu)}{4\pi}\varphi(x,\nu,\omega,t)d\omega d\nu\right)C_1(x,t)$$

$$=C_2(x,t)\left(A_{21}+C_{21}n_e(x)+B_{21}\int_I\int_\Omega\frac{\kappa(\nu)}{4\pi}\varphi(x,\nu,\omega,t)d\omega d\nu\right), \tag{2}$$

$$C_1(x,t)+C_2(x,t)=f(x) \tag{3}$$

is considered to determine the intensity of radiation $\varphi(x,\nu,\omega,t)$, which propagates in the direction $\omega$ with frequency $\nu$, and the concentrations $C_1(x,t)$ and $C_2(x,t)$ of atoms in the ground and excited states, respectively. System (1)–(3) corresponds to the model of a two-level atom with the complete redistribution of radiation in frequency [3,4].

Here $x=\{x_1,x_2,x_3\}\in G\subset \mathrm{R}^3$, $\omega\in\Omega=\{\omega=\{\omega_1,\omega_2,\omega_3\}\in\mathrm{R}^3:\omega_1^2+\omega_2^2+\omega_3^2=1\}$; $\nu\in I=[0,\nu_0]$; $t\in[0,T]$, $\nu_0>0$, $T>0$; $G$ is a convex bounded set with a diameter $d>0$ and with a smooth boundary $\partial G$, at each point $x\in\partial G$ there is a unit vector of the external normal $n(x)$. The system (1)–(3) is considered under the boundary and initial conditions

$$\varphi(x,\nu,\omega,t)=0,\quad x\in\partial G,\ (\omega,n(x))<0,\ \omega\in\Omega,\ \nu\in I,\ t\in[0,T], \tag{4}$$

$$\varphi(x,\nu,\omega,0)=\varphi_0(x,\nu,\omega),\ x\in G,\ \nu\in I,\ \omega\in\Omega. \tag{5}$$

The boundary condition (4) means that there is no external radiation flux incident on the boundary of the region.

In (1)–(3) $h$ is Planck's constant, $c$ is the speed of light, $\nu_{12}$ is the frequency of transition between levels, $A_{21}$, $B_{12}$, $B_{21}$, $C_{12}$, $C_{21}$ are Einstein coefficients, $n_e(x)$ is the electron concentration, $f(x)$, $x\in G$, is the density of atoms, $\kappa(\nu)$, $\nu\in I$, is the scattering indicatrix [2,3]. All of the above are constants are positive and

$$B_{12}C_{21}-B_{21}C_{12}>0; \tag{6}$$

the functions $n_e$, $f$, $\kappa$ are measurable and non-negative almost everywhere in their domains,

$$\operatorname{esssup} n_e(x)=n_e^*<\infty,\ \operatorname{esssup} f(x)=f^*<\infty,$$

$$\operatorname{esssup}\kappa(\nu)=\kappa^*<\infty,\ \int_I\kappa(\nu)d\nu=1. \tag{7}$$

The family of characteristics $\{l_\omega\}$ of a differential operator $\partial/c\partial t + (\omega, \nabla)$ is determined by a system of equations

$$c\,dt = \frac{dx_1}{\omega_1} = \frac{dx_2}{\omega_2} = \frac{dx_3}{\omega_3} = \frac{d\omega_1}{0} = \frac{d\omega_2}{0} = \frac{d\omega_3}{0} = \frac{d\nu}{0},$$

$1/c\,(d/d\tau)_\omega$ denotes the operator of differentiation along the characteristic $\{l_\omega\}$:

$$\frac{1}{c}\left(\frac{d}{d\tau}\right)_\omega \varphi(x,\nu,\omega,t) = \frac{1}{c}\left.\frac{d\varphi(x+c\omega(\tau-t),\nu,\omega,\tau)}{d\tau}\right|_{\tau=t}.$$

Let $D = G \times I \times \Omega$, $K_\infty(\Pi)$ denotes the cone of non-negative functions from $L_\infty(\Pi)$, where $\Pi \subset \mathrm{R}^m$ is an arbitrary measurable set; $\mathcal{D}_\infty(D \times [0,T])$ is the space of functions $\psi \in L_\infty(D \times [0,T])$ that absolutely continuous along almost every characteristic $\{l_\omega\}$ and

$$\frac{1}{c}\left(\frac{d}{d\tau}\right)_\omega \psi \in L_\infty(D \times [0,T]),$$

$$\mathcal{M}_T = \mathcal{D}_\infty(D \times [0,T]) \times L_\infty(G \times [0,T]) \times L_\infty(G \times [0,T])),$$
$$\mathcal{K}_T = K_\infty(D \times [0,T]) \times K_\infty(G \times [0,T]) \times K_\infty(G \times [0,T]).$$

Similar classes were first introduced in the work of V.S. Vladimirov [6] for stationary problems of transfer theory, in [27,30] these functional spaces were used in the study of non-stationary problems.

Let $\varphi_0 \in K_\infty(D)$. The solution of problem (1)–(5) is the function $\Phi = \{\varphi, C_1, C_2\} \in \mathcal{M}_T$, satisfying system (1)–(3) and initial and boundary conditions (4), (5) almost everywhere, the differential operator in (1) is considered as the operator of differentiation along the characteristics of $\{l_\omega\}$.

**Theorem 1.** *Let the conditions (6), (7) be fulfilled. Then for any $\varphi_0 \in K_\infty(D)$ problem (1)–(5) has a unique solution $\Phi \in M_T$. It continuously depends on $\varphi_0$ and there is the inclusion $\Phi \in \mathcal{M}_T \cap \mathcal{K}_T$.*

Theorem 1 is proved in [30].

We use the following notations, similar to those were introduced in [15,27] when studying other classes of nonstationary problems of the theory of radiation transfer.

$$R_{12}(\psi)(x,t) = C_{12}n_e(x) + B_{12}J(\psi)(x,t),$$
$$R_{21}(\psi)(x,t) = A_{21} + C_{21}n_e(x) + B_{21}J(\psi)(x,t),$$
$$R(\psi) = R_{12}(\psi)(x,t) + R_{21}(\psi)(x,t),$$
$$J(\psi)(x,t) = \int_I\int_\Omega \frac{\kappa(\nu)}{4\pi}\psi(x,\nu,\omega,t)\,d\omega d\nu,$$
$$F(C)(x,t) = h\nu_{12}(B_{12}C_1(x,t) - B_{21}C_2(x,t)),$$
$$P(C)(x,t) = h\nu_{12}A_{21}C_2(x,t).$$

System (1)–(3) takes the form

$$\frac{1}{c}\left(\frac{d}{d\tau}\right)_{\omega}\varphi(x,\nu,\omega,t)+\frac{\kappa(\nu)}{4\pi}F(C)(x,t)\varphi(x,\nu,\omega,t)=\frac{\kappa(\nu)}{4\pi}P(C)(x,t), \tag{8}$$

$$C_1(x,t)=f(x)\frac{R_{21}(\varphi)(x,t)}{R(\varphi)(x,t)},\ C_2(x,t)=f(x)\frac{R_{12}(\varphi)(x,t)}{R(\varphi)(x,t)}. \tag{9}$$

The function $\varphi\in D_\infty(\mathcal{D}\times[0,T])$ is the solution of problem (8), (4), (5), where $C_1$, $C_2$ are determined by (9), if and only if $\Phi=\{\varphi,C_1,C_2\}\in M_T$ is a solution to problem (1)–(5).

To solve problem (1)–(5), the following linearizing iterative algorithm is proposed. Let $\varphi^0\in K_\infty(D)$. For $k=0,1,...$ we define $C_1^k$, $C_2^k$ by (9), where $\psi=\varphi^k$, that is

$$C_1^k(x,t)=\frac{A_{21}+C_{21}n_e(x)+B_{21}J(\varphi^k)(x,t)}{A_{21}+(C_{12}+C_{21})n_e(x)+(B_{12}+B_{21})J(\varphi^k)(x,t)}f(x), \tag{10}$$

$$C_2^k(x,t)=\frac{C_{12}n_e(x)+B_{12}J(\varphi^k)(x,t)}{A_{21}+(C_{12}+C_{21})n_e(x)+(B_{12}+B_{21})J(\varphi^k)(x,t)}f(x); \tag{11}$$

then function $\varphi^{k+1}\in\mathcal{D}_\infty(D\times[0,T])$ is defined as the solution of the problem

$$\frac{1}{c}\left(\frac{d}{d\tau}\right)_{\omega}\varphi^{k+1}(x,\nu,\omega,t)+h\nu_{12}\frac{\kappa(\nu)}{4\pi}[B_{12}C_1^k(x,t)$$

$$-B_{21}C_2^k(x,t)]\varphi^{k+1}(x,\nu,\omega,t)=h\nu_{12}\frac{\kappa(\nu)}{4\pi}A_{21}C_2^k(x,t), \tag{12}$$

$$\varphi^{k+1}(x,\nu,\omega,0)=\varphi_0(x,\nu,\omega).$$

**Theorem 2.** *For any $\varphi^0\in K_\infty(D\times[0,T])$ and $1\le p<\infty$ the iterative process (12) converges in $L_p(D\times[0,T])$ to some function $\varphi$, and $\{\varphi,C_1,C_2\}\in\mathcal{M}_T\cap\mathcal{K}_T$, where $C_1$, $C_2$ determined by (9), is a solution of problem (1)–(5).*

*Proof.* Denote for $\psi\in K_\infty(D\times[0,T])$

$$F(\psi)(x,t)=h\nu_{12}f(x)\frac{A_{21}B_{12}+(B_{12}C_{21}-B_{21}C_{12})n_e(x)}{A_{21}+(C_{12}+C_{21})n_e(x)+(B_{12}+B_{21})J(\psi)(x,t)}, \tag{13}$$

$$P(\psi)(x,t)=h\nu_{12}f(x)A_{21}\frac{C_{12}n_e(x)+B_{12}J(\psi)(x,t)}{A_{21}+(C_{12}+C_{21})n_e(z)+(B_{12}+B_{21})J(\psi)(x,t)}. \tag{14}$$

Consider the problem of determining the function $\varphi\in\mathcal{D}_\infty(D\times[0,T])$, satisfying the equation

$$\frac{1}{c}\left(\frac{d}{d\tau}\right)_{\omega}\varphi(x,\nu,\omega,t)+\frac{\kappa(\nu)}{4\pi}F(\psi)(x,t)\varphi(x,\nu,\omega,t)=\frac{\kappa(\nu)}{4\pi}P(\psi)(x,t) \tag{15}$$

and the initial condition

$$\varphi(x,\nu,\omega,0)=\varphi_0(x,\nu,\omega),\ (x,\nu,\omega)\in D. \tag{16}$$

Let $\{l_\omega\}=\{(x+c\omega(\tau-t),\nu,\omega,\tau),\tau\in \mathrm{R}\}$ is the characteristic passing at the moment $\tau=t$ through the point $(x,\nu,\omega)$. Denote by $t_\omega^-(x,t)$ the first moment when the characteristic intersects the boundary of the region $\mathcal{D}\times[0,T]$. Then either $t_\omega^-(x,t)=0$, either this intersection occurs in the part of $\partial G\times I\times\Omega\times[0,T]$, where $(\omega,n)<0$. Let

$$g_0(x,\nu,\omega,t)=\varphi(x+c\omega(t_\omega^-(x,t)-t),\nu,\omega,t_\omega^-(x,t))$$
$$\equiv\begin{cases}0, & t_\omega^-(x,t)>0\\ \varphi_0(x-c\omega t,\nu,\omega), & t_\omega^-(x,t)=0\end{cases},$$

then the solution of Eq. (15) has the form

$$\varphi(x,\nu,\omega,t)=g_0(x,\nu,\omega,t)\exp\left\{-\int_{t_\omega^-(x,t)}^{t}\left(c\frac{\kappa(\nu)}{4\pi}F(\psi)(x+c\omega(s-t),s)\right)ds\right\}$$
$$+\int_{t_\omega^-(x,t)}^{t}c\frac{\kappa(\nu)}{4\pi}P(\psi)(x+c\omega(\tau-t),\tau)$$
$$\times\exp\left\{\int_t^{\tau}\left(c\frac{\kappa(\nu)}{4\pi}F(\psi)(x+c\omega(s-t),s)\right)ds\right\}d\tau. \tag{17}$$

It follows from (17) that $\varphi\in\mathcal{D}_\infty(D\times[0,T])\cap K_\infty(D\times[0,T])$. Let $M=\max\{A_{21}C_{12}/(B_{12}C_{21}-B_{21}C_{12}),\|\varphi_0\|_{L_\infty(D)}\}$. Then if $\|\psi\|_{L_\infty(D\times[0,T])}\le M$, for almost all $(x,\nu,\omega,t)\in D\times[0,T]$

$$\frac{P(\psi)(x,t)}{F(\psi(x,t))}=\frac{A_{21}(C_{12}n_e(x)+B_{12}J(\psi)(x,t))}{A_{21}B_{12}+(B_{12}C_{21}-B_{21}C_{12})n_e(x)}\le M,$$

$$\varphi(x,\nu,\omega,t)\le M\exp\left\{-\int_{t_\omega^-(x,t)}^{t}\left(c\frac{\kappa(\nu)}{4\pi}F(\psi)(x+c\omega(s-t),s)\right)ds\right\}$$
$$+M\int_{t_\omega^-(x,t)}^{t}c\frac{\kappa(\nu)}{4\pi}F(\psi)(x+c\omega(\tau-t),\tau)$$
$$\times\exp\left\{\int_t^{\tau}\left(c\frac{\kappa(\nu)}{4\pi}F(\psi)(x+c\omega(s-t),s)\right)ds\right\}d\tau=M. \tag{18}$$

Define the operator $A:K_\infty(D\times[0,T])\to K_\infty(D\times[0,T])$ as follows: for $\psi\in K_\infty(D\times[0,T])$ let $A\psi=\varphi$ is the solution of problem (15), (16). Then the solution of problem (8), (4), (5) is the fixed point of the operator $A$.

Let $\psi_1\succ\psi_2$ for $\psi_1,\psi_2\in K_\infty(D\times[0,T])$ if and only if almost everywhere in $D\times[0,T]$ $\psi_1(x,\nu,\omega,t)\ge\psi_2(x,\nu,\omega,t)$. Then $K_\infty(D\times[0,T])$ with a partial order relation $\succ$ is a conditionally complete lattice [31] and

$$K_0=\{\psi\subset K_\infty(D\times[0,T]):\psi\prec M\},$$

is a complete sublattice. It follows from (18) that $A\psi\in K_0$ for all $\psi\in K_0$.

Let $\psi_1$, $\psi_2 \in K_0$, $\psi_1 \succ \psi_2$, $\varphi_1 = A\psi_1$, $\varphi_2 = A\psi_2$. We denote $\psi = \psi_1 - \psi_2$, $\varphi = \varphi_1 - \varphi_2$. Then

$$\frac{1}{c}\left(\frac{d}{d\tau}\right)_\omega \varphi + \frac{\kappa(\nu)}{4\pi}F(\psi_1)\varphi = \frac{\kappa(\nu)}{4\pi}\left(P(\psi_1) - P(\psi_2) + (F(\psi_2) - F(\psi_1))\varphi_2\right), \tag{19}$$

$$\varphi(x,\nu,\omega,0) = 0,\ (x,\nu,\omega) \in D. \tag{20}$$

Thus,

$$\varphi(x,\nu,\omega,t)$$
$$= \int_{t_\omega^-(x,t)}^{t} c\frac{\kappa(\nu)}{4\pi}S(x + c\omega(\tau - t),\nu,\omega,\tau)\cdot$$
$$\exp\left\{\int_t^\tau \left(c\frac{\kappa(\nu)}{4\pi}F(\psi_1)(x + c\omega(s-t),s)\right)ds\right\}d\tau,$$

where

$$S(x,\nu,\omega,t) = (P(\psi_1)(x,t) - P(\psi_2)(x,t) + (F(\psi_2)(x,t) - F(\psi_1)(x,t))\varphi_2(x,\nu,\omega,t).$$

Therefore, $A\psi_1 \succ A\psi_2$, that is, the operator $A$ is isotone.

For almost all $(x,\nu,\omega,t) \in D \times [0,T]$

$$S(x,\nu,\omega,t)$$
$$\leq h\nu_{12}f^*\frac{(A_{21}B_{12} + (B_{12}C_{21} - B_{21}C_{12})n_e^*)(A_{21} + (B_{12} + B_{21})M)}{A_{21}^2}J(\psi)(x,t)$$
$$\leq C\int_I\int_\Omega \psi(x,\nu,\omega,t)d\omega d\nu,$$

$$\varphi(x,\nu,\omega,t) \leq C\frac{c\kappa(\nu)}{4\pi}\int_{t_\omega^-(x,t)}^{t}\int_I\int_\Omega \psi(x + c\omega(\tau - t),\nu_1,\omega_1,\tau)d\omega_1 d\nu_1 d\tau,$$

$$\int_G \varphi(x,\nu,\omega,t)dx \leq C\frac{c\kappa(\nu)}{4\pi}\int_I\int_\Omega\int_G\int_0^T \psi(x,\nu_1,\omega_1,\tau)d\tau dx d\omega_1 d\nu_1,$$

$$\|A\psi_1 - A\psi_2\|_{L_1(D\times[0,T])} \leq cTC\|\psi_1 - \psi_2\|_{L_1(D\times[0,T])}. \tag{21}$$

Obviously, by the definition of the operator $A$, the iterative method, formulated in Theorem, converges if and only if the successive approximations

$$\varphi^{k+1} = A\varphi^k \tag{22}$$

converge for any $\varphi^0 \in K_0$.

We define the sequences $\{\xi_n\}_{n=0}^\infty$, $\{\eta_n\}_{n=0}^\infty \subset K_0$, where $\xi_0 = 0$, $\xi_{n+1} = A\xi_n$, $\eta_0 = M$, $\eta_{n+1} = A\eta_n$, $n = 0, 1, \ldots$. Obviously,

$$\xi_0 \prec \xi_1 \prec \ldots,\ \eta_0 \succ \eta_1 \succ \ldots.$$

Let $\xi = \sup\{\xi_n\}$, $\eta = \inf\{\eta_n\}$. Then $\|\xi - \xi_n\|_{L_p(D)} \to 0$, $\|\eta - \eta_n\|_{L_p(D)} \to 0$, because convergence by the norm in $L_p(D \times [0,T])$, $p \geq 1$, follows from the order convergence [32].

It follows from the inequality (18) that

$$\|A\xi - A\xi_n\|_{L_1(D\times[0,T])} \to 0, \; \|A\eta - A\eta_n\|_{L_1(D\times[0,T])} \to 0.$$

Due to the monotony of the norm $\|\cdot\|_{L_1(D)}$,

$$\|A\xi - \xi\|_{L_1(D\times[0,T])} \leq \|A\xi - A\xi_n\|_{L_1(D\times[0,T])} \to 0,$$

$$\|A\eta - \eta\|_{L_1(D\times[0,T])} \leq \|A\eta - A\eta_n\|_{L_1(D\times[0,T])} \to 0.$$

Thus $\xi = \eta$ is the unique fixed point of $A$, $\lim_{n\to\infty} \|\xi_n - \eta_n\|_{L_p(D\times[0,T])} = 0$.

Let us now take the sequence $\{\varphi^k\}_{k=0}^{\infty}$ defined by (22) for $\varphi^0 \in K_0$. Obviously, $\xi_n \prec A^{(n+1)}\varphi^0 \prec \eta_n$ and

$$\|A^{(n+1)}\varphi^0 - \xi\|_{L_p(D)} \leq \|A^{(n+1)}\varphi^0 - \xi_n\|_{L_p(D\times[0,T])} + \|\xi - \xi_n\|_{L_p(D\times[0,T]))}$$

$$\leq 2\|\xi_n - \eta_n\|_{L_p(D\times[0,T])}.$$

Note that the evaluation (21), which is valid for all $\psi_1$, $\psi_2 \in K_0$, implies that $A$ is an ordinal continuous operator [31,32] and the iterative process (22) convergences. Theorem is proved.

## 3 Linearizing Algorithm. Numerical Analysis

This section presents the results of numerical implementation of the proposed algorithm in the form of software modules within the framework of the finite element method [23] and the finite difference method [19]. The integro-interpolation method is used for finite-difference approximation by spatial variables, linear basis functions are used in the finite element method.

A rectangular coordinate system is used for the spatial description of the computational domain, and a spherical one is used to define a single vector of particle flight. When approximating the system of integro-differential equations (1)–(5) by an angular variable, the method of discrete ordinates [19] is used. The unit vector of the direction of flight of particles $\omega$ is represented by a pair of numbers $(\mu, \phi)$, where (see Fig. 1)

$$\omega_1 = \xi = \sqrt{1-\mu^2}\cos\phi, \; \omega_2 = \eta = \sqrt{1-\mu^2}\sin\phi, \; \omega_3 = \mu = \cos\theta.$$

In the method of discrete ordinates, a discrete description of the angular dependence is carried out. Let $n$ be allocated for particle propagation flows in the directions $(\mu_j, \phi_{lj})$, where $j = 1, ..., M$, $l = 1, ..., L$ and for the integral of the particle flux density, a quadrature formula, that is a formula for approximate calculation of a certain integral by the values of the integrand function in a finite

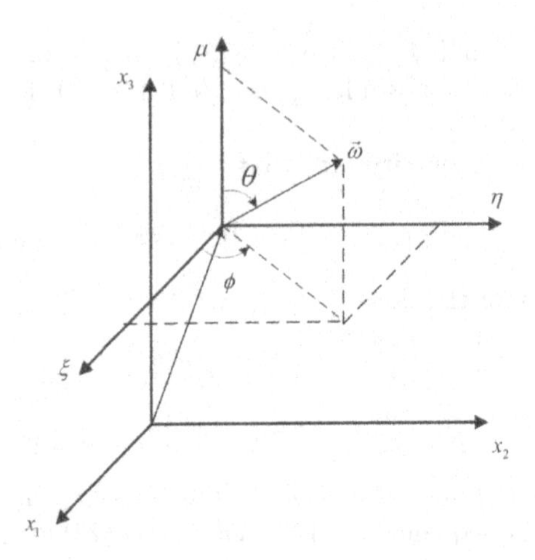


**Fig. 1.** Setting the direction of flight of particles

number of points, is known. Then the original transfer Eq. (1) can be replaced on iteration $k+1$ by a system of $L$ equations

$$\frac{1}{c}\frac{\partial \varphi^{k+1}}{\partial t} + \xi_{jl}\frac{\partial}{\partial x_1}\varphi^{k+1} + \eta_{jl}\frac{\partial}{\partial x_2}\varphi^{k+1} + a^k\varphi^{k+1} = b^k, \tag{23}$$

$$a^k(x,\nu,t) = h\nu_{12}\frac{\kappa(\nu)}{4\pi}[B_{12}C_1^k(x,t) - B_{21}C_2^k(x,t)],$$

$$b^k(x,\nu,t) = h\nu_{12}\frac{\kappa(\nu)}{4\pi}A_{21}C_2^k(x,t),$$

where $C_1^k(x,t)$ and $C_2^k(x,t)$ are calculated by formulas (10) and (11). The integral of the particle flux density takes the form

$$\int_I\int_\Omega \frac{\kappa(\nu)}{4\pi}\varphi(x,\nu,\omega,t)d\omega d\nu$$

$$= \int_I\int_{-1}^{1}\int_0^{2\pi} \frac{\kappa(\nu)}{4\pi}\varphi(x,\nu,\mu,\phi,t)d\mu d\phi d\nu$$

$$\approx \sum_I\sum_{i=1}^{L}\sum_{j=1}^{M_L} a_{lj}\varphi(x,\nu,\mu_j,\phi_{lj},t). \tag{24}$$

The issues of choosing a quadrature formula and weights $a_{lj}$ for it are discussed in [19,33–35].

The following are convenient calculation formulas for computation the coefficients of interaction of radiation with matter. The values $A_{21}, B_{12}, B_{21}$ are the Einstein coefficients of spontaneous and forced radiation and absorption, respectively. They are interconnected by well-known relations [2]

$$A_{21} = \frac{2h\nu_{12}^3}{c^2} B_{21}, \tag{25}$$

$$B_{21} = \frac{g_1}{g_2} B_{12}, \tag{26}$$

where $g_1$, $g_2$ are the statistical weights of the first and second levels, respectively.

The coefficient of spontaneous transition $A_{21}$ or the strength of the oscillator $f_{21}$ is determined by the relation

$$A_{21} = \frac{g_1}{g_2} \frac{8\pi e^2 \nu_{12}^2}{m_e c^3}. \tag{27}$$

Using the ratio

$$\nu_{12} = \frac{c}{\lambda_{12}}, \tag{28}$$

we get the expression

$$A_{21} = \frac{g_1}{g_2} \frac{8\pi e^2}{\lambda_{12}^2 m_e c}. \tag{29}$$

The last expression can be rewritten as

$$A_{21} = 1.499 \cdot 10^8 \frac{g_1}{g_2} \frac{f_{21}}{\lambda_{12}^2}. \tag{30}$$

The values for $g_1, g_2, \lambda_{12}, f_{21}$ are determined from [36,37]. For a known $A_{21}$, the coefficients $B_{12}$ and $B_{21}$ are calculated using formulas (25), (26).

Let's pay attention to the calculation of $C_{12}$ and $C_{21}$, where $C_{ik}$ is the number of transitions $i \to k$ under the action of electronic shocks in 1 s per atom at the $i$th level and a single electron concentration. If the electron velocity distribution is Maxwellian with the electron temperature $T_e$, the coefficients $C_{12}$ and $C_{21}$ are expressed through each other [2]:

$$C_{12} = \frac{g_1}{g_2} e^{-\frac{h\nu_{12}}{T_e}} C_{21}.$$

Coefficient value is

$$C_{12} = \frac{e^2 c^3 m_e}{\sqrt{3} h \nu_{12}^2} \frac{A_{21}}{\sqrt{2\pi m_e k T_e}} \frac{g_1}{g_2} e^{-\frac{h\nu_{12}}{T_e}} P(\frac{h\nu_{12}}{kT_e}),$$

where $P(X)$ is determined by linear interpolation from the table values [2].

## 4 Computational Results

To numerically illustrate the operability and efficiency of the proposed iterative algorithm, model problems were considered for media with different optical densities of matter. In the region $N = \{(x_1, x_2) \in \mathrm{R}^2 : 0 \leq x_1 \leq 0.3,\ 0 \leq x_2 \leq 0.3\}$

(dimensions are given in centimeters), a system of transport equations and statistical equilibrium with zero boundary conditions is solved, provided there is no incoming flow. As a result of calculations, the total number of particles

$$\tilde{\varphi} = \int_{-1}^{1} \int_{0}^{2\pi} \varphi d\mu d\phi$$

was considered, the initial value of which was assumed to be equal to $\tilde{\varphi} = 1$. The energy during the transition from one level to another in this problem was equal to $h\nu_{12} = 3.2[eV]$. The calculations were carried out on a regular spatial grid with a uniform step over the space $h_1 = h_2 = 0.01$. According to the angular variable, 144 directions were taken. Time variable step $\tau = 0.001$. The criterion for stopping the iterative process has the form

$$|\tilde{\varphi}^{k+1} - \tilde{\varphi}^{k}| \leq \varepsilon_0 \tilde{\varphi}^{k+1} + \varepsilon_1, \tag{31}$$

where $\varepsilon_0$, $\varepsilon_1$ are given constants, we take $\varepsilon_0 = 10^{-4}$, $\varepsilon_1 = 10^{-16}$.

A series of calculations was carried out using the finite element method (FEM) and the finite difference method (FDM) in which the composition of the region, namely its optical properties, determined by absorption and scattering coefficients, was changed.

As a first example, the area was filled with helium and was optically transparent. Figure 2 shows the spatial distribution of the scalar particle flux at time $t = 0.1$. The picture on the right is the distribution obtained using FDM, the picture on the left is the distribution obtained using FEM.

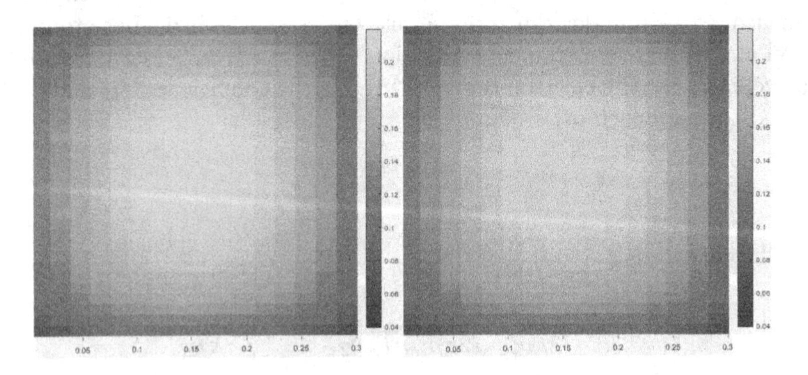

**Fig. 2.** Spatial distribution of the scalar flow at time $t = 0.1$. Left - FEM. Right - FDM

Figure 2 demonstrates that the results are in good agreement with each other. The iterative process converged in 354 iterations with a given accuracy according to criterion (31), which corresponds to 3–4 iterations per step.

An optically dense problem was chosen as the next example. The geometry of the area and the parameters of the account(spatial and angular grids, the size of the step in space and time) fully correspond to the first example. The results obtained by the FMD and by the FEM are in good agreement. The iterative process in optically dense regions converges much slower, this is due to the values of the coefficients of interaction of radiation with matter. The iterative process converged in 813 iterations with a given accuracy according to criterion (31), which corresponds to 8–9 iterations per step.

As a third example, the region was divided into three subdomains, different in physical properties. In this problem, the first and third regions are optically dense, and the second is transparent. Table 1 shows the values of the interaction coefficients, namely the Einstein coefficients for spontaneous and forced radiation and absorption, respectively, for each of the regions.

**Table 1.** Parameters of areas in a task.

| Area | Coefficient $A_{21}$ | Coefficients $B_{12}$ and $B_{21}$ |
|---|---|---|
| $1 : 0 \leq x_1 \leq 0.1,\ 0 \leq x_2 \leq 0.3$ | $\sim 0.48 \cdot 10^7 [s^{-1}]$ | $\sim 10^{14} [J^{-1} cm^3 s^{-2}]$ |
| $2 : 0.1 \leq x_1 \leq 0.2,\ 0 \leq x_2 \leq 0.3$ | $\sim 0.21 \cdot 10^8 [s^{-1}]$ | $\sim 10^{13} [J^{-1} cm^3 s^{-2}]$ |
| $3 : 0.2 \leq x_1 \leq 0.3,\ 0 \leq x_2 \leq 0.3$ | $\sim 0.48 \cdot 10^7 [s^{-1}]$ | $\sim 10^{14} [J^{-1} cm^3 s^{-2}]$ |

Table 2 shows the main characteristics of the calculation, namely, the total number of particles in the system at the time $t = 0.1$ and data on the number of iterations.

**Table 2.** Characteristics of calculations

| Characteristics | FEM | FDM |
|---|---|---|
| Number of particles in the system | 0.0121 | 0.0120 |
| Number of iterations | 2132 | 2452 |

Based on the results given in Table 2, it can be concluded that the relative error obtained in the calculations for the FEM in comparison with the FDM was about 0.8 percent. The difference in the number of iterations is due to the fact that a second-order precision difference scheme was used in the FDM, which in general does not have the monotonicity property [11]. To improve this property, monotonization algorithms are used, which require additional calculations.

Figures 3 and 4 show the computational grid used in the calculations and the spatial distribution of the density of the scalar particle flux at the time $t = 0.1$, obtained from the FDM.

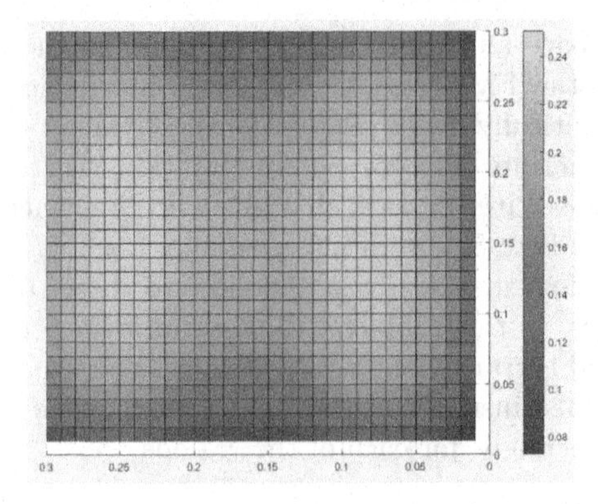

**Fig. 3.** Problem geometry

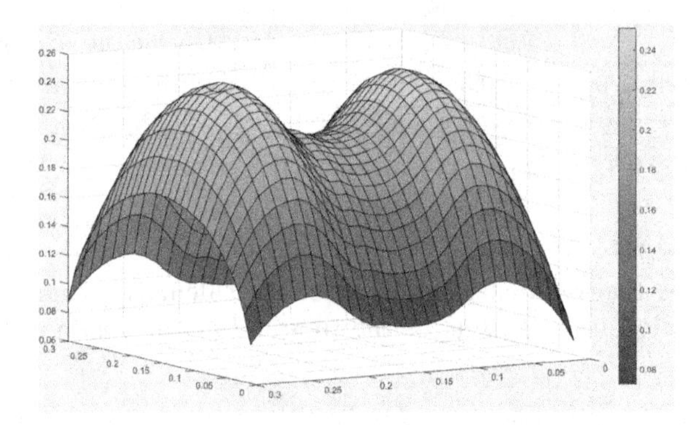

**Fig. 4.** Spatial distribution of the scalar flow at time $t = 0.1$

In this paper we propose and numerically implement an iterative algorithm for solving a nonlinear nonstationary system of radiative transfer equations in the kinetic approximation and statistical equilibrium corresponding to the model of a two-level atom for a spatially multidimensional case. To illustrate the effectiveness of the proposed algorithm, three model problems of radiation propagation were solved, which differ in the optical properties of the medium. Numerical results obtained in the work allow us to conclude that the iteration process is convergent, the results obtained using the finite element method are in good agreement with the results using the finite difference method, the iterative process converges much faster in optically transparent media.

## References

1. Zel'dovich, Ya.B., Raizer, Y.P.: Physics of Shock Waves and High-Temperature Hydrodynamic Phenomena. Academic Press, Cambridge (1967)
2. Ivanov, V.V.: Transfer Theory and the Spectra of Calestial Objects. Nauka, Moscow (1969). (in Russian)

3. Mihalas, D.: Stellar Atmospheres. Freeman, San Francisco (1978)
4. Chetverushkin, B.N.: Mathematical Modeling of Problems in the Dynamics of a Radiating Gas. Nauka, Moscow (1985). (in Russian)
5. Bell, G.I., Glasstone, S.: Nuclear Reactor Theory. Van Nostrand Reinhold Co., New York (1970)
6. Vladimirov, V.S.: Mathematical problems in the one-velocity theory of particle transport. Tr. MIAN SSSR 61. Publ. house of the Academy of Sciences of the USSR, Moscow (1961)
7. Agoshkov, V.I.: Some questions of the theory and approximate solution of particle transfer problems. Department of Comp. Math. USSR Academy of Sciences, Moscow (1983)
8. Germogenova, T.A.: Local Properties of the Transport Equation Solutions. Nauka, Moscow (1986). (in Russian)
9. Marchuk, G.I., Lebedev, V.I.: Numerical Methods in the Theory of Neutron Transport. Harwood Academic Publication (1986)
10. Prilepko, A.I., Volkov, N.P.: Inverse problems for determining the parameters of the nonstationary kinetic transport equation from additional information on the traces of the unknown function. Differ. Equ. **24**(1), 107–115 (1988)
11. Sushkevich, T.A.: Mathematical Models of Radiation Transfer. BINOM, Moscow (2005). (in Russian)
12. Anikonov, D.S., Kovtanyuk, A.E., Konovalova, D.S., Nazarov, V.G., Prokhorov, I.V., Yarovenko, I.P.: Radiation tomography and transport equation. Dal'nevost. Mat. Zh. **8**(1), 5–18 (2008)
13. Amosov, A.A.: Initial-boundary value problem for the nonstationary radiative transfer equation with diffuse reflection and refraction conditions. J. Math. Sci. **235**, 117–137 (2018)
14. Prokhorov, I.V., Sushchenko, A.A.: The Cauchy problem for the radiative transfer equation in an unbounded medium. Dal'nevost. Mat. Zh. **18**(1), 101–111 (2018). (in Russian)
15. Morozov, S.F., Sumin, V.I.: Nonlinear integro-differential systems of equations of nonstationary transport. Sibirsk. Mat. Zh. **19**(4), 842–848 (1978)
16. Makin, R.S.: On the existence of solutions of a nonlinear integro-differential system of transport equations. Math. Notes **90**(1), 102–124 (2011)
17. Chetverushkin, B.N., Olkhovskaya, O.G., Gasilov, V.A.: Three-layer scheme for solving the radiation diffusion equation. Dokl. Ross. Akad. Nauk. Math., Inf. Proc. Upr. **512**, 89–95 (2023)
18. Gol'din, V.Ya.: A quasi-diffusion method of solving the kinetic equation. Comp. Math. Math. Phys. **4**(6), 136–149 (1964)
19. Bass, L.P., Voloshchenko, A.M., Germogenova, T.A.: Methods of discrete ordinates in problems of radiation transfer. IPM im. Keldysha AN SSSR, Moscow (1986). (in Russian)
20. Cheremisin, F.G.: A conservative method for calculation of the Boltzmann collision integral. Dokl. Phys. **42**(11), 607–610 (1997)
21. Tcheremissine, F.: Direct numerical solution of the Boltzmann equation. Rarefied gas dynamics. In: 24th International Symposium on Rarefied Gas Dynamics. AIP Conference Proceedings, vol. 762, pp. 667–685. Melville (2005)
22. Anikin, Y.A.: Numerical study of radiometric forces via the direct solution of the Boltzmann kinetic equation. Comput. Math. Math. Phys. **51**(7), 1251–1266 (2011)
23. Sychugova, E.P.: Discontinuous finite element method for solving the transport equation on an unstructured grid of triangular cells. IPM im. Keldysha **85** (2013)

24. Kalinin, A.V., Morozov, S.F.: The method of linearization for a nonlinear integro-differential system of transport equations. Soviet Math. **27**(12), 23–32 (1983)
25. Kalinin, A.V., Morozov, S.F.: Solvability "in the large" of a nonlinear problem of radiative transfer. Differ. Uravn. **21**(3), 484–494 (1985)
26. Kalinin, A.V., Morozov, S.F.: Stabilization of the solution of a nonlinear system of radiation transport in a two-level approximation. Dokl. Math. **35**(3), 239–241 (1990)
27. Kalinin, A.V., Morozov, S.F.: A mixed problem for a nonstationary system of nonlinear integro-differential equations. Sib. Math. J. **40**(5), 887–900 (1999)
28. Kalinin, A.V., Morozov, S.F.: A non-linear boundary-value problem in the theory of radiation transfer. Comput. Math. Math. Phys. **30**(4), 76–83 (1990)
29. Kalinin, A.V., Tyukhtina, A.A., Busalov, A.A.: An iterative method for solving a nonlinear system of the theory of radiation transfer and statistical equilibrium in a plane-parallel layer. In: Balandin, D., et al. (eds.) MMST 2022. CCIS, vol. 1750, pp. 106–120. Springer, Cham (2022). https://doi.org/10.1007/978-3-031-24145-1_9
30. Kalinin, A.V., Tyukhtina, A.A.: On a nonlinear problem for a system of integro-differential equations of radiative transfer theory. Comput. Math. Math. Phys. **62**(6), 933–944 (2022)
31. Kantorovich, L.V.: The method of successive approximations for functional analysis. Acta Math. **71**, 63–97 (1939)
32. Krasnoselskiy, M.A.: Positive Solutions of Operator Equations. Fizmatgiz, Moscow (1962)
33. Fornberg, B., Martel, J.M.: On spherical harmonics based numerical quadrature over the surface of a sphere. Adv. Comput. Math. **40**, 1169–1184 (2014)
34. Carlson, B.G.: Transport Theory: Discrete Ordinates Quadrature over the Unit. Los Alamos Scientific Laboratory Report, LA-4554 (1970)
35. Awono, O., Tagoudjeu, J.: A splitting iterative method for solving the neutron transport equation. Math. Model. Anal. **14**(3), 271–289 (2009)
36. Elyashevich, M.A.: Atomic and Molecular Spectroscopy. Fizmatlit, Moscow (1962). (in Russian)
37. Allen, C.W.: Astrophysical Quantities. The Athlone Press, London (1973)

# Empirical Analysis of the Dynamic Processes of Node Appearance and Disappearance in Temporal Networks

Timofei D. Emelianov(✉)

Saratov State University, Saratov 410012, Russia
emelianov.timofey1337@gmail.com

**Abstract.** The paper presents an analysis of real scale-free networks, which incorporates a mechanism for the disappearance of edges and nodes over time. Constantly growing scale-free networks and many of their properties are well studied. However, these models and the real network dynamics analysis typically only consider growing networks, without any mechanism for removing edges or vertices. Therefore, the paper focuses on the analysis of the network reduction process, paying close attention to changes in various local characteristics over time and determining whether the power law is satisfied for such networks. The study also reveals general patterns of the disappearance of edges in real networks, which can be used to create models that account for the network reduction process.

**Keywords:** Complex scale-free networks · Temporal graphs · Network analysis

## 1 Introduction

In the world there are a large number of different types of scale-free networks that change in time [1–6]: social (community and subscriptions), collaborative (authors, actors), economic (firms and their interactions), biological (food chains). In reality, such networks, in addition to the emergence of new nodes/edges, can also lose existing ones: people can unsubscribe from each other, authors can stop writing articles in co-authorship, firms can close, and certain animal species die out. However, this process is currently little studied, since most papers consider only the mechanism of the appearance of new nodes and edges in such networks and also model only the process of network growth (for example, the Barabasi-Albert models [7], triadic closure [8], as well as their varieties [9–13]).

Perhaps one of the few papers that explores the node removal mechanism, albeit in a truncated form, is the paper [14]. The authors use a list of active nodes, the removal of which deactivates the node and excludes the possibility of new

The work was supported by the Russian Science Foundation, project 23-21-00148.


D. Balandin et al. (Eds.): MMST 2023, CCIS 1914, pp. 59–71, 2024.
https://doi.org/10.1007/978-3-031-52470-7_5

nodes joining it in the process of network growth. Perhaps the first study that proposes a new mechanism for the formation of scale-free networks by removing nodes is the paper [15]. Two nodes form one node when merged, the neighbours of which are the neighbours of the parent nodes.

The main question in this study was the following: do real scale-free networks cease to satisfy the power law of degree distribution if edges and nodes can disappear from the graph? And also in the paper, we will try to understand whether different real networks have common patterns when reducing.

To answer the questions posed, the article analyses the characteristics of three real temporal networks:

- Co-authors network;
- Actors network;
- Keywords network.

The paper will show that even if the disappearance of edges and nodes takes place, all the considered networks continue to satisfy the power law of the degree distribution. In addition, some general patterns have been demonstrated in how edges disappear for various networks, which in the future may form the basis of models that can be used to simulate the process of network reduction.

## 2 Data and Methodology

### 2.1 Notations and Definitions

This work considers only an unweighted undirected temporal graph $G(t)$.

The set $A$ is called the set of *time-edges* [16] and it refers to the structure of the temporal graph at a particular time. In particular, $A(t) = \{(u, v) : (u, v, t) \in A\}$ is the set of all edges that appear in the temporal graph at time $t$. In turn, $A(t)$ can be used to define a snapshot of the temporal graph $G$ at time $t$, which is usually called the *t-th instance* of $G$, and is the static graph $G(t) = (V, A(t))$, which means that a temporal graph can be viewed as a sequence of static graphs $G = (G_1, G_2, \ldots, G_N)$.

Let $d_u(t)$ denote the number of edges connected to node $u$ at a given time $t$, i.e. the degree of node $u$.

### 2.2 Datasets

The paper considers 3 datasets:

- Dataset of articles and authors (DBLP V12) [17];
- Dataset of movies and actors (IMDB)[1];
- Dataset of article keywords from the scientific electronic library "Cyber-Leninka"[2].

All of these datasets use years as time points.

[1] https://datasets.imdbws.com/.
[2] https://cyberleninka.ru/.

**DBLP V12.** The dataset is a network of co-authors and has the following characteristics:

- 4,900,000 articles;
- 4,100,000 authors;
- 55 years: from 1966 to 2020.

When constructing the graph, we will use the authors as nodes and the presence of a common publication as an edge. Thus, we get an undirected graph. Since the year of publication of each article is known, such a graph can be considered temporal.

**IMDB.** The dataset is a network of actors and has the following characteristics:

- 9,700,000 movies/series;
- 2,320,000 actors;
- 113 years: from 1911 to 2023.

When constructing the graph, we will use actors as nodes and the fact that two actors played their roles in the same movie/series as an edge. By analogy with the network of co-authors, we have information about the year of release of each movie, so such a graph can also be considered temporal.

**CyberLeninka.** The dataset is a network of article keywords and has the following characteristics:

- 31,300 articles;
- 245,700 keywords;
- 23 years: from 1994 to 2016.

When constructing the graph, we will use keywords as nodes and the fact that two keywords appeared in the same article as an edge. Temporality is achieved by analogy with previous networks due to the fact that we know the year the article was published.

### 2.3 Data Preparation

Let us describe the data preparation process for a network of co-authors. For other datasets, the preparation is completely similar.

The DBLP V12 dataset contains a list of authors and publication years for each article. Obviously, we cannot use adjacency matrices to store this information since for each moment of time (a year, in our case), it was necessary to store a matrix of dimensions $N \times N$, where $N$ is the number of authors, i.e., 4.1 million $\times$ 4.1 million. Instead, we will store a table like this:

**author #1, author #2, year.**

```
"id" : "53e99784b7602d9701f3e133",
"title" : "The relationship between canopy...",
"authors" : [0, 1, 2],
"year" : 2011,
```

| | author_1 | author_2 | year |
|---|---|---|---|
| 0 | 0 | 1 | 2011 |
| 1 | 0 | 2 | 2011 |
| 2 | 1 | 2 | 2011 |

**Fig. 1.** The data preparation example

Two authors refer to the table if they have a common article, and the publication year of this article is recorded in the appropriate column.

Thus, if an article has $K$ authors, then we get $C_K^2$ rows in the final table for this article since we need to store all pairs of co-authors.

The Fig. 1 shows an example of preparing data for the co-authors network.

The main rules that were used to clear the resulting table:

- Do not use articles with only one author;
- Store only unique triples (author #1, author #2, year);
- The rows (author #1, author #2, year) and (author #2, author #1, year) are replaced by a single row (author #1, author #2, year) because the graph is undirected;
- Do not use corrupted or unreliable data: articles without a publication year or with a publication year later than the current date.

The Table 1 shows the characteristics of networks after data preparation.

**Table 1.** The networks and their main characteristics

| Network | $\|V\|$ | $\|E\|$ | Time points |
|---|---|---|---|
| Co-authors | 4,000,000 | 19,200,000 | 55 |
| Actors | 2,000,000 | 17,400,000 | 113 |
| Keywords | 245,000 | 1,034,000 | 23 |

### 2.4 Network Reduction Mechanism

In the considered networks, there is generally no natural mechanism for removing nodes or edges. That is, if an edge between nodes $u$ and $v$ appeared at time $t$, then at time $t' \geq t$ the edge between nodes $u$ and $v$ still exists. To fix this, let us define our own criteria for removing edges and nodes in the temporal graph, which will be based on how long ago two nodew interacted with each other.

We say that two nodes $u$ and $v$ in graph $G$ *interact* at time $t$ if $(u, v, t) \in A(t)$, where $A(t)$ is the time-edges set of graph $G$ at time $t$.

In all networks that are considered in the article, we will remove edges between two nodes if there was no interaction between them during the last $W$ moments of time. And we remove the nodes if they do not have any common edges with other nodes. This algorithm is presented in Algorithm 1.

---

**Algorithm 1:** The network reduction algorithm

---

**Input**: A temporal graph $G$ which is defined by time-edges sets $(A(1), A(2), \ldots, A(N))$ and the time window $W$ during which we consider that the edge between two nodes still exists.

1 **for** $n \leftarrow 1$ **to** $N$ **do**

2   Consider an edge $e = (u, v, n)$ *removed* if

$$\forall t \in [n - W + 1, n] \; (u, v, t) \notin A(t),$$

$$(u, v, n - W) \in A(n - W).$$

3   Consider a node $u$ **removed** if all the edges of this node have been **removed**.

4 **end**

---

The Fig. 2 shows an example of how the network changes over time via the mechanism of disappearance of edges and nodes described in Algorithm 1 for a given time window $W = 3$. Let us describe each moment of time t:

1. The nodes $a$ and $b$ interact with each other, so we connect them with an edge;
2. The nodes $b$ and $c$ interact with each other, so we connect them with an edge. We do not remove the edge between the nodes $a$ and $b$, because at the previous moment of time (falls into our window $W$) they interacted;
3. There are no interactions between the nodes, but we do not remove the edges since the nodes interacted during the $W$ last moments of time;
4. There are no interactions between nodes. The last time the nodes $a$ and $b$ interacted was at time $t = 1$, which does not fall within the given window $W$, which means that we remove the edge $(a, b)$. Since node $a$ no longer has edges between other nodes, we remove $a$ from the graph;
5. The nodes $b$ and $c$ interact again, which means that we do not remove the edge between them.

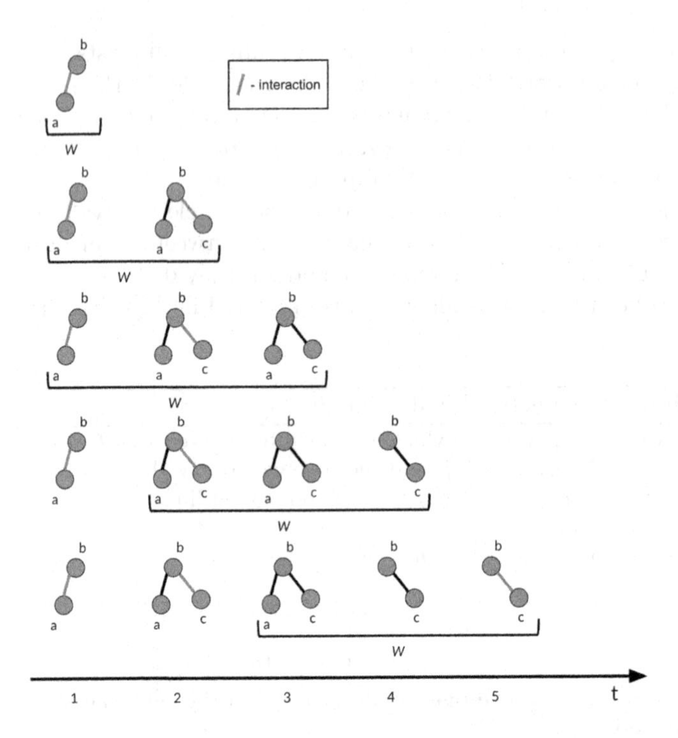


**Fig. 2.** An example of the appearance and disappearance of nodes and edges in the network for $W = 3$

## 3 Empirical Results

If the value of $W$ is not explicitly specified below, then we assume $W = 3$.

Since we are dealing with temporal graphs, the number of nodes $|V(t)|$, edges $|E(t)|$ and the graph density

$$density(t) = \frac{|E(t)|}{|V(t)|(|V(t)| - 1)}$$

can change at any given time. The Fig. 2 shows how these characteristics change over time.

It can be seen that the reduction mechanism sometimes leads to the fact that the number of nodes and edges at the next time point is less than at the previous one (Fig. 3).

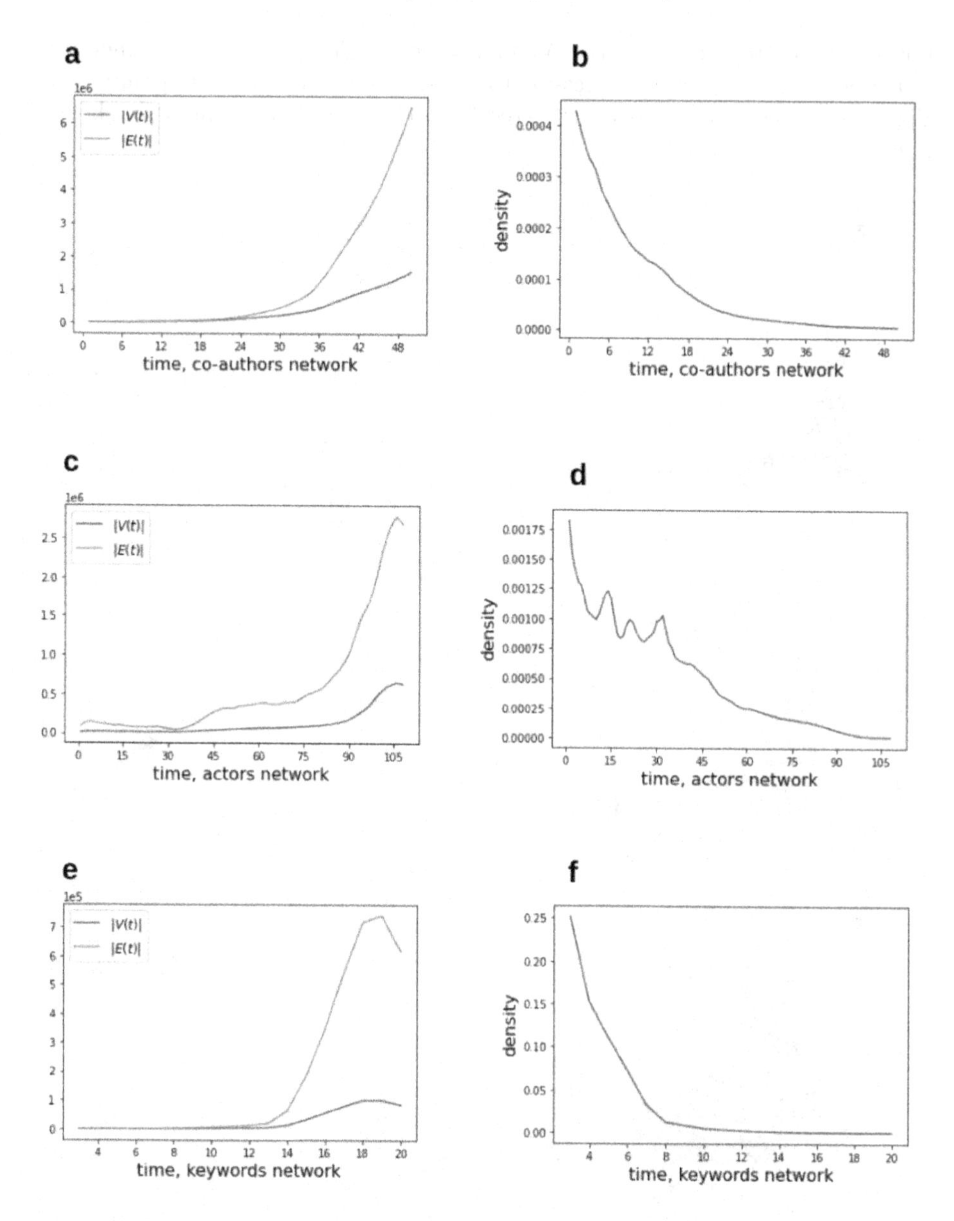


**Fig. 3.** The figures **a–f** show the change in the number of nodes and edges over time

Now let us find out if the networks satisfy the power law of the degree distribution if the network reduction mechanism is added to them.

First, we show the principle of calculating the power law exponent $\gamma$ for one randomly chosen moment in time. We build histograms to observe the node degree distribution. For each network we use linear and log-log plots, the latter is required to show that the power law holds. The power law exponents $\gamma$ were obtained fitting linear regression to logarithmically binned data points. The Fig. 4 shows the histograms constructed for each network at a randomly chosen point in time.

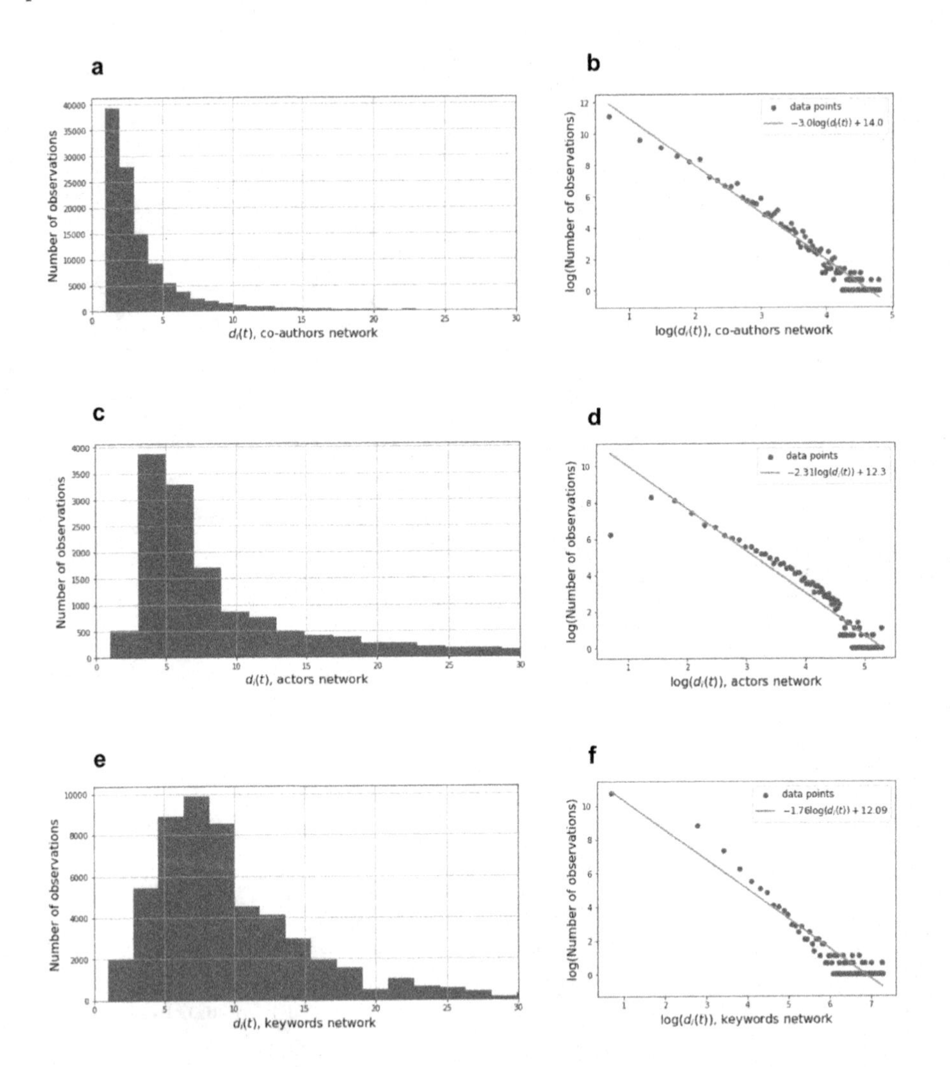


**Fig. 4.** The figures **a–f** show empirical histograms $d_i(N)$, their variants on a logarithmic scale, and the value of $\gamma$ for one randomly chosen time $N$

In the same way that we calculated the value of $\gamma$ for one moment in time $N$, let us calculate it for all other points in time to see how it changes over time.

The Fig. 5 shows the change in the value of $\gamma$ over time in each network. It is noteworthy that over time, in all networks, the value of $\gamma$ turns out to be in the range of $[-2; -3]$, which is typical for many free-scale networks [1–6].

Thus, we can conclude that even despite the presence of a mechanism for removing nodes and edges, the considered real networks are free-scale networks (that is, the degree distribution follows a power law).

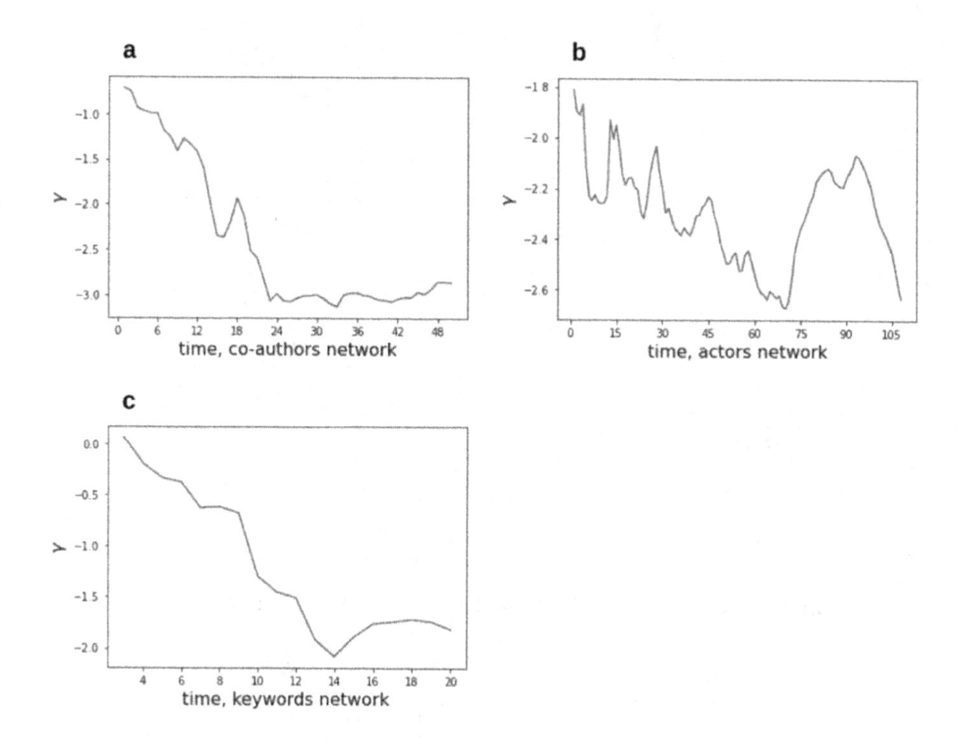


**Fig. 5.** The figures **a**–**c** show how the power law exponent $\gamma$ changes over time

Now let us try to find patterns in how edges disappear in real networks.

First, consider how the degree of the node $d_i(t)$ changes at the next time $d_i(t+1)$. An obvious assumption is that, on average, the degree at time $t+1$ practically does not differ from the degree at time $t$, and confirmation of this can be seen in the Fig. 6.

Let $r_i(t)$ denote *the fraction of removed edges* at the node $i$ at time $t$.

The Figs. 7, 8, 9 show the dependence of the proportion of removed edges at a node at time $t+1$ $r_i(t+1)$ on the degree of the node at time $t$ $d_i(t)$ for various values $W$.

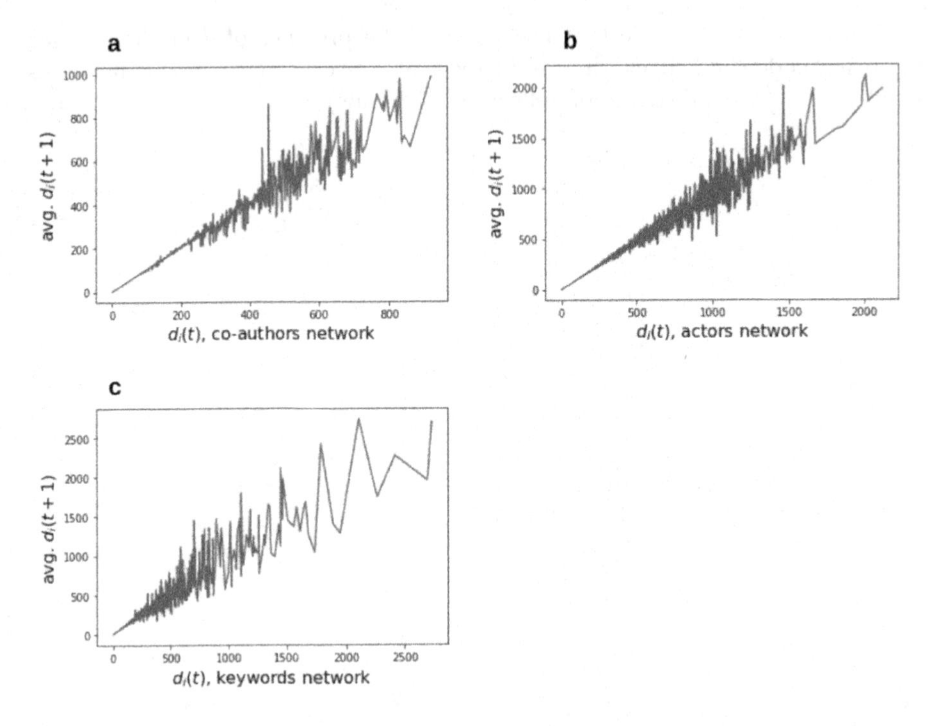


**Fig. 6.** The figures **a–c** show the dependence of $d_i(t+1)$ on $d_i(t)$

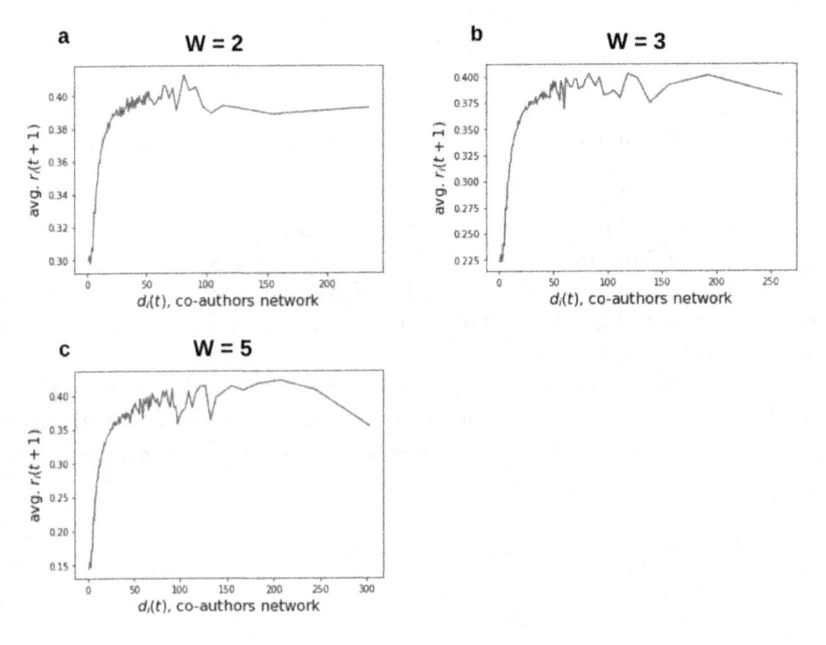


**Fig. 7.** The figures **a–c** show the dependence of $r_i(t+1)$ on $d_i(t)$ for the co-authors network

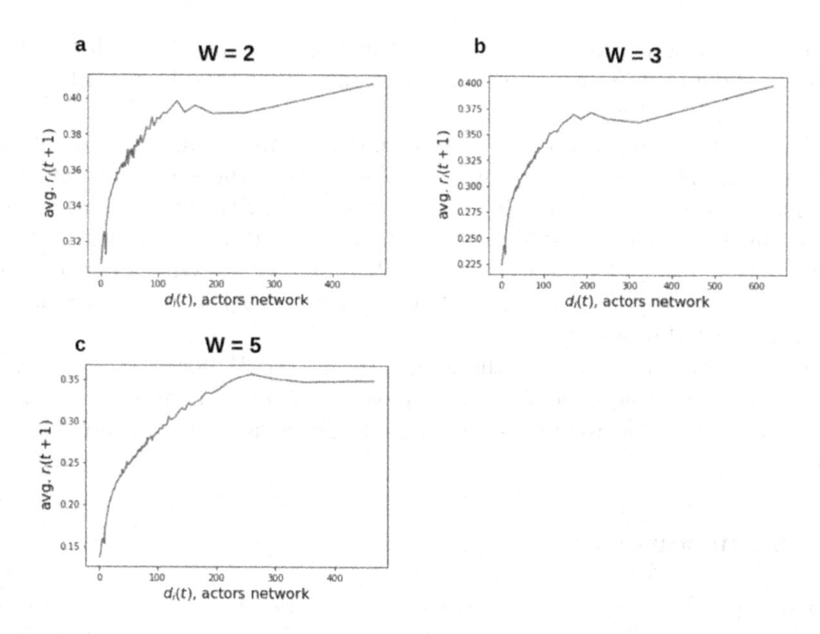


**Fig. 8.** The figures **a**–**c** show the dependence of $r_i(t+1)$ on $d_i(t)$ for the actors network

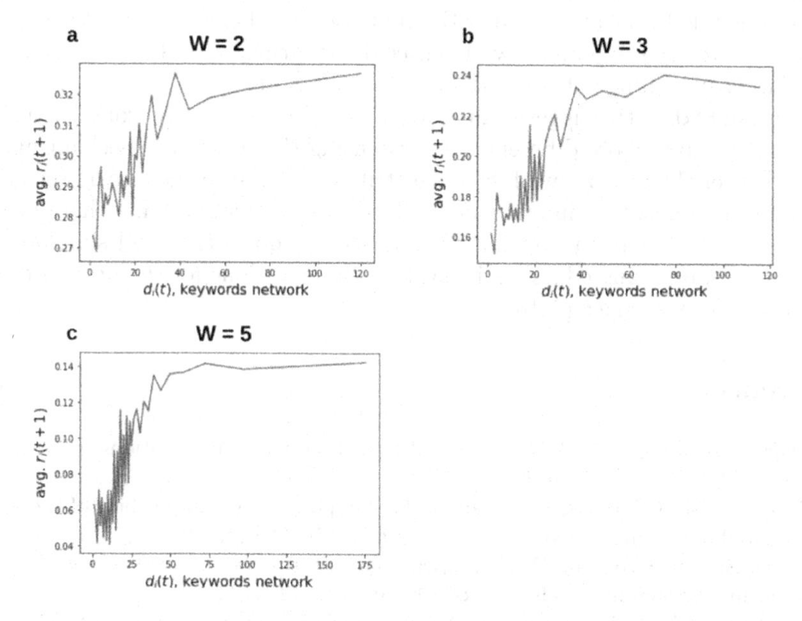


**Fig. 9.** The figures **a**–**c** show the dependence of $r_i(t+1)$ on $d_i(t)$ for the keywords network

It can be seen that the percentage of removed edges for low-degree nodes is almost 2 times less (for $W = 3$) than for high-degree nodes. But we also saw earlier that, on average, the degree of a node remains practically the same at the next moment of time compared to the current one. This means that high-degree nodes are more likely to create new connections rather than keep existing ones than low-degree nodes.

That is, if you are a popular author and have many different co-authors, then, most likely, next year you will also have about the same number of co-authors, but ≈40% of them will be new co-authors with whom you have not had common publications over the past three years. But authors with a small number of co-authors have only ≈20% probability not to write an article with a co-author with whom they already have a common article in the current year within the next three years.

Also, the Figs. 7, 8, 9 show the longer the period $W$ for which we believe that the connection between the nodes is lost the smaller the proportion of lost connections for low-degree nodes. For high-degree nodes it practically does not change.

## 4 Conclusions

The paper presents an analysis of real scale-free networks, which incorporates a mechanism for the disappearance of edges and nodes, the main characteristics of networks and their change over time are studied. Based on the results, it can be concluded that after adding the reduction mechanism, the networks still continue to satisfy the power law of degree distribution. Also, high-degree nodes lose a larger percentage of edges than low-degree nodes.

Our results directly suggest a number of questions for future work. First, it is necessary to study more different networks using the reduction mechanism and check whether the power law of degree distribution holds in them. And secondly, if it turns out that edges in other networks disappear more often at high-degree nodes, then this property can form the basis of models that will simulate the process of disappearing edges (as it has already been done for the process of the appearance of new edges [7–13]).

## References

1. Clauset, A., Shalizi, C.R., Newman, M.E.J.: Power-law distributions in empirical data. SIAM Rev. **51**(4), 661–703 (2009)
2. Faloutsos, M., Faloutsos, P., Faloutsos, C.: On power-law relationships of the internet topology. Comput. Commun. Rev. **29**(4), 251–262 (1999)
3. Klaus, A., Yu, S., Plenz, D.: Statistical analyses support power law distributions found in neuronal avalanches. PLoS One **6**(5), 1–12 (2011)
4. Newman, M.E.J.: The structure and function of complex networks. SIAM Rev. **45**(2), 167–256 (2003)
5. De Solla, P.D.: Networks of scientific papers. Science **149**, 292–306 (1976)
6. Albert, R., Barabási, A.-L.: Statistical mechanics of complex networks. Rev. Mod. Phys. **74**(1), 47–97 (2002)
7. Barabási, A.-L., Albert, R.: Emergence of scaling in random networks. Science **286**(5439), 509–512 (1999)

8. Holme, P., Kim, B.J.: Growing scale-free networks with tunable clustering. Phys. Rev. E **65**(2), 026107 (2002)
9. Sidorov, S., Mironov, S.: Growth network models with random number of attached links. Physica A **576**, 126041 (2021)
10. Krapivsky, P.L., Redner, S., Leyraz, F.: Connectivity of growing random networks. Phys. Rev. Lett. **85**(21), 4629–4632 (2000)
11. Pal, S., Makowski, A.M.: Asymptotic degree distributions in large (homogeneous) random networks: a little theory and a counterexample. IEEE Trans. Netw. Sci. Eng. **7**(3), 1531–1544 (2020)
12. Brot, H., Muchnik, L., Louzoun, Y.: Directed triadic closure and edge deletion mechanism induce asymmetry in directed edge properties. Eur. Phys. J. B **88**, 12 (2015)
13. Itzhack, R., Muchnik, L., Erez, T.: Empirical extraction of mechanisms underlying real world network generation. Physica A **389**(22), 5308–5318 (2010)
14. Klemm, K., Eguíluz, V.M.: Growing scale-free networks with small-world behavior. Phys. Rev. E **65**(5), 057102 (2002)
15. Naglić, L., Šubelj, L.: War pact model of shrinking networks. PLOS ONE **14**(10), 1–14 (2019)
16. Erlebach, T., Hoffmann, M., Kammer, F.: On temporal graph exploration. J. Comput. Syst. Sci. **119**, 1–18 (2021)
17. Tang, J., Zhang, J., Yao, L., Li, J., Zhang, L., Su, Z.: ArnetMiner: extraction and mining of academic social networks. In: Proceedings of the Fourteenth ACM SIGKDD International Conference on Knowledge Discovery and Data Mining (SIGKDD 2008), pp. 990–998 (2008)

# Dynamics of a Multi-piston Vibropercussion Mechanism Equipped with a Reaction Weight

Vladimir S. Metrikin and Irina V. Nikiforova(✉)

Lobachevsky University, Nizhny Novgorod, Gagarin Avenue, 23 bld, 603022 Nizhny Novgorod, Russia
tsii@list.ru, irina.nikiforova@itmm.unn.ru

**Abstract.** Dynamics of a multi-weight vibrating rammer with the physical diagram consisting of an anvil, which consolidates the treated medium via a plate, is considered. The mechanism is fixed on supports bearing a crank drive with phase-shifted crank shafts. The shafts are connected to the main and auxiliary connecting rods. Piston hammers and an additional reaction weight with ram heads, which is spring-biased relative to the supports towards the plate, are fixed on the free ends of the rods. The reaction weight is equipped with an anvil fixed with an adjusting bolt, while the crank drive is equipped with equal-sized crank shafts, which are phase-shifted to 180 degrees, and opposing piston hammers located within the mechanism body. Unlike the known vibrating rammers, the studied mechanism is characterized by sufficiently high consolidation efficiency and stable operation in a wide frequency range.

**Keywords:** vibrating rammer · piston hammers · strongly nonlinear dynamic system

## 1 Introduction

Vibropercussion machines and mechanisms are used widely in various industries, mining, everyday living, etc. Currently, a whole class of various types of vibration and vibropercussion machines exists, which are used for widely varying applications [1,2].

Due to the fact that complex phenomena in the soil are caused by vibrations, the choice of optimal parameters of vibrodrivers is complicated. Therefore, the development of new designs of vibration and vibropercussion machines require theoretical studies of the process of vibration driving [3–12].

Frequently, when one studies systems with impacts, the issue of determining the stress states of the colliding elements of the system is secondary, while the more significant problem is the study of the motion allowing for the presence of

Supported by the Ministry of Science and Higher Education of the Russian Federation, project no. FSWR-2023-0034, and by the Research and Education Mathematical Center "Mathematics for Future Technologies".


D. Balandin et al. (Eds.): MMST 2023, CCIS 1914, pp. 72–85, 2024.
https://doi.org/10.1007/978-3-031-52470-7_6

the impacts. In such cases, the impact is described conventionally in accordance with Newton's hypothesis [13,14]. According to this classical theory, the impact duration is assumed to be equal to zero, and the associated forces are infinite. Additionally, it is assumed that the relative velocity, at which the contact areas of the colliding bodies approach, reverses its sign in the impact processes, and its absolute value decreases by $R$ times ($0 < R < 1$), where $R$ is the coefficient of the velocity return, which is determined by the material and shape of the contact areas. Along with the equation of the system momentum conservation during the impact allows one to determine unambiguously the dynamic state of the system after the impact.

In this paper, we study the mathematical models of two design versions of vibrating rammers.

The first version comprises eccentric vibropercussion mechanisms with crank-type vibrators [15,16]. This design concept is based on the inverted vibrator principle, where the actuator being a balance weight swings on an eccentric shaft and is brought to equilibrium by rotation of the balance weight. The power momentum transferred to the surface (soil, pile, etc.) is the result of both the thrust of the arm of the eccentric shaft, and the kinetic drop energy of the actuator. The efficiency of the tamping and driving machines depends significantly not so much on the amount of the energy transferred to the processed medium, as on the character of the transmission of this energy, i.e., the driver signature, which should be varied by redistributing individual dynamic factors of the single driving loop. This multiple driving method can be implemented using multi-hammer eccentric vibropercussion mechanisms with crank oscillation exciters, whose design allows one to adjust the operation regimes by varying the geometry of kinematic connections and complete the tasks of soil tamping in the confined industrial and civil engineering areas.

The second type is the vibrating rammer with an anvil, whose supports carry the crank drive of the percussive mechanism with phase-shifted equal-sized cranks connected to the main and auxiliary connections rods. The piston hammers fixed at the free ends of the rods are directed toward the plate. A drawback of the known vibrating rammers is insufficient tamping efficiency, a relatively narrow range of stable operation, and insufficient impact rate, since the duration of the free run of the percussion element between the impacts is great, and it increases with the growing resistance of the medium due to the increase in the climb height. Since the reaction of the medium affected by the shock-and-vibration actuators determines considerably the duration and value of the force impact in each impact cycle, the efficiency of the actuators can be increased by applying additional forces that ensure forcing of the impact weight to the limiter at the instants of their interaction, which results in a greater energy doing useful work [17,18].

It is known that the impact momentum depends on the linear momentum transferred by the impact element to the soil. Therefore, the transfer of the same linear momentum by a lighter weight can be ensured by increasing its applied velocity, while decreasing the specific content of metal in the actuator.

This objective is achieved by complementing the vibrating rammer with a plate which tamps the processed medium. The supports of the plate bear the crank shaft of the impact mechanism with two phase-shifted crank shafts connected to the main and auxiliary connecting rods. The piston hammers and an additional reaction weight with ram heads, which is spring-biased relative to the supports towards the plate, are fixed on the free ends of the connecting rods. The reaction weight is equipped with an anvil fixed with an adjustment screw, and the crank drive contains equal-size crank shafts phase-shifted to the angle and opposing piston hammers, which are installed within the body, each in its own guide aligned with the vertical axis.

## 2 Vibropercussion Mechanism with a Crank-Type Vibrator and a Reaction Weight

The physical scheme of the mechanism can be presented as follows (Fig. 1).

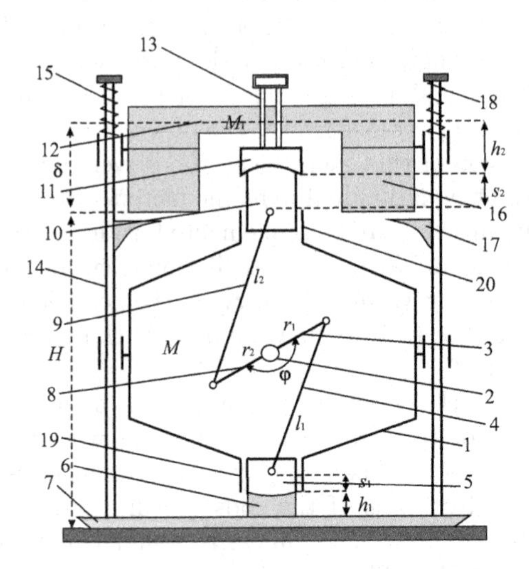


**Fig. 1.** Structure of the vibrating rammer with a reaction weight.

The driving torque is transferred from the motor to crank shaft 2. Then, piston hammers 5 and 10 shuttle relative to body 1 in guides 19 and 20 in the opposite directions, interacting periodically with anvils 6 and 11. As a result, body 1, tamping plate 7, and reactive weight 12 start shuttling vertically relative to each other. With proper adjustment of the structure parameters, weight 12 can enhance significantly the quality of material densification.

### 2.1 Mathematical Model

Let $y$ be the coordinate of the center of mass $M$ of the body, $y_1$ be the coordinate of the center of mass $M_1$ of the reaction weight counting from the tamping plate, and $y_{p_1}$ and $y_{p_2}$ be the values characterizing the deflection of the bases of the $i$th piston hammer.

Under the condition $y_1 - \delta > H$, the equation for the vibration of the reactive weight $M_1$ on the springs has the form

$$M_1 \frac{\mathrm{d}^2 y_1}{\mathrm{d}t^2} + c y_1 + M_1 g = 0.$$

The impact of mass $M_1$ with anvil 17 $y_1 - \delta = H$, $\frac{\mathrm{d}y_1}{\mathrm{d}t} < 0$ is described by the relationship

$$\left.\frac{\mathrm{d}y_1}{\mathrm{d}t}\right|_+ = -R \left.\frac{\mathrm{d}y_1}{\mathrm{d}t}\right|_- .$$

The impact of piston hammer 10 having the mass $m$ with auxiliary anvil 11 occurs if the following conditions are fulfilled:

$$\begin{cases} y_1 - \delta > H, y_{p_2} = y_1 - h_2 \\ \left.\frac{\mathrm{d}y_1}{\mathrm{d}t}\right|_+ - \left.\frac{\mathrm{d}y_{p_2}}{\mathrm{d}t}\right|_+ = -R\left(\left.\frac{\mathrm{d}y_1}{\mathrm{d}t}\right|_- - \left.\frac{\mathrm{d}y_{p_2}}{\mathrm{d}t}\right|_-\right), \\ M_1 \left.\frac{\mathrm{d}y_1}{\mathrm{d}t}\right|_+ + m \left.\frac{\mathrm{d}y_{p_2}}{\mathrm{d}t}\right|_+ = M_1 \left.\frac{\mathrm{d}y_1}{\mathrm{d}t}\right|_- + m \left.\frac{\mathrm{d}y_{p_2}}{\mathrm{d}t}\right|_- . \end{cases}$$

The free-motion equation for the weight $M$ occurs at $y_{p_1} > h_1, y_{p_2} < y_1 - h_2$ can be written as follows:

$$M \frac{\mathrm{d}^2 y}{\mathrm{d}t^2} = -Mg.$$

The impact of piston 5 with anvil 6, if the conditions $y_{p_1} = h_1$, $\frac{\mathrm{d}y_{p_1}}{\mathrm{d}t} < 0$ are fulfilled, is

$$\left.\frac{\mathrm{d}y_{p_1}}{\mathrm{d}t}\right|_+ = -R \left.\frac{\mathrm{d}y_{p_1}}{\mathrm{d}t}\right|_- .$$

The positions of the eccentricities having the lengths $r_1$ and $r_2$ will be measured with the angles $\theta = \omega t$ and $\alpha = \varphi - \theta$ counted from the vertical axis, respectively. Then, one can write down the apparent relationships (see Fig. 2).

$$y_{p_1} = y - s_1 + r_1 \cos\theta - \sqrt{l_1^2 - r_1^2 \sin^2\theta},$$

$$y_{p_2} = y + s_2 + r_2 \cos(\theta - \varphi) + \sqrt{l_1^2 - r_1^2 \sin^2(\theta - \varphi)}.$$

Let us take into account that $l_1 \approx l_2 = l, r_i \ll l (i = 1, 2)$. Then, the two latter equations are written as

$$y_{p_1} = y - s_1 + r_1 \cos\theta - l,$$
$$y_{p_2} = y + s_2 + r_2 \cos(\theta - \varphi) + l.$$

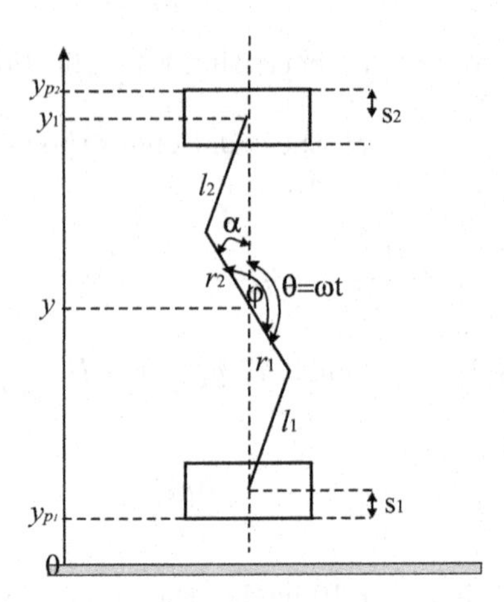


**Fig. 2.** Structure of the vibrating rammer with a reaction weight.

Thus, the motion equations of the mechanism in the dimensional form are written as

$$\begin{cases} y_1 - \delta > H \\ M_1 \dfrac{\mathrm{d}^2 y_1}{\mathrm{d}t^2} = -cy_1 - M_1 g. \end{cases} \tag{1}$$

$$\begin{cases} y_1 - \delta = H, \dfrac{\mathrm{d}y_1}{\mathrm{d}t} < 0 \\ \left.\dfrac{\mathrm{d}y_1}{\mathrm{d}t}\right|_+ = -R \left.\dfrac{\mathrm{d}y_1}{\mathrm{d}t}\right|_- . \end{cases} \tag{2}$$

$$\begin{cases} y_1 - \delta > H, y_{p_2} = y_1 - h_2, \\ \left.\dfrac{\mathrm{d}y_1}{\mathrm{d}t}\right|_+ - \left.\dfrac{\mathrm{d}y_{p_2}}{\mathrm{d}t}\right|_+ = -R\left( \left.\dfrac{\mathrm{d}y_1}{\mathrm{d}t}\right|_- - \left.\dfrac{\mathrm{d}y_{p_2}}{\mathrm{d}t}\right|_- \right) \\ M_1 \left.\dfrac{\mathrm{d}y_1}{\mathrm{d}t}\right|_+ + m \left.\dfrac{\mathrm{d}y_{p_2}}{\mathrm{d}t}\right|_+ = M_1 \left.\dfrac{\mathrm{d}y_1}{\mathrm{d}t}\right|_- + m \left.\dfrac{\mathrm{d}y_{p_2}}{\mathrm{d}t}\right|_- . \end{cases} \tag{3}$$

$$\begin{cases} y_{p_1} = h_1, \dfrac{\mathrm{d}y_1}{\mathrm{d}t} < 0 \\ \left.\dfrac{\mathrm{d}y_{p_1}}{\mathrm{d}t}\right|_+ = -R \left.\dfrac{\mathrm{d}y_{p_1}}{\mathrm{d}t}\right|_- . \end{cases} \tag{4}$$

$$\begin{cases} y_{p_1} > h_1, \\ y_{p_2} < y_1 - h_2, \\ \dfrac{\mathrm{d}^2 y}{\mathrm{d}t^2} = -g. \end{cases} \tag{5}$$

## 3 Vibropercussion Mechanism with a Crank-Type Vibrator Without a Reaction Weight

The physical structure of the considered vibropercussion mechanism without a reaction weight can be represented as shown in Fig. 3. It consists of body 1 having the mass $M$, and shaft 2 with flywheels at its ends is installed in the bearings of the body. Each of the eccentric mechanisms mounted on this shaft consists of two eccentrics 3 and 4 inserted into each other and providing the possibility to change the positions of the washers. This allows one to adjust the eccentricities $r_i$ and phase shifts $\varphi_i$ between them (here, $i = 1, 2, 3$). Pivoted piston hammers are installed on the free ends of the connecting rods. Together with the connecting rods and the piston hammers, the eccentric mechanisms convert the motion of the shaft, which is rotatory with the constant cyclic frequency $\omega$ of the flywheel rotation, into the shuttling motion of the body relative to supports 7. The piston hammers are located one inside the other. Each of them impacts with the respective anvil 6, and the vibropercussion action produced in this way is transferred to the tamped material via anvil block 8.

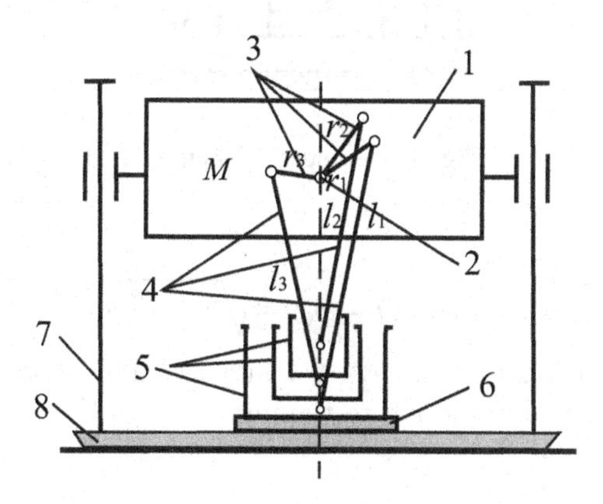


**Fig. 3.** Structure of the vibropercussion mechanism.

### 3.1 Mathematical Model

We denote the coordinate of the center of mass $M$ of the body by $y$, and the coordinates of the masses $m_i$ by $y_{p_i}$ (where $i = 1, 2, 3$). Then, we form the formulas for the kinetic $V$ and potential $\Pi$ energies:

$$T = \frac{1}{2}M\dot{y}^2 + \frac{1}{2}\sum_{i=1}^{3} m_i \dot{y}_{p_i}^2; \Pi = Mgy + g\sum_{i=1}^{3} m_i y_{p_i}.$$

and represent the Lagrangian function as

$$L = \frac{1}{2}M\dot{y}^2 + \frac{1}{2}\sum_{i=1}^{3} m_i \dot{y}_{p_i}^2 - Mgy - g\sum_{i=1}^{3} m_i y_{p_i}$$

Using the Lagrangian equations of the second kind [13] and the relation between the coordinates of the body and the piston hammers (Fig. 4).

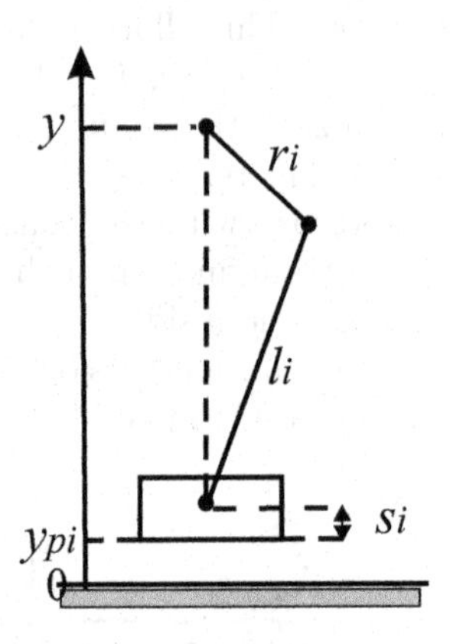


**Fig. 4.** Geometric scheme.

$$y_{p_i} = y - s_i + r_i \cos(\omega t - \varphi_i) - \sqrt{l_i^2 - r_i^2 \sin^2(\omega t - \varphi_i)},$$

and allowing for the relationship $r_i \ll l$ and assuming $l_i \approx l, i = 1, 2, 3$, we transform the Lagrangian function to the form

$$L = \frac{1}{2}M\dot{y}^2 + \frac{1}{2}\sum_{i=1}^{3} m_i(\dot{y} - r_i\omega \sin(\omega t - \varphi_i))^2 - Mgy$$

$$-\sum_{i=1}^{3} m_i g(y - s_i + r_i \cos(\omega t - \varphi_i) - l).$$

Then, by passing over to the dimensionless coordinate $x = \dfrac{y - s_2 - l}{l}$ and the time $\tau = \omega t$, the motion equations of the mechanism can be written as follows:

$$\begin{cases} \dfrac{\mathrm{d}^2 x}{\mathrm{d}\tau^2} - A\cos(\tau - \Phi) + p = 0, (x > f(\tau)) \\ \left.\dfrac{\mathrm{d}x}{\mathrm{d}\tau}\right|_+ = -R\left.\dfrac{\mathrm{d}x}{\mathrm{d}\tau}\right|_- + (1+R)\dfrac{\mathrm{d}f(\tau)}{\mathrm{d}\tau}, (x = f(\tau), \dot{x} - \dfrac{\mathrm{d}f}{\mathrm{d}\tau} < 0). \end{cases} \tag{6}$$

$f(\tau) = \max_{\tau}(f_1(\tau), f_2(\tau), f_3(\tau))$, $f_i(\tau) = \varepsilon_i - \mu\gamma_i \cos(\tau - \varphi_i), i = 1, 2, 3$, where $\mu = r_1/l, \gamma_i = r_i/r_1, \epsilon_i = (s_i - s_2)/l, p = g/\omega^2 l$, $a = \mu \sum_{i=1}^{3} \gamma_i \lambda_i \sin \varphi_i$, $b = \mu \sum_{i=1}^{3} \gamma_i \lambda_i \cos \varphi_i$, $A = \sqrt{a^2 + b^2}$, $tg\Phi = a/b$, $\varphi_1 = 0$, $\gamma_1 = 1$, $\varepsilon_2 = 0$.

The phase space of the system $\Phi(x \geq f(\tau), \dot{x} < +\infty)$ in the coordinates $x, \dot{x}, \tau$ is truncated with respect to $x$. The surface $S(x = f(\tau))$ is a corrugated cylindrical surface formed by crossing of three surfaces, $x = f_i(\tau)$. All the phase trajectories are located either on the surface $S$ or above it. The case $x > f(\tau)$ corresponds to the free motion of the mechanism, and $x = f(\tau)$ corresponds to the impact interaction of one of the pistons with the anvil block. The qualitative form of the phase trajectories is shown in Fig. 5.

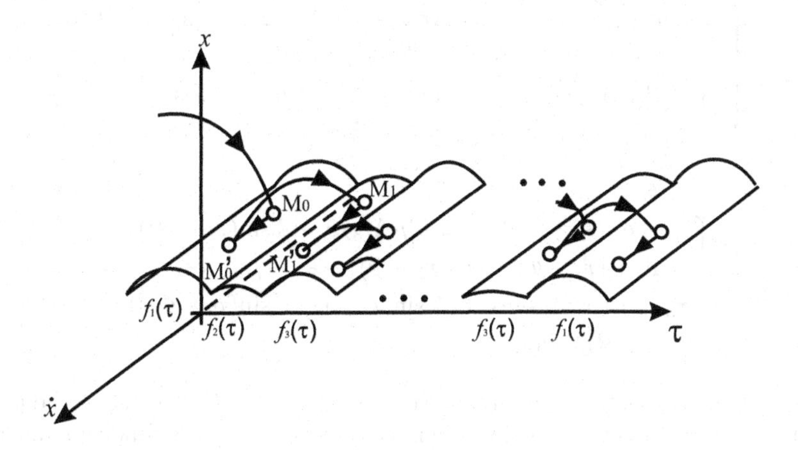


**Fig. 5.** Phase space of system.

It should be noted that the shape of the surface $S$ shown in Fig. 5 can take place only when the surfaces $f_1(\tau), f_2(\tau)$ and $f_3(\tau)$ intersect. The conditions for the intersection of these surfaces can be written as

$$\varepsilon_1^2 \leq \mu^2(\gamma_2^2 + 1 - 2\gamma_2 \cos \varphi_2),$$

$$\varepsilon_3^2 \leq \mu^2(\gamma_3^2 + \gamma_2^2 - 2\gamma_2\gamma_3 \cos(\varphi_3 - \varphi_2)).$$

Taking into account the structure of the phase space and the behavior of the phase trajectories in it, the dynamics of the vibropercussion mechanism was studied further by investigating the properties of the point transformation [19] of the surface $S(x = f(\tau))$.

Let

$$M_0(\tau = \tau_0, x = f_1(\tau_0), \dot{x} = \dot{x_0}) \in S_1(x = f_1(\tau)),$$

$$M_1(\tau = \tau_1, x = f_2(\tau_1), \dot{x} = \dot{x_1}) \in S_2(x = f_2(\tau)),$$

$$M_2(\tau = \tau_2, x = f_3(\tau_2), \dot{x} = \dot{x_2}) \in S_3(x = f_3(\tau)),$$
$$M_3(\tau = \tau_3, x = f_1(\tau_3), \dot{x} = \dot{x_3}) \in S_1(x = f_1(\tau)).$$

be the four sequential points that belong to the surface $x = f(\tau)$.

Then, the conversion $T = T_3T_2T_1$ of the points $M_0 \xrightarrow{T_1} M_1 \xrightarrow{T_2} M_2 \xrightarrow{T_3} M_3$ can be written as

$$T_1 = \begin{cases} -\mu\gamma_2 \cos(\tau_1 - \varphi_2) = \Delta\tau_1(\dot{x_0} - p\Delta\tau_1/2 - A\sin(\tau_0 - \Phi)) \\ -A(\cos(\tau_1 - \Phi) - \cos(\tau_0 - \Phi)) + \varepsilon_1 - \mu\cos\tau_0, \\ \dot{x_1} = R(p\Delta\tau_1 - \dot{x_0} + A(\sin(\tau_1 - \Phi) - \sin(\tau_0 - \Phi))) \\ +(1+R)\mu\sin\tau_1. \end{cases} \quad (7)$$

$$T_2 = \begin{cases} \varepsilon_3 - \mu\gamma_3 \cos(\tau_2 - \varphi_3) = \Delta\tau_1(\dot{x_1} - p\Delta\tau_2/2 - A\sin(\tau_1 - \Phi)) \\ -A(\cos(\tau_2 - \Phi) - \cos(\tau_1 - \Phi)) - \mu\gamma_2\cos(\tau_1 - \varphi_2), \\ \dot{x_2} = R(p\Delta\tau_2 - \dot{x_1} + A(\sin(\tau_2 - \Phi) - \sin(\tau_1 - \Phi))) \\ +(1+R)\mu\gamma_2\sin(\tau_2 - \varphi_2). \end{cases} \quad (8)$$

$$T_3 = \begin{cases} \varepsilon_1 - \mu\cos\tau_3 = \Delta\tau_3(\dot{x_2} - p\Delta\tau_3/2 - A\sin(\tau_2 - \Phi)) \\ -A(\cos(\tau_3 - \Phi) - \cos(\tau_2 - \Phi)) + \varepsilon_3 - \mu\gamma_3\cos(\tau_2 - \varphi_3), \\ \dot{x_3} = R(p\Delta\tau_3 - \dot{x_2} + A(\sin(\tau_3 - \Phi) - \sin(\tau_2 - \Phi))) \\ +(1+R)\mu\sin\tau_3. \end{cases} \quad (9)$$

Due to the truncated character of the phase space with respect to the phase coordinate $x$, the point conversion $T$ will be determined, if the following inequalities are fulfilled:

$$\begin{aligned} &\varepsilon_1 - \mu\cos\tau_0 > (-\mu\gamma_2\cos(\tau_0 - \varphi_2)), \varepsilon_3 - \mu\gamma_3\cos(\tau_0 - \varphi_3)) \\ &- \mu\gamma_2\cos(\tau_1 - \varphi_2) > (\varepsilon_1 - \mu\cos\tau_1, \varepsilon_3 - \mu\gamma_3\cos(\tau_1 - \varphi_3)) \\ &\varepsilon_3 - \mu\gamma_3\cos(\tau_2 - \varphi_3) > (\varepsilon_1 - \mu\cos\tau_2 \geq -\mu\gamma_2\cos(\tau_2 - \varphi_2)) \\ &\varepsilon_1 - \mu\cos\tau_3 > (-\mu\gamma_2\cos(\tau_3 - \varphi_2)) \geq \varepsilon_3\gamma_3\cos(\tau_3 - \varphi_3) \\ &x(\tau) > f(\tau), \tau_i < \tau < \tau_{i+1}, i = 0, 1, 2 \end{aligned} \quad (10)$$

The first three inequalities of system (13) mean that at the time instants $\tau = \tau_i, i = 0, 1, 2$ the representative point belongs to the surface $S(x = f_i(\tau), i = 1, 2, 3)$ (these conditions correspond to the impact interaction of the $i$th piston with the anvil block), while the fourth inequality means that at the time instant $\tau = \tau_3$ the representative point belongs again to the surface $S(x = f_1(\tau))$ (this condition corresponds to the impact interaction of the first piston with the anvil block). The fifth inequality reflects the self-evident fact that the representative point belongs to the subspace $x > f(\tau)$ (free motion of the mechanism).

The equations that yield the coordinates of the stationary points of the point transformation $T$, which correspond to periodic motions with successive hits performed by each piston hammer, are obtained by supplementing Eqs. (7)–(9) with the periodicity conditions [20]

$$\dot{x_3} = \dot{x_0} = \dot{x}, \tau_3 = \tau_0 + 2\pi n, (n = 1, 2, ...). \tag{11}$$

Solving system (7)–(9), (11) with regard to $\dot{x}, \dot{x_1}, \dot{x_2}$, we find their dependences on the time $\tau = \tau_i, i = 0, 1, 2$ and the parameters, whose formulas are omitted because of their awkwardness. Substituting the expressions for the coordinates of the stationary points to Eqs. (7)–(9), respectively, and performing a series of transformations, we obtain a system of three transcendental equations for the times $\tau_0, \tau_1, \tau_2$. An original numerical analytical method was developed for solving the above-mentioned equations [16].

It is known [20] that stability in small of a stationary point is determined by the value of the roots of the characteristic equation $\chi$. In our case, it has the form

$$\chi(z) = A_1 z^2 + B_1 z + C_1 = 0. \tag{12}$$

The coefficients $A_1, B_1, C_1$ are calculated after linearization of the point transformation Eqs. (7)–(9) in the vicinity of the found stationary points.

Let us denote the region of existence and stability of the periodic motions in the parametric space by $D(m_1, m_2, m_3)$, where $m_i$ is the number of impacts of the $i$th piston with the anvil block.

It is known [20] that the boundary of the region of existence and stability of the stationary points of the point representation $T$ is determined by the surfaces $N_+, N_-, N_\phi$, the equations for which are obtained by substituting $z = 1, z = -1, z = \exp(\pm i\varphi), 0 \le \varphi \le 2\pi$, respectively, to Eq. (12). The equations for these boundaries have the form

$$\begin{aligned} N_+ &: A_1 + B_1 + C_1 = 0, \\ N_- &: A_1 - B_1 + C_1 = 0, \\ N_\varphi &: A_1 - C_1 = 0. \end{aligned} \tag{13}$$

### 3.2 Numerical Study of Mechanism Dynamics

The dynamics of the three-piston mechanism was studied further using numerical calculations performed by the software suite developed in the Borland Developer Studio 2006.

Figures 6, 7 shows two bifurcation diagrams for the three-piston mechanism, which were plotted for the sets of parameters

$$\mu = 0.12, \varepsilon_1 = 0.018, \varepsilon_3 = 0.02, \gamma_2 = 3, \gamma_3 = 3, \varphi_2 = 0.2, \varphi_3 = 1.1,$$

$$\lambda_1 = 0.1, \lambda_2 = 0.2, \lambda_3 = 0.3.$$

and various R equal to 0.2 (Fig. 6) and 0.4 (Fig. 7).

In the diagram, the frequency parameter $p$ is plotted on the $x$ axis, and the after-impact velocities of the piston hammers, on the $y$ axis. It is seen in Fig. 6 that at the frequencies $0.13 \leq p \leq 0.15$ and $0.16 \leq p \leq 0.28$ there exists a periodic motion with $m_i = 1, i = 1, 2, 3$, whereas at the frequencies $0.11 \leq p \leq 0.12$ and $0.15 \leq p \leq 0.16$, the chaotic motion regime is observed. Here, $p = 0.13$ is the bifurcation value of the parameter, at which one observes the duplication of the number of impacts performed by the first, second, and third pistons.

It is seen in Fig. 7 that the periodic motion $m_i = 1, i = 1, 2, 3$ is observed at the frequencies $0.2 \leq p \leq 0.3$, whereas at $0.18 \leq p \leq 0.2$, there exists the motion with two impacts of the first piston, one impact of the second piston, and two impacts of the third piston. At $0.14 \leq p \leq 0.18$, one observes the motion with two impacts of the first piston, three impacts of the second and two impacts of the third piston. The chaotic regime is observed at $0.11 \leq p \leq 0.14$.

Thus, when comparing Fig. 6 and Fig. 7, it is possible to conclude that, as $R$ increases, the periodic motion regime $D(1, 1, 1)$ shifts toward greater values of the frequency parameter $p$.

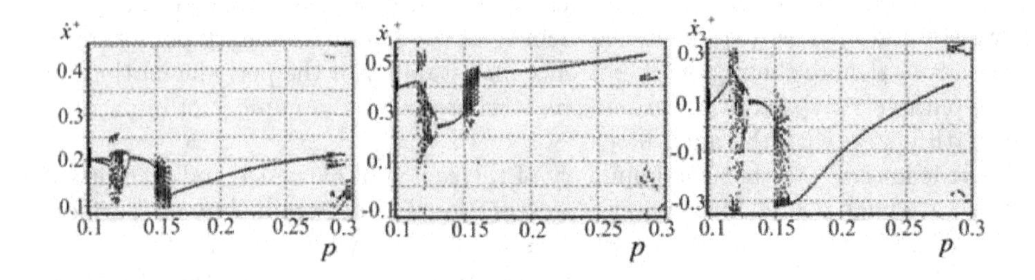


**Fig. 6.** Bifurcation diagrams with respect to the frequency parameter p for $R = 0.2$

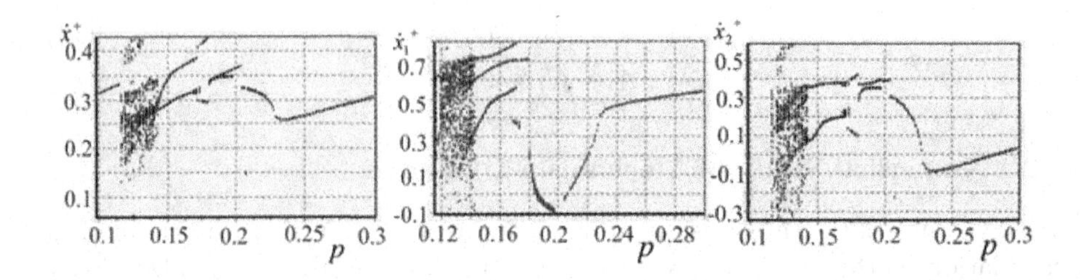


**Fig. 7.** Bifurcation diagrams with respect to the frequency parameter p for $R = 0.4$

Figures 8 and 9 present the bifurcation diagrams with respect to the frequency parameter $p$ for the following sets of parameters:

$$\mu = 0.12, \varepsilon_1 = 0.018, \varepsilon_3 = 0.02, \gamma_2 = 3, \gamma_3 = 3, \lambda_1 = 0.1, \lambda_2 = 0.2,$$

$$\lambda_3 = 0.3, R = 0.6$$

and various $\varphi_2, \varphi_3$ equal to 0.2, 1.1 (Fig. 8) and 0.2 and 0.8 (Fig. 9), respectively.

Figure 8 shows that the regime with one impact of the first piston and the chaotic regime for the second and third piston are observed, at $0.12 \leq p \leq 0.14$ the chaotic regime is observed for the all three pistons, at $0.14 \leq p \leq 0.17$ one observes the regime with two impacts of the first piston, three impacts of the second piston and two impacts of the third piston, at $0.17 \leq p \leq 0.2$ the regime with two impacts of the first and third pistons and one impact of the second piston is observed, and at $0.2 \leq p \leq 0.3$, there exists the periodic motion with $m_i, i = 1, 2, 3$.

In Fig. 9, the chaotic regime is observed at $0.1 \leq p \leq 0.18$ for all three pistons. At $0.18 \leq p \leq 0.19$, the regime with four impacts of the first piston, three impacts of the second piston and three impacts of the third piston is observed, at $0.19 \leq p \leq 0.2$ the regime with four impacts of the first and second pistons and two impacts of the third piston is observed, at $0.2 \leq p \leq 0.21$ one observes the regime with two impacts of the first and second piston and one impact of the first piston, the chaotic regime is observed for all the three pistons at $0.21 \leq p \leq 0.225$, and $0.225 \leq p \leq 0.3$ there exists the regime with three impacts of the first piston and two impacts of the second and third pistons. Comparing Figs. 8 and 9, one can see that a decrease in $\varphi_3$ leads to disappearance of the periodic regime with alternating collisions of the pistons and an increase in the range of the values of the parameter $p$, at which the chaotic regime is observed.

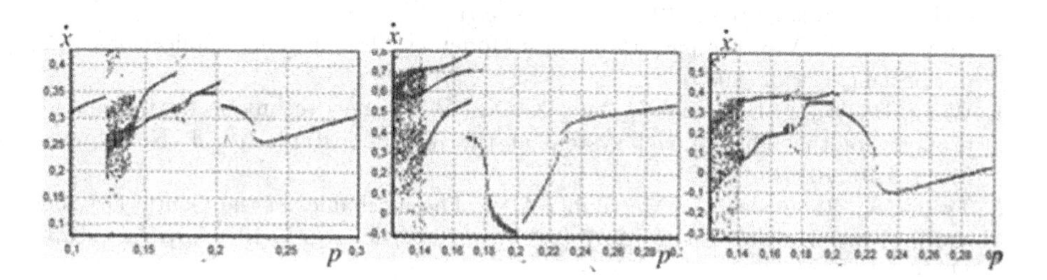

**Fig. 8.** Bifurcation diagrams with respect to the frequency parameter p for $\varphi_2 = 0.2, \varphi_3 = 1.1$

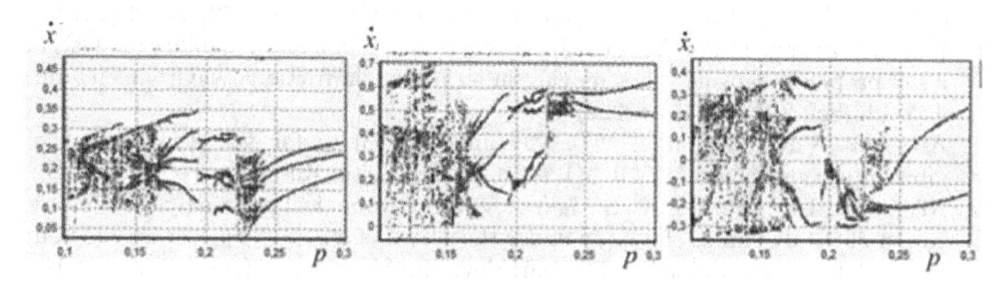

**Fig. 9.** Bifurcation diagrams with respect to the frequency parameter p for $\varphi_2 = 0.2, \varphi_3 = 0.8$

# 4 Conclusion

- A new mathematical model of a vibropercussion mechanism with a reactive mass has been developed, which differs from the known ones in that when various media are compacted, the reactive mass enhances the impact and improves the structure and density of the treated medium.
- Using the developed original numerical method for studying the dynamics of essentially nonlinear systems, numerical experiments were carried out, the results of which are presented in the form of bifurcation diagrams for the main parameters of the system, which made it possible to study the processes of rearranging the operating modes of mechanisms depending on the parameters of the dynamic system.
- The ranges of parameter values are found for which there are regimes with a finite number of impacts of the mechanism and chaotic regimes.

# References

1. Blekhman, I.I., Dzhanelidze, Yu.G.: Vibratory movement. Nauka, 410 p. (1964). (in Russian)
2. Blekhman, I.I.: Synchronization of dynamic systems. Fizmatgiz, 894 p. (1971). (in Russian)
3. Babitsky, V.I., Krupenin, V.L.: Vibration of Strongly Nonlinear Discontinuous Systems, 404 p. Springer, Berlin (2001)
4. Kobrinskiy, A.E., Kobrinskiy, A.A.: Vibropercussion mechanism (Dynamics and stability). Nauka, 591 p. (1973). (in Russian)
5. Ibrahim, R.A.: Vibro-Impact Dynamics: Modeling, Mapping and Applications, 312 p. Springer, Berlin (2009)
6. Wang, B., Wang, L., Peng, J., Yue, X., Xu, W.: A new technique for the global property of the vibro-impact system at the impact instant. Int. J. Non-Linear Mech. **140**, 103914 (2022)
7. Tsetas, A., Tsouvalas, A., Metrikine, A.V.: The mechanics of the Gentle Driving of Piles. Int. J. Solids Struct. **207** (2023). https://doi.org/10.1016/j.ijsolstr.2023.112466
8. Liu, Y., Wiercigroch, M., Pavlovskaia, E., Hongnian, Y.: Modelling of a vibro-impact capsule system. Int. J. Mech. Sci. **66**, 2–11 (2013)
9. Liu, Y., Pavlovskaia, E., Hendry, D., Wiercigroch, M.: Vibro-impact responses of capsule system with various friction models. Int. J. Mech. Sci. **72**, 39–54 (2013)
10. Liao, M., Zhang, J., Liu, Y., Zhu, D.: Speed optimisation and reliability analysis of a self-propelled capsule robot moving in an uncertain frictional environment. Int. J. Mech. Sci. **221**, 107156 (2022)
11. Astashev, V.K., Krupenin, V.L.: Dynamic principles for the development of vibroimpact machines. VNTR **5**(129), 3–10 (2018). (in Russian)
12. Nikiforova, I.V., Metrikin, V.S., Igumnov, L.A.: Numerical and analytical investigation of the dynamics of a body under the action of a periodic piecewise constant external force. In: Balandin, D., Barkalov, K., Meyerov, I. (eds.) MMST 2022. CCIS, vol. 1750, pp. 67–81. Springer, Cham (2022). https://doi.org/10.1007/978-3-031-24145-1_6

13. Butenin, N.V., Lunts, Ya.L., Merkin, D.R.: Theoretical mechanics course. V 2 t. T.2. Dynamics. Nauka, 496 p. (1985). (in Russian)
14. Panovko, Ya.G.: Introduction to the theory of mechanical shock. Nauka, 224 p. (1977). (in Russian)
15. Igumnov, L.A., Metrikin, V.S., Nikiforova, I.V.: The dynamics of eccentric vibration mechanism (Part 1). J. Vibroengineering **19**(7) (2017). ISSN 1392-8716
16. Nikiforova, I.V., Metrikin, V.S., Igumnov, L.A.: Mathematical modeling of multidimensional strongly nonlinear dynamic systems. In: Balandin, D., Barkalov, K., Gergel, V., Meyerov, I. (eds.) MMST 2020. CCIS, vol. 1413, pp. 63–76. Springer, Cham (2021). https://doi.org/10.1007/978-3-030-78759-2_5
17. Tyuremnov, I.S.: Determination of the coefficients of elastic and viscous soil resistance during its vibrational compaction. MMTT **8**(90), 46–49 (2016). (in Russian)
18. Cytovich, N.A.: Soil mechanics. LIBROKOM, 271 p. (2011). (in Russian)
19. Neymark, Yu.I.: Method of point mappings in the theory of nonlinear oscillations. LIBROKOM, 472 p. (2010). (in Russian)
20. Feygin, M.I.: Forced vibrations of systems with discontinuous nonlinearities. Nauka, 288 p. (1994). (in Russian)

# Parametric Perturbations of a Duffing–Type Equation with Nonmonotonic Rotation

K. E. Morozov(✉) and A. D. Morozov

Lobachevsky State University of Nizniy Novgorod, Nizniy Novgorod, Russia
kirill.morozov@itmm.unn.ru, morozov@mm.unn.ru

**Abstract.** Quasiperiodic parametric perturbations of a Duffing–type equation with nonmonotonic rotation are studied. It is assumed that the perturbations are nonconservative. The solutions behavior in the neighborhood of nearly degenerate resonance levels of energy is described and conditions for the existence of new resonance three–dimensional invariant tori in the extended phase space are found. Bifurcations that lead to the appearance of these solutions are also studied.

**Keywords:** Averaging · Quasiperiodic · Parametric · Perturbation · Duffing equation

## 1 Introduction

When it comes to nonatonomous perturbations of Hamiltonian systems, it can be noticed that the most studied case is the case of periodic perturbations (systems with 3/2 degrees of freedom). The most noticeable advance in the theory of resonance in such systems was achieved in 1983 by L. P. Shilnikov and A. D. Morozov [1]. Their research was continued in many papers (e.g., see monograph [2] and references therein).

While studying periodic and quasiperiodic perturbations of two–dimensional Hamiltonian systems, resonances play a key role. If the unperturbed system has a cell filled with closed trajectories, resonances occur when the natural frequency becomes commensurable with frequencies of the perturbation. If the natural frequency is a nonmonotonic function of energy, so–called degenerate resonances may take place (see, for instance, [3–6]). The study of such resonances in Hamiltonian systems can be sometimes reduced to the study of nonmonotonic twist maps (see [8] and references there). Among applied works on degenerate resonances, let us note the paper [7].

In the present paper, we consider two–frequency *parametric* perturbations of the asymmetric Duffing equation with nonmonotonic rotation. Although there is a number of studies on degenerate resonances in nearly Hamiltonian systems, the case of parametric perturbation was left out of consideration. The objective of the article is to fill this gap. We will find conditions for the existence of new


D. Balandin et al. (Eds.): MMST 2023, CCIS 1914, pp. 86–97, 2024.
https://doi.org/10.1007/978-3-031-52470-7_7

three–dimensional invariant tori in the resonance zone and show that these tori is a peculiarity of parametric perturbations. As well, we will briefly dwell upon bifurcations of these solutions when parameters change.

So, let us consider the equation

$$\ddot{x} + x + ax^2 + x^3 = \varepsilon f(x, \dot{x}, \omega_1 t, \omega_2 t), \tag{1}$$

where $f = (1 - p_1 x^2 + p_2 x b(t))\dot{x} + p_3 b(t)$, $p_i$ are parameters, $b(t) = \sin\omega_1 t \sin\omega_2 t$, $|a| \in (0, 2)$ is fixed, $\omega_1/\omega_2$ is an irrational number. The time–dependent term $x\dot{x}b(t)$ is called parametric. Note that this equation at $a = 0$ (i.e. when the natural frequency is a monotonic function of energy) was considered in [9].

## 2 Preliminaries

The unperturbed equation ($\varepsilon = 0$) is Hamiltonian with

$$H(x, y) = y^2/2 + U(x),$$

where $y \equiv \dot{x}$, $U(x) = x^2/2 + ax^3/3 + x^4/4$ is potential energy. Since $U(-x) \neq U(x)$ if $a \neq 0$, the equation is called the asymmetric Duffing equation. When $|a| < 2$, it has the only equilibrium of the center type, which is surrounded by closed phase curves $H(x, \dot{x}) = h > 0$. The period of motion on these curves is given by

$$T(h) = 2\pi/\omega(h) = \sqrt{2}\int_{x_1(h)}^{x_2(h)} \frac{dx}{\sqrt{h - U(x)}},$$

where $x_1 < x_2$ are the real roots of the equation $h - U(x) = 0, h > 0$. It can be shown that if $|a| \in (0; 2)$, then the natural frequency $\omega(h) = 2\pi/T(h)$ has a simple minimum, i.e. there exists an energy level $H(x, \dot{x}) = h_0$ such that $\omega'(h_0) = 0, \omega''(h_0) > 0$. Energy levels satisfying the condition $\omega'(h) = 0$ are called degenerate. The graph of $T(h)$ and the phase portrait at $a = -1.6$, where the degenerate trajectory is highlighted in bold, are shown in Fig. 1(a) and (b).

It is well-known that at $a = 0$ solutions to the unperturbed equation are given by

$$x(h, \omega t) = x_0 \mathrm{cn}\left(\frac{2\omega\mathbf{K}}{\pi}t\right),$$

where $x_0 = \sqrt{-1 + \sqrt{4h + 1}}$, $\omega = \omega(h) = \pi(1 + 4h)^{1/4}/(2\mathbf{K}(k))$, $\mathbf{K}(k)$ is the complete elliptic integral of the first kind, $k = x_1/(\sqrt{2}(1 + 4h)^{1/4})$. When $a \neq 0$, solutions are also can be found explicitly. They have the form [10]:

$$x(h, \omega t) = \frac{\beta x_2 + \alpha x_1 + (\alpha x_1 - \beta x_2)\mathrm{cn}(\frac{2\omega\mathbf{K}}{\pi}t)}{\alpha + \beta + (\alpha - \beta)\mathrm{cn}(\frac{2\omega\mathbf{K}}{\pi}t)}, \tag{2}$$

where $\alpha = \sqrt{(m - x_2)^2 + n^2}, \beta = \sqrt{(m - x_1)^2 + n^2}$, $x_{3,4} = m \pm in$ are the complex roots of the equation $h - U(x) = 0$, $\omega = \pi\sqrt{\frac{\alpha\beta}{2}}/\mathbf{K}(k)$.

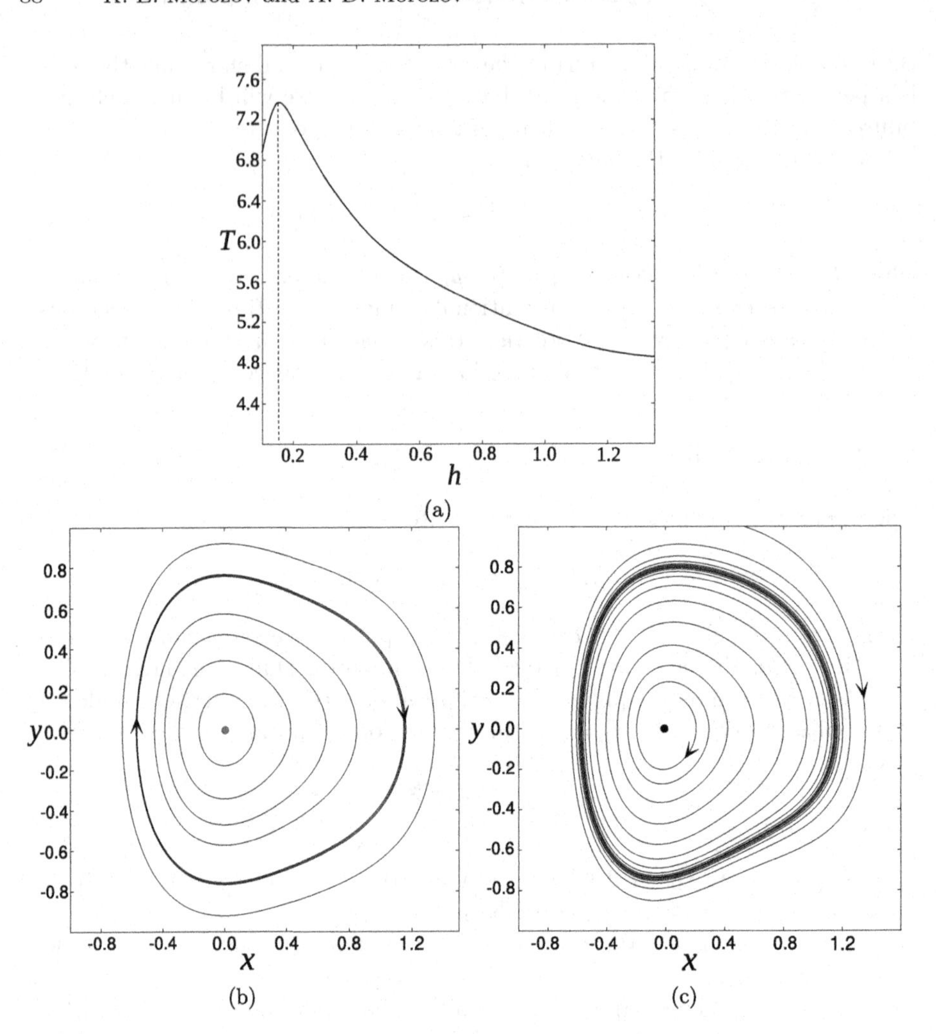


**Fig. 1.** The period of motion $T(h)$ (a); the phase portrait of Eq. (1) at $a = -1.6, \varepsilon = 0$ (b) and at $a = -1.6, \varepsilon = 0.1, p_1 = 4, p_2 = p_3 = 0$ (c). (Color figure online)

Now suppose that $\varepsilon \neq 0$ and consider the perturbed autonomous equation obtained by averaging of the right–hand parts of the initial equation over $t$:

$$\ddot{x} + x + ax^2 + x^3 = \varepsilon(1 - p_1x^2)\dot{x}. \tag{3}$$

When $|a| < 2$, $p_1 > 0$ and $\varepsilon > 0$ is small enough, Eq. (3) has a unique stable limit cycle [5]. Figure 1(c) shows the phase portrait of this equation, where the cycle is highlighted by red. Limit cycles of Eq. (3) play an important role in the behavior of solutions to Eq. (1).

Now, let us turn to Eq. (1). Consider a compact region $D \subset R^2$ filled with closed phase trajectories $H(x; y) = h, 0 < h_1 \leq h \leq h_2$ (thus, $D$ is an annulus). Passing in $D$ to action $I$–angle $\theta$ variables, one can rewrite Eq. (1) in the form:

$$\begin{aligned} \dot{I} &= \varepsilon G_1(I, \theta, \theta_1, \theta_2) \\ \dot{\theta} &= \omega(I) + \varepsilon G_2(I, \theta, \theta_1, \theta_2) \\ \dot{\theta}_k &= \omega_k, \quad k = 1, 2, \end{aligned} \tag{4}$$

where

$$\begin{aligned} G_1 &= f(x(I,\theta), \dot{x}(I,\theta), \theta_1, \theta_2) x'_\theta(I,\theta) \\ G_2 &= -f(x(I,\theta), \dot{x}(I,\theta), \theta_1, \theta_2) x'_I(I,\theta), \end{aligned}$$

the function $x(I, \theta) \equiv x(h(I), \theta)$ is given by Eq. (2) and determines the change of variables[1]. Let us now give a definition of a degenerate phase curve.

**Definition 1.** *If*

$$\omega^{(i)}(I_0) = 0,\ i = \overline{1, j-1};\ \omega^{(j)}(I_0) \neq 0, \quad j > 1,$$

*then a closed phase trajectory $H(x, y) = h(I_0)$ of the unperturbed equation is called degenerate, $j$ is the order of degeneracy. If $j = 1$, it is said that the phase trajectory is non–degenerate.*

In the case of Eq. (1), we have a degenerate phase curve with the order of degeneracy $j = 2$.

At $\varepsilon = 0$ the four–dimensional phase space of the system given by Eq. (4) is comprised of three–dimensional invariant tori $T^3$, the dynamics on which is conditionally periodic with the frequencies $\omega, \omega_1, \omega_2$. When $\varepsilon \neq 0$, most of these tori are destroyed because the perturbation is nonconservative and resonances arise.

**Definition 2.** *A closed phase trajectory $H(x, y) = h(I_{n\mathbf{k}})$ of the unperturbed equation is referred to as a resonance one if there exist setwise coprime integers $(n, k_1, k_2)$ such that*

$$n\omega(I_{n\mathbf{k}}) = k_1\omega_1 + k_2\omega_2. \tag{5}$$

If $\omega(I)$ is a nonmonotonic function, several resonance phase curves may have the same frequency $\omega_{n\mathbf{k}} = k_1\omega_1/n + k_2\omega_2/n$ and, therefore, the same set of numbers $(n, k_1, k_2)$. On the other hand, for each resonance curve, the numbers $n, k_1, k_2$ are determined uniquely. When the resonance phase curve $H(x, y) = h(I_{n\mathbf{k}})$ is degenerate, a degenerate resonance is said to occur.

[1] Details on action–angle variables can be found in [11].

## 3 Resonances. Averaged System

Let us suppose that the degenerate level $H(x, y) = h(I_0)$ satisfies the resonance condition given by (5) (i.e. $I_0 \equiv I_{n\mathbf{k}}$) and consider the neighborhoods $U_\varepsilon = \{(I, \theta) : I_0 - C\varepsilon^s < I < I_0 + C\varepsilon^s, \theta \in [0, 2\pi)\}$, where $C = const > 0, s = 1/3$. Reducing the system given by Eq. (4) in the neighborhood $U_\varepsilon$ to the standard form of the averaging method by the change

$$\theta = v + k_1\theta_1/n + k_2\theta_2/n, \quad I = I_0 + \varepsilon^{1/3}u$$

and applying the averaging procedure, we obtain the following system (up to terms of order $O(\varepsilon^{4/3})$):

$$\begin{aligned} \dot{u} &= \varepsilon^{2/3}A(v, I_0) + \varepsilon P(v, I_0)u \\ \dot{v} &= \varepsilon^{2/3}b_2u^2 + \varepsilon(b_3u^3 + Q(v, I_0)), \end{aligned} \tag{6}$$

where $b_i = \omega^{(i)}(I_0)/i!, i = 1, 2,$

$$\begin{aligned} A =& \frac{1}{(2\pi n)^2}\int_0^{2\pi n}\int_0^{2\pi n} G_1(I_0, v + \frac{1}{n}\sum_{j=1}^{2} k_j\theta_j, \theta_1, \theta_2)x'_\theta d\theta_1 d\theta_2, \\ P =& \frac{1}{(2\pi n)^2}\int_0^{2\pi n}\int_0^{2\pi n} \frac{\partial}{\partial I}G_1(I_0, v + \frac{1}{n}\sum_{j=1}^{2} k_j\theta_j, \theta_1, \theta_2)x'_\theta d\theta_1 d\theta_2 = \frac{\partial A}{\partial I}, \\ Q =& \frac{1}{(2\pi n)^2}\int_0^{2\pi n}\int_0^{2\pi n} G_2(I_0, v + \frac{1}{n}\sum_{j=1}^{2} k_j\theta_j, \theta_1, \theta_2)x'_I d\theta_1 d\theta_2. \end{aligned} \tag{7}$$

It is not hard to see that the functions $A, P, Q$ have the fundamental period $T = 2\pi/n$. Therefore, the phase space of the averaged system is a cylinder $C = S^1 \times R^1 = (v \bmod \frac{2\pi}{n}, u)$. Since the eigenvalues of equilibrium states have zero real parts, the Bogolubov theorem justifying the averaging procedure is inapplicable to the system [12]. Therefore, we consider deformations of the vector field of the system given by Eq. (6). Our objective here is to study the dynamics in the neighborhood $U_\varepsilon$ when the resonance curve approaches a degenerate one. So, let us introduce a parameter of deformation $\gamma$ that defines the distance between $I_0$ and $I_{n\mathbf{k}}$:

$$\begin{aligned} \dot{u} &= \varepsilon^{2/3}A(v, I_0) + \varepsilon P(v, I_0)u, \\ \dot{v} &= \varepsilon^{2/3}(b_2u^2 + \gamma u) + \varepsilon(b_3u^3 + Q(v, I_0)). \end{aligned} \tag{8}$$

Without loss of generality, one can put $b_2 = 1$. The divergence of the system given by Eq. (8) is $\varepsilon\sigma$, where

$$\sigma(v, I_0) = P(v, I_0) + dQ(v, I_0)/dv. \tag{9}$$

It can be shown that

$$\sigma(v, I_0) = \frac{1}{(2\pi n)^2}\int_0^{2\pi n}\int_0^{2\pi n} f'_{\dot{x}}(x, \dot{x}, \theta_1, \theta_2)d\theta_1 d\theta_2$$

where $x = x(I_0, v + k_1\theta_1/n + k_2\theta_2/n), \dot{x} = \dot{x}(I_0, v + k_1\theta_1/n + k_2\theta_2/n)$. As in the case of periodic perturbations [5], the system given by Eq. (8) is reduced to the form (keeping the notations):

$$\begin{aligned} \dot{u} &= A(v, I_0) + \varepsilon^{1/3}\sigma(v, I_0)u, \\ \dot{v} &= \gamma u + u^2, \end{aligned} \tag{10}$$

where the slow time $\tau = \varepsilon^{2/3}t$ is introduced. Now, let us compute the functions $A(v, I_0)$ and $\sigma(v, I_0)$. For the sake of certainty, we consider only resonances with $k_1, k_2 \in N$. Resonances with negative values of $k_1$ or $k_2$ can be studied similarly. Straightforward calculations give us the following expressions:

$$A(v, I_0) = B_0 + \begin{cases} p_2 C_1 \cos nv + p_3 C_2 \sin nv, & k_1 = k_2 = 1, n \text{ is odd}, \\ 0, & \text{otherwise}, \end{cases} \tag{11}$$

$$\sigma(v, I_0) = B_1 + \begin{cases} p_2 C_3 \cos nv, & k_1 = k_2 = 1, n \text{ is odd}, \\ 0, & \text{otherwise}. \end{cases} \tag{12}$$

Note that if $a = 0$, the inequality $n < \omega_1 + \omega_2$ holds, since $\omega(h) > 1$ for $h > 0$. For $a \neq 0$, this condition may be violated due to the non-monotonicity of $\omega(h)$. The constants $B_0, B_1, C_1, C_2, C_3$ are given by

$$B_0 = \frac{\omega}{2\pi}\int_0^{2\pi}(1 - p_1 x^2(\theta))x_\theta'^2(\theta)d\theta, \quad B_1 = \frac{1}{2\pi}\int_0^{2\pi}(1 - p_1 x^2(\theta))d\theta,$$

$$C_1 = -\frac{\omega}{4\pi}\int_0^{2\pi}\cos(n\theta)x(\theta)x'^2(\theta)d\theta, \quad C_2 = -\frac{1}{4\pi}\int_0^{2\pi}\sin(n\theta)x'(\theta)d\theta,$$

$$C_3 = -\frac{1}{4\pi}\int_0^{2\pi}\cos(n\theta)x(\theta)d\theta = \frac{1}{4\pi}\int_0^{2\pi}\sin(n\theta)x'(\theta)d\theta = -C_2,$$

where $x(\theta) = x(I_0, \theta)$ is the unperturbed solution on the phase curve $H(x, \dot{x}) = h(I_0)$. One can see that $B_0 = B_0(I_0)$ is the generating Poincaré–Pontryagin function for the perturbed autonomous equation given by Eq. (3). From the Pontryagin theorem it follows that if $B_0(I_0) = 0, B_1(I_0) = B_0'(I_0) < 0$, this equation has a stable limit cycle in the neighborhood of $H(x, \dot{x}) = h(I_0)$ for small enough $\varepsilon > 0$ (see details in [5]).

*Remark 1.* Resonances, for which the condition $B_0(I_{n\mathbf{k}}) = 0$ holds, are called impassable.

For the rest of the paper, we will suppose that the considered resonance is impassable. From the condition $B_0(I_0) = 0$, we find that

$$p_1^* = \frac{\int_0^{2\pi} x_\theta'^2(\theta)d\theta}{\int_0^{2\pi} x^2(\theta)x_\theta'^2(\theta)d\theta},$$

$$B_1 = 1 - \frac{1}{2\pi}\frac{\int_0^{2\pi} x^2(\theta)d\theta \int_0^{2\pi} x_\theta'^2(\theta)d\theta}{\int_0^{2\pi} x^2(\theta)x_\theta'^2(\theta)d\theta}.$$

At $a = -1.6$ numerical calculations give us the following values:

$$C_1 \approx 0.01, \quad C_2 = -C_3 \approx -0.16, \quad p_1^* \approx 7.56 \quad B_1 \approx -1.2.$$

So, when $k_1 = k_2 = 1$ and $n$ is odd, we obtain

$$\begin{aligned} \dot{u} &= B_0 + p_2 C_1 \cos nv + p_3 C_2 \sin nv + \varepsilon^{1/3}(B_1 - p_2 C_2 \cos nv)u, \\ \dot{v} &= \gamma u + u^2. \end{aligned} \tag{13}$$

By changing the scale of the variable $v$, we can get $n = 1$.

There is a peculiarity of parametric perturbations: if $|p_2 C_2| > |B_1|$, then $\sigma(v, I_0)$ is a sign–alternating function. As a result, contractible limit cycles that do not encircle the phase cylinder can exist in the averaged system. It leads to the existence of new resonance invariant tori in the initial system that have not been studied yet.

### 3.1 First Approximation

Consider the system of the first approximation

$$\begin{aligned} \dot{u} &= p_2 C_1 \cos v + p_3 C_2 \sin v \\ \dot{v} &= \gamma u + u^2, \end{aligned} \tag{14}$$

which is Hamiltonian with

$$H_1(v, u) = \gamma u^2/2 + u^3/3 + U_1(v), \quad U_1(v) = -p_3 C_1 \sin v + p_2 C_2 \cos v.$$

When $\gamma \neq 0$, the system has four simple equilibrium states on the phase cylinder:

$$O_1(v_1, 0), O_2(v_2, 0), O_3(v_1, -\gamma), O_4(v_2, -\gamma),$$

where $v_1, v_2 \in [-\pi, \pi)$ are the roots of the equation:

$$p_2 C_1 \cos v + p_3 C_2 \sin v = 0.$$

It is easy to see that these equilibria are saddles and centers. For instance, if $\gamma(-p_3 C_1 \sin v_1 + p_2 C_2 \cos v_1) < 0$, then $O_1, O_4$ are centres, $O_2, O_3$ are saddles. Possible phase portraits under variation of $\gamma$ are shown in Fig. 2. The bifurcation value $\gamma_{cr}$ (see Fig. 2(b)) can be found from the condition $h_1 = h_2$, where $h_1$ and $h_2$ are the values of the first integral $H_1(v, u)$ near the saddles. Thus, we obtain

$$\gamma_{cr} = \pm \left(6|U_1(v_2) - U_1(v_1)|\right)^{1/3}.$$

When decreasing $|\gamma|$, the saddles and the centers come closer and finally coincide forming complex equilibrium states at $\gamma = 0$.

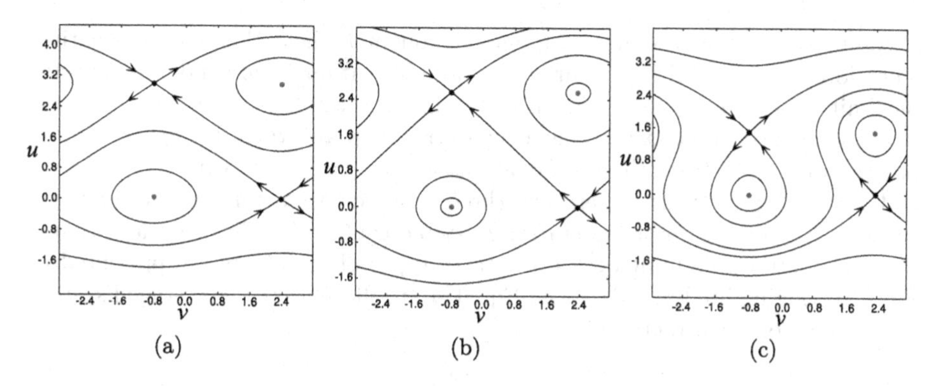


**Fig. 2.** Phase portraits of the first approximation system under variation of $\gamma$ ($p_2C_1 = 1, p_3C_2 = 1$ and $\gamma = -4$ (a); $\gamma = \gamma_{cr} = -(12\sqrt{2})^{1/3} \approx -2.57$ (b); $\gamma = -1.5$ (c).

### 3.2 Second Approximation

Now let us study the system of the second approximation, which is given by Eq. (13). When $\varepsilon \neq 0$ is small enough, the equilibrium states of the center type become foci. The stability of a focus $(u_0, v_0)$ is determined by the sign of $\sigma(v_0, I_0)$. If $\sigma(v_0, I_0) < 0$, the corresponding focus is asymptotically stable. The following theorem holds.

**Theorem 1.** *Let the system given by Eq. (13) have a simple equilibrium state* $(v_0, u_0)$ *and* $\sigma(v_0, I_0) \neq 0$. *Then, if* $\varepsilon > 0$ *is small enough, there exists a quasiperiodic solution* $(x_{n\mathbf{k}}(t), y_{n\mathbf{k}}(t))$ *to the initial system given by Eq. (1) with the frequency basis* $(\omega_1/n, \omega_2/n)$. *This solution is asymptotically stable if* $\sigma(v_0, I_0) < 0$. *Accordingly, the system given by Eq. (4) has a two-dimensional invariant torus* $T^2$, *which is asymptotically stable if* $\sigma(v_0, I_0) < 0$.

*Proof.* Under the conditions of the theorem, the averaged system determined by Eq. (13) has a simple equilibrium state, for which the roots of the characteristic equation have non-zero real parts. Then, the first Bogolubov theorem holds (see [12], p. 379). According to this theorem, the system determined by Eq. (1) has the specified quasiperiodic solution.

Thus, for $\gamma \neq 0$, the original system has four quasiperiodic solutions in the resonance zone with the frequencies $(\omega_1/n, \omega_2/n)$, two of which are saddle solutions.

*Remark 2.* When the degeneracy order equals $j$, there appear $2j$ quasiperiodic solutions with the frequencies $(\omega_1/n, \omega_2/n)$. In this case, $j$ of these solutions are saddle solutions.

From the condition $\sigma(v_0, I_0) = 0$, we find the values of parameters, at which a complex focus is formed:

$$p_3 = \pm \frac{p_2 B_1 C_1}{C_2 \sqrt{p_2^2 C_2^2 - B_1^2}}. \tag{15}$$

It can be shown that the first Lyapunov exponent $L_1$ near the focus is negative. Therefore, the supercritical Andronov–Hopf bifurcation occurs and a unique stable limit cycle appears. When $|p_3|$ is increasing, the amplitude of the cycle grows. At some critical moment it forms a separatrix loop and then disappears. Phase portraits that illustrate these bifurcations are shown in Fig. 3 ($0 < |\gamma| < |\gamma_{cr}|$) and Fig. 4 ($|\gamma| > |\gamma_{cr}|$). There are three types of attractors in the averaged system: 1) stable equilibrium states; 2) stable contractible limit cycles; 3) stable non–contractible limit cycles. All three types of attractors are presented in Fig. 4(b) (highlighted in red). From Eq. (15), it follows that the averaged system has at most one contractible limit cycle.

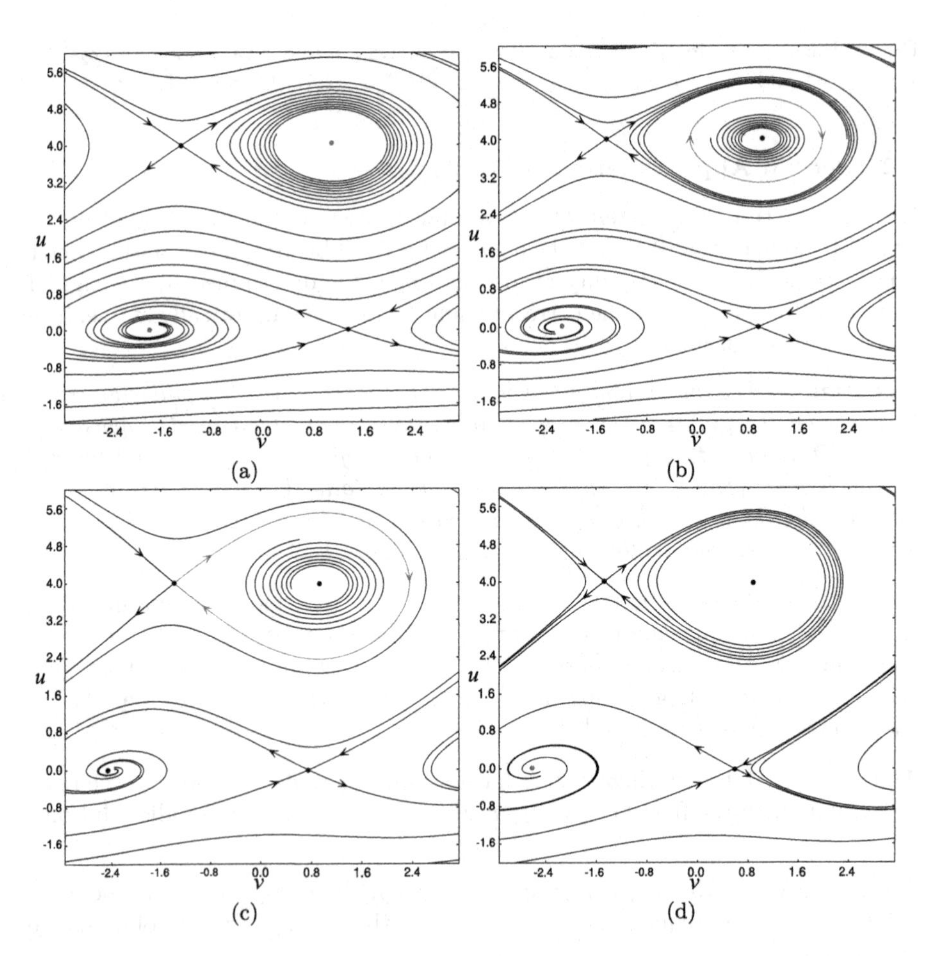


**Fig. 3.** Phase portraits of the second approximation system for $\gamma > \gamma_{cr}$ ($B_1 = -1, C_1 = 0.5, C_2 = 1, \varepsilon^{1/3} = 0.1$, $p_2 = 2, p_3$ is varied).

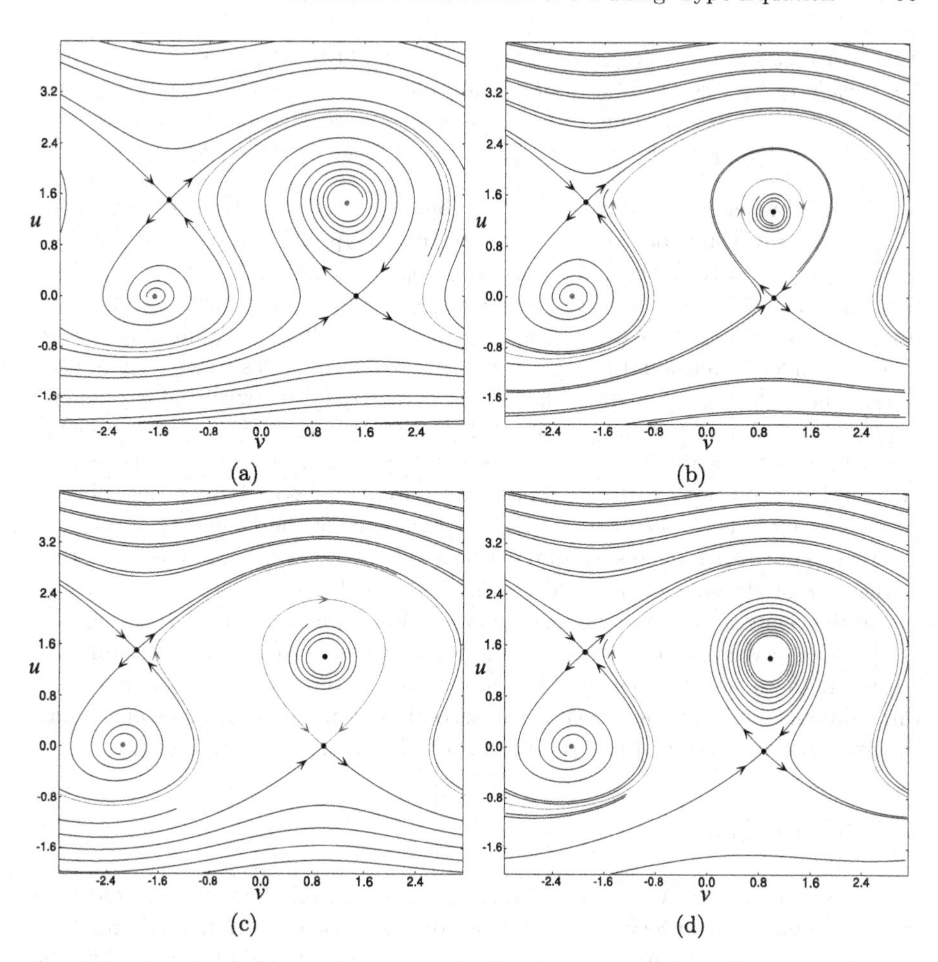


**Fig. 4.** Phase portraits of the second approximation averaged system $0 < \gamma < \gamma_{cr}$ ($B_1 = -1, C_1 = 0.5, C_2 = 1, \varepsilon^{1/3} = 0.1$, $p_2 = 2, p_3$ is varied). (Color figure online)

In Fig. 4, there exists a non–contractible limit cycle. According to [9], this cycle corresponds to a three–dimensional invariant torus in the system determined by Eq. (4). For such a torus, there exists a generating "Kolmogorov" invariant torus $I = I(h_0)$ in the unperturbed system given by Eq. (4) at $\varepsilon = 0$. We call these invariant three–dimensional tori *tori of the first type*. Along with tori of the first type, other invariant three–dimensional tori can exist in the original system. It follows from the second Bogolubov theorem that a contractible limit cycle of the averaged system corresponds to a three-dimensional invariant torus in the system given by Eq. (4), for which there is no generating one in the unperturbed system. These tori are called *tori of the second type*. More accurately, we come to the following theorem.

**Theorem 2.** *Let the averaged system given by Eq.* (13) *have a rough contractible limit cycle in the region of oscillatory motion and assume that* $\gamma \neq 0$, $\gamma \neq \gamma_{cr}$. *Then, if* $\varepsilon > 0$ *is small enough, there exists a three–dimensional invariant torus* $T^3$ *in the system given by Eq. (4). The torus is stable if the cycle is stable.*

Three–dimensional invariant tori of the second type bifurcate from two–dimensional invariant tori corresponding to simple equilibrium states of the averaged system as a result of the Andronov–Hopf bifurcation (see Figs. 3 and 4). One can also see that the initial system has at most six invariant tori in the resonance zone: four two–dimensional tori corresponding to equilibrium states and two three–dimensional tori corresponding to limit cycles (one torus of each type). The motion on two–dimensional tori is quasiperiodic with the frequencies $(\omega_1/n, \omega_2/n)$. The dynamics on three–dimensional tori is the following: either any solution is quasiperiodic with three fundamental frequencies or there exist quasiperiodic solutions with two fundamental frequencies and any other solution on the torus tends to one of them. For the existence of the second type tori, the divergence of the averaged system must be a sign-alternating function. Note that in the case of the so–called forced resonance ($p_2 = 0, p_3 \neq 0$), the divergence of the initial system does not depend on time and, consequently, the divergence of the averaged system is constant. In this case, according to the Bendixson criterion, there are no contractible limit cycles. Therefore, there cannot exist three-dimensional invariant tori of the second type in the initial system. Thus, the existence of these tori is a peculiarity of parametric perturbations.

## 4 Conclusion

In the present paper, we study a Duffing–like equation that is subjected to a two–frequency nonconservative perturbation. The perturbation is assumed to contain the so–called parametric terms. It is shown that these terms lead to the existence of new three–dimensional invariant tori in the extended phase space.

The unperturbed equation has a center equilibrium, which is surrounded by closed phase trajectories $H(x, y) = h, h > 0$. One can see that the frequency of motion on these trajectories $\omega = \omega(h)$ is a nonmonotonic function: it has a simple minimum $h_0$. The corresponding closed phase curve $H(x, y) = h_0$ is called degenerate. If $\omega(h_0)$ is commensurable with the frequencies of the perturbation, we say that degenerate resonance arises. Our main objective is to study the behavior of solutions in the neighborhood of a resonance level when it is nearly degenerate. We introduce a deformation parameter that determines the deviation of the resonance level from the degenerate one. Based on the analysis of the averaged system and the Bogolubov theorems, we find conditions for the existence of new resonance three–dimensional invariant tori in the extended phase space. It is indicated that the presence of such tori is a feature of parametric perturbations. As well, we investigate bifurcations that result in the emergence of such solutions.

**Acknowledgements.** Authors acknowledge a financial support from the Russian Science Foundation [grant 24–21–00050]. Numerical simulations of the paper were supported by the Ministry of Science and Higher Education of the Russian Federation [grant FSWR-2020-0036]. K.E. Morozov was partially supported by the RSciF [grant number 19–11–00280] (Section 3, Theorem 2).

## References

1. Morozov, A.D., Shil'nikov, L.P.: On nonconservative periodic systems similar to two-dimensional Hamiltonian ones. Pricl. Mat. i Mekh. **47**(1), 385–394 (1983)
2. Morozov, A.D.: Resonances, cycles and chaos in quasi-conservative systems. Regular and Chaotic Dynamics, Moscow-Izhevsk (2005)
3. Morozov, A.D., Boykova, S.A.: On investigation of the degenerate resonances. Regular Chaotic Dyn. **4**(4), 70–82 (1999)
4. Morozov, A.D.: Degenerate resonances in Hamiltonian systems with 3/2 degrees of freedom. Chaos **12**(3), 539–548 (2002)
5. Morozov, A.D.: On degenerate resonances and "vortex pairs". Regular Chaotic Dyn. **13**(1), 27–36 (2008)
6. Morozov, A.D., Morozov, K.E.: Synchronization of quasiperiodic oscillations in nearly Hamiltonian systems: the degenerate case. Chaos **31**, 083109 (2021)
7. Soskin, S.M., Luchinsky, D.G., Mannella, R., Neiman, A.B., McClintoc, P.V.: Zero-dispersion nonlinear resonance. Int. J. Bifurcation Chaos **7**(4), 923–936 (1997)
8. Howard, J.E., Humpherys, J.: Nonmotonic twist maps. Physica D **80**, 256–276 (1995)
9. Morozov, A.D., Morozov, K.E.: On quasi-periodic parametric perturbations of Hamiltonian systems. Russian J. Nonlinear Dyn. **16**(2), 369–378 (2020)
10. Morozov, A.D., Morozov, K.E.: Quasiperiodic perturbations of two-dimensional Hamiltonian systems with nonmonotone rotation. J. Math. Sci. **255**(6), 741–752 (2021)
11. Arnold, V.I.: Mathematical Methods of Classical Mechanics. Springer, New York (1978). https://doi.org/10.1007/978-1-4757-1693-1
12. Bogolubov, N.N., Mitropolskiy, J.A.: Asimptotic Methods on the Theory of Nonlinear Oscillations. Fizmatgiz, Moscow (1958). (in Russian)

# Competitiveness Function for the Generalized Abrams-Strogatti Model in the Case of Non-constant Community Size

Alexander Medvedev(✉) and Oleg Kuzenkov

National Research Lobachevsky State Nizhny University, Nizhny Novgorod, Russia
a.medvedev.unn@gmail.com
http://www.unn.ru

**Abstract.** The *purpose* of this work is to construct competitiveness function for the modified Abrams-Strogatti model describing the competition of several languages. *Materials and methods.* The modified model differs from the original Abrams-Strogatti model in that: volatility may be different for each language; the number of competing languages in a community is not limited (unlike the bilingual Abrams-Strogatti community); changes in the size of the community are allowed. Languages are characterized by parameters of prestige, volatility and the initial number of native speakers. The problem of determining the results of language competition by their characteristic parameters is considered. *Results.* A new method of solving the problem of the results of language competition is proposed. For this purpose, a new concept is introduced in language dynamics: the competitiveness function. To restore the competitiveness function, a ranking method is used. The competitiveness function is sought in the form of a power function depending on the language parameters. The values of the competitiveness function are analyzed, and on its basis a forecast is made for the further development of dynamics. At the first stage, the methodology is tested on a model that has an analytical solution. At the second stage, its application is demonstrated in the case when finding a solution in analytical form is difficult. *Conclusion.* The proposed methodology for constructing the competitiveness function is quite general and can be applied to a wide range of models describing the dynamics of populations. The forecast made on the basis of the constructed competitiveness functions is in good agreement with empirical data.

**Keywords:** Language competition · Language dynamics · Selection · Language preservation · Competitiveness function · Fitness function · Selection criterion · Selection processes · Mathematical model · Ordinary differential equations

## 1 Introduction

Mathematical modeling is widely used to study language dynamics [1–7]. Abrams and Strogatti laid the foundation for mathematical modeling of language


D. Balandin et al. (Eds.): MMST 2023, CCIS 1914, pp. 98–111, 2024.
https://doi.org/10.1007/978-3-031-52470-7_8

dynamics. They proposed a simple model of language competition, which allows us to explain the historical data on the decline of Welsh, Scottish, Gaelic and other endangered languages [1]. The Abrams-Strogatti (AS) model assumes a homogeneous population, all members of which speak one of two languages. This hypothesis assumes that even a member of a community knows two languages prefers only one in life, and we can talk not about language proficiency, but about its actualization at a given point in time. Abrams and Strogatti introduced the concepts of language prestige, its attractiveness to those who do not speak it, and language volatility as the readiness of native speakers to change it. The attractiveness of a language in the considered community increases with the number of its speakers in proportion to its prestige. The number of community members is assumed to be constant. The AS model shows that one language is always replaced by another over time. The latter in the AS model of language dynamics is called language death [1].

When studying language competition, the most important thing is not the competitive dynamics deployed in time, but its result (selection). The presence of such a result makes it possible to identify trends in which one or another language will displace others or disappear. Since the concept of language death is a general and critical trend in language dynamics, the central task for the researcher is to identify the reasons why one language displaces the others, as well as to find solutions to change it. The forecast is the result of selection, which is determined on the basis of the characteristics of the language and the initial number of its speakers.

From the point of view of qualitative analysis of a dynamic system, selection is the movement of a phase trajectory towards a certain state of equilibrium located on the coordinate axis. In this case, the phase space is divided into basins of attraction of stable equilibrium states. Each basin of attraction corresponds to the survival of one language. To predict the results of selection, it is necessary to express the equations of the boundaries separating the basins of attraction. In some models, this can be done by the classical method of studying dynamic systems, but often in models with complex dynamics, determining the basins of attraction becomes a much more complex and non-trivial mathematical task, because the analytical expression of the curves separating the basins of attraction is very difficult. We propose to do this by constructing a competitiveness function (CF) for each language. Comparing the values of the competitiveness function for languages allows us to decide which language will be displaced or remain.

The competitiveness function is understood as a function that depends on the parameters of the language: volatility, prestige and the number of native speakers at a time. If the $i$-th language eventually displaces the $j$-th language, then the value of the competitiveness function for the $i$-th language should be greater than the value for the $j$-th language. Restoring the competitiveness function is a special case of the more general problem of finding comparison functions in ranking problems. In some cases, it is constructed analytically [8–19], but for the model we are considering, the classical research method does not allow us to explicitly express the competitiveness function, so we have to use numerical methods [20–23], in particular machine learning methods. As an approximation of the competitiveness function, a polynomial of the $n$-th degree is taken, where $n$ is the order of approximation. The problems of linear approximation and binary classification are solved, as a result of which the coefficients of the polynomial are determined.

The purpose of this work is to construct competitiveness function for the modified Abrams-Strogatti model in the case of non-constant community size, as well as different volatilities for languages. The constructed competitiveness function are used to predict the development of the dynamics of language competition in the considered communities.

## 2 Materials and Methods

### 2.1 Mathematical Model

The principle of interaction between native speakers in a community generalizes the well-known hypothesis of effective meetings [24], which was used in the Volterra model.

Let's take the following input data to build the model:

- $N$ is the number of languages used in the community;
- $x_i$ is the number of individuals using $i$-th language at a time $t$;
- $s_i$ is the prestige of the $i$-th language;
- $c$ is the tendency of community members to change languages;;
- the proportion of individuals who do not speak any language is negligible;
- it is assumed that each individual at a given time speaks only one of the $N$ languages;
- indicators of the prestige of all languages of the community are standartized, i.e. the sum of the prestige of $N$ languages is limited to one - $\sum_{i=1}^{N} s_i \leq 1$;
- the number of members of the community is not fixed;
- language acquisition is considered to be consistent with AS hypotheses and there is individual volatility for each $\alpha_i$ language.
- the number of speakers of the $i$-th language increases in proportion to the coefficient $k_i$;
- the number of speakers of the $i$-th language decreases as a result of language competition in proportion to the coefficient $k$.

Taking into account the accepted hypotheses, we find that the distribution of languages in the community characterizes the state of the following system:

$$\begin{cases} \dot{x_i} = \left(\sum_{j\neq i}^{N} x_j\right) c s_i x_i^{\alpha_i} + k_i x_i - x_i c\left(\sum_{j\neq i}^{N} s_j x_j^{\alpha_j}\right) - k x_i \sum_{j\neq i}^{N} x_j, \\ \sum_{i=1}^{N} s_i \leq 1, \\ 1 < \alpha_i < 2. \end{cases} \tag{1}$$

We investigate a special case of this model, when $k_i = 0$, $k = 0$. This case is important to consider because for $k_i = 0$, $k = 0$, the separating boundaries can be expressed analytically, which makes it possible to check the effectiveness of the method. The phase space for the system (1), when $k_i = 0$, $k = 0$, is the locus of points with non-negative coordinate values. Model (1) has stable equilibrium states of the degenerate node type, which are located on the coordinate axes. The coordinates of unstable equilibrium states are determined by the equations of surfaces, each point of which is an unstable degenerate node. Their equations are as follows:

$$x_i = \sqrt[\alpha_j - 1]{\frac{\sum_{j \neq i}^{N} s_j x_j^{\alpha_j - 1}}{s_i \sum_{j \neq i}^{N} x_j}}, i = \overline{1, N}.$$

Since stable states are located only on the coordinate axes, the phase space is divided into basins of attraction in such a way that the phase trajectories of these basins of attraction converge over time to the corresponding coordinate axis. For $N = 2$ the phase plane of the system is shown (in Fig. 1). Based on this, it can be argued that selection processes are carried out in this system. From the point of view of qualitative analysis of a dynamic system, selection is the movement of a phase trajectory towards a certain state of equilibrium located on the coordinate axis. In this regard, the following problem arises - how to determine which language will be preserved and which will be lost by the characteristics of the language. In [25], an analytical solution to this problem was found and the language CF was constructed for a special case of this model - when the volatility parameters coincided $\alpha_i = \alpha_j$, and the size of the community was constant. It had a very convenient form for calculations and was characterized by the time-average specific growth rate of the number of native speakers. In this model, using the equations of the curves separating the basins

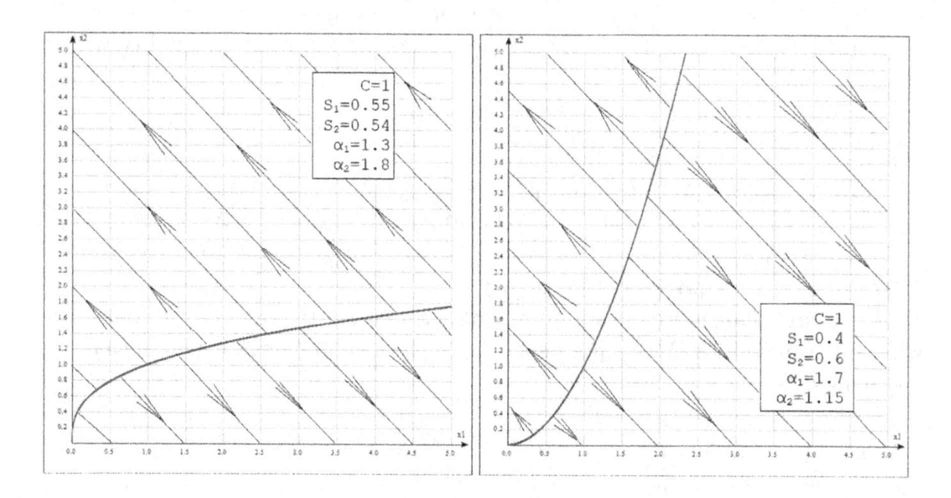


**Fig. 1.** Phase plane of system (1) for N=2.

of attraction, it is possible to check the correctness of the obtained CF, because at the points of this curve the values of the CF for the first and second languages must coincide.

### 2.2 Construction of the Competitiveness Function

Let us characterize $i$-th language, at an arbitrary point in time, by the following set of parameters:

$$v_i = \{\alpha_i, s_i, x_{i0}\}.$$

**Definition 1.** *We will say that the i-th language is better than the j-th if the phase trajectory, with initial conditions determined from $v_i$ and $v_j$, eventually tends to the $x_i$ coordinate axis, i.e. $x_j, j = \overline{1,N}$ values tends to zero.*

It is intended to express the CF as a power function of the following form: $J(x_{i0}, s_i, \alpha_i) = \lambda_0 x_{i0} + \lambda_1 x_{i0} s_i + \lambda_2 x_{i0} \alpha_i + \lambda_3 x_{i0} s_i \alpha_i + \lambda_4 x_{i0}^2 + \lambda_5 x_{i0}^2 \alpha_i + \lambda_6 x_{i0}^2 s_i + + \lambda_7 x_{i0}^2 s_i^2 + \lambda_8 x_{i0}^2 \alpha_i^2 + \lambda_9 x_{i0}^2 \alpha_i s_i + \lambda_1 0 x_{i0}^2 \alpha_i^2 s_i^2 + \lambda_1 1 x_{i0}^2 \alpha_i^2 s_i + \lambda_1 0 x_{i0}^2 \alpha_i s_i^2$.

**Definition 2.** *The functional $J(v_i)$, which determines the order of advantage within the set $\{v_i\}$, will be called a competitiveness function if from $J(v_i) > J(v_j)$ it should follow that the i-th language is preferable to the j-th at a time t.*

To determine the coefficients, we assign the following vector to each $v_i$:

$$M(v_i) = \{x_{i0}, x_{i0}s_i, x_{i0}\alpha_i, x_{i0}\alpha_i s_i, x_{i0}^2, x_{i0}^2 s_i, x_{i0}^2\alpha_i, x_{i0}^2 s_i^2, x_{i0}^2\alpha_i^2, x_{i0}^2\alpha_i s_i,$$

$$x_{i0}^2\alpha_i^2 s_i^2, x_{i0}^2\alpha_i^2 s_i, x_{i0}^2\alpha_i s_i^2\},$$

and to the pair $(v_i, v_j)$ we assign a point $(M(v_i) - M(v_j))$ and to the pair $(v_j, v_i)$ - point $(M(v_j) - M(v_i))$ in the 15-dimensional parameter space $MV$. If the $i$-th language is better than the $j$-th language, then the following inequality will be true:

$$J(v_i) > J(v_j),$$

which can be expanded as follows:

$\lambda_0 x_{i0} + \lambda_1 x_{i0} s_i + \lambda_2 x_{i0} \alpha_i + \lambda_3 x_{i0} s_i \alpha_i + \lambda_4 x_{i0}^2 + \lambda_5 x_{i0}^2 \alpha_i + \lambda_6 x_{i0}^2 s_i + + \lambda_7 x_{i0}^2 s_i^2 + \lambda_8 x_{i0}^2 \alpha_i^2 + \lambda_9 x_{i0}^2 \alpha_i s_i + \lambda_1 0 x_{i0}^2 \alpha_i^2 s_i^2 + \lambda_1 1 x_{i0}^2 \alpha_i^2 s_i + \lambda_1 0 x_{i0}^2 \alpha_i s_i^2 > \lambda_0 x_{j0} + \lambda_1 x_{j0} s_j + \lambda_2 x_{j0} \alpha_j + \lambda_3 x_{j0} s_j \alpha_j + \lambda_4 x_{j0}^2 + \lambda_5 x_{j0}^2 \alpha_j + \lambda_6 x_{j0}^2 s_j + \lambda_7 x_{j0}^2 s_j^2 + \lambda_8 x_{j0}^2 \alpha_j^2 + \lambda_9 x_{j0}^2 \alpha_j s_j + \lambda_1 0 x_{j0}^2 \alpha_j^2 s_j^2 + \lambda_1 1 x_{j0}^2 \alpha_j^2 s_j + \lambda_1 2 x_{j0}^2 \alpha_j s_j^2$.

After cancellation we obtain an inequality defining the area of the space $MV$ which contains all the$M(v_i)$ points at which the $i$-th language is better than the $j$-th language:

$\lambda_0(x_{i0} - x_{j0}) + \lambda_1(x_{i0}s_i - x_{j0}s_j) + \lambda_2(x_{i0}\alpha_i - x_{j0}\alpha_j) + \lambda_3(x_{i0}s_i\alpha_i - x_{j0}s_j\alpha_j) + \lambda_4(x_{i0}^2 - x_{j0}^2) + \lambda_5(x_{i0}^2\alpha_i - x_{j0}^2\alpha_j) + \lambda_6(x_{i0}^2 s_i - x_{j0}^2 s_j) + \lambda_7(x_{i0}^2 s_i^2 - x_{j0}^2 s_j^2) + \lambda_8(x_{i0}^2\alpha_i^2 - x_{j0}^2\alpha_j^2) + \lambda_9(x_{i0}^2\alpha_i s_i - x_{j0}^2\alpha_j s_j) + \lambda_{10}(x_{i0}^2\alpha_i^2 s_i^2 - x_{j0}^2\alpha_j^2 s_j^2) + \lambda_{11}(x_{i0}^2\alpha_i^2 s_i - x_{j0}^2\alpha_j^2 s_j) + \lambda_{12}(x_{i0}^2\alpha_i s_i^2 - x_{j0}^2\alpha_j s_j^2)$.

If last expression is turned into equality, then the hyperplane equation will be obtained, which will divide the space $MV$ into two parts. In one of them, the $i$-th language will be better than the $j$-th language, and vice versa in the other. The area in which the $i$-th language will be better than all other languages at once is determined by the system of equations:

$$\begin{cases} \lambda_0(x_{i0} - x_{j0}) + \lambda_1(x_{i0}s_i - x_{j0}s_j) + ... + \lambda_1 2x_{j0}^2\alpha_j s_j^2 > 0, \\ i = \overline{1, N}, \\ j \neq i. \end{cases} \tag{2}$$

Thus, the construction of the competitiveness function is reduced to determining the normal of the hyperplane, which is given by the vector of the $\lambda_i$ coefficients:

$$\overline{N} = (\lambda_0, \lambda_1, \lambda_2, ..., \lambda_{12}).$$

The CF formula is obtained as a power function. By substituting the language parameters into the CF, we obtain the values of its competitiveness. By comparing the resulting competitiveness of languages, we can say that over time, the language that has the greatest competitiveness will remain. The boundary dividing the phase space into basins of attraction will pass through the points at which the CF values for the languages are equal.

## 3 Results

### 3.1 Construction of a Training Sample

On the phase plane, 90 arbitrary points were selected with different initial conditions for the languages from which the sets $v_1 = \{s_1, \alpha_1, x_{10}\}$ and $v_1 = \{s_2, \alpha_2, x_{20}\}$ were formed. An example of such samples is shown (in Fig. 2).

For various parameters of languages $\{s_1, s_2, \alpha_1, \alpha_2\}$, by solving a system of differential equations, it was established which language would win as a result of competition for all $v_1$ and $v_2$ sets. The set of points $(M(v_1) - M(v_2))$ and $(M(v_2) - M(v_1))$ was calculated. This set of points was classified according to the survival of one of the two languages. Using the method of support vectors, the dividing line equation for these two classes was obtained. From the equation of the dividing line, the values of the $\lambda_i$ coefficients were obtained. The results of the procedures for restoring the coefficients of the competitiveness function and constructing the functions themselves are presented in Table 1.

CF of the $i$-th language:

$J(x_{i0}) = 6047,907x_{i0} - 12173,6x_{i0}s_i - 3447,73x_{i0}\alpha_i + 6865,399x_{i0}\alpha_i s_i + 112729,2x_{i0}^2 - 297,779x_{i0}^2\alpha_i - 454935,3x_{i0}^2 s_i + 460092,9x_{i0}^2 s_i^2 - 0,000022x_{i0}^2\alpha_i^2 + 62,452x_{i0}^2\alpha_i s_i - 0,00005x_{i0}^2\alpha_i^2 s_i^2 + 10,159x_{i0}^2\alpha_i^2 s_i - 0,000035x_{i0}^2\alpha_i s_i^2$

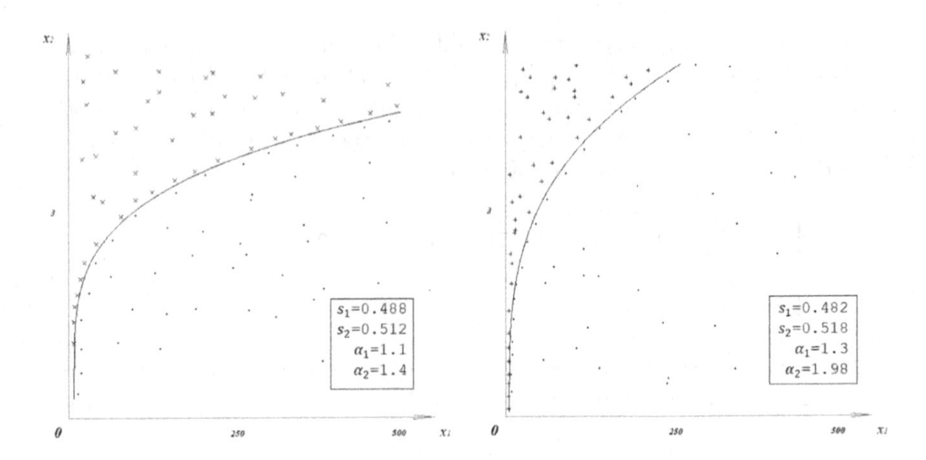


**Fig. 2.** Fragments of $v_1$ and $v_2$ training samples for identifying the parameters of the competitiveness function. Dots mark the initial conditions under which the first language dominates, crosses - the second language dominates.

**Table 1.** Competitiveness function coefficients.

| $\lambda_0$ | $\lambda_1$ | $\lambda_2$ | $\lambda_3$ | $\lambda_4$ | $\lambda_5$ | $\lambda_6$ |
|---|---|---|---|---|---|---|
| 6047,907 | -12173,6 | -3447,73 | 6865,399 | 112729,2 | -298,779 | -454935,3 |

| $\lambda_7$ | $\lambda_8$ | $\lambda_9$ | $\lambda_{10}$ | $\lambda_{11}$ | $\lambda_{12}$ |
|---|---|---|---|---|---|
| 460092,9 | -0,000022 | 62,452 | -0,00005 | 10,159 | -0,000035 |

### 3.2 Restoring the Competitiveness Function

Using the least squares method, the parameters of model (1) were identified, the values of the coefficients are given in Tables 2, 3. Statistical data on the shares of Welsh and English languages for 1901–2001, as well as Gaelic and English for 1891–1971 in England are considered [26,27]. The statistical data of the multilingual community in Brussels (Belgium) for 1842–1947 are considered [28]. In this community, two main languages are distinguished (French and Dutch), and the remaining languages are combined into the "other languages" group, for which average values of characteristics (prestige and volatility) are derived. Based on the data given in Table 1, competitiveness formulas were obtained for language pairs from Tables 2, 3. Graphs of competitiveness functions for the Welsh and English languages (Fig. 3), Gaelic and English (Fig. 4) (Table 4).

**Table 2.** Model coefficients for two languages.

| Language group | $s_1$ | $s_2$ | $\alpha_1$ | $\alpha_2$ | $c$ |
|---|---|---|---|---|---|
| Welish and English | 0,488 | 0,512 | 1,1 | 1,4 | 2,1 |
| Gaelic and English | 0,482 | 0,518 | 1,3 | 1,98 | 1,8 |

**Table 3.** Model coefficients for three languages.

| Language group | $s_1$ | $s_2$ | $s_3^*$ | $\alpha_1$ | $\alpha_2$ | $\alpha_3^*$ | $c$ |
|---|---|---|---|---|---|---|---|
| French, Dutch and other | 0,284 | 0,315 | 0,4 | 1,749 | 1,63 | 1,65 | 1,5 |

* cumulative average values of prestige and volatility of other languages.

**Table 4.** Competitiveness functions of Welsh and English languages.

| | |
|---|---|
| CF of Welish | $J(x_{10}) = 0,0334x_{10} + 0,0031x_{10}^2$ |
| CF of English | $J(x_{20}) = 49,5892x_{20}^2 - 90,6802x_{20}$ |

A comparison of the CF for languages is shown in Fig. 5. The gray areas in the graph on the left correspond to the initial conditions for languages, in which the CF values of the Welsh language are greater than the CF values of the English language. In the gray areas of the graph on the right, the Gaelic CF values exceed the English CF values. In white areas, English CF values exceed the values of the Welsh and Gaelic respectively (see Table 5).

The CF values for the Welsh and English languages as of 2001 were calculated (Table 7). The CF values for Gaelic and English as of 1971 were calculated (Table 7). The CF values for French, Dutch and other languages as of 1947 were calculated (Table 8) (see Table 6).

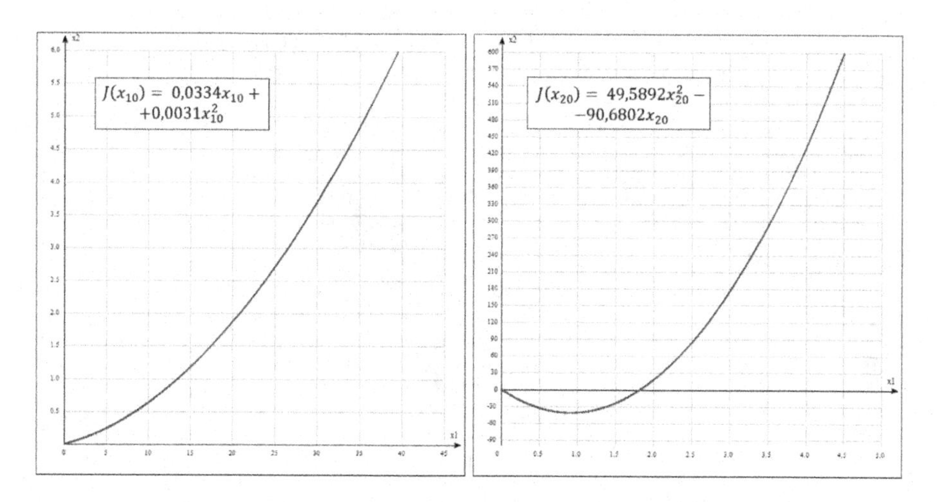


**Fig. 3.** A graph of the Welsh language competitiveness function on the left, a graph of the English language competitiveness function on the right.

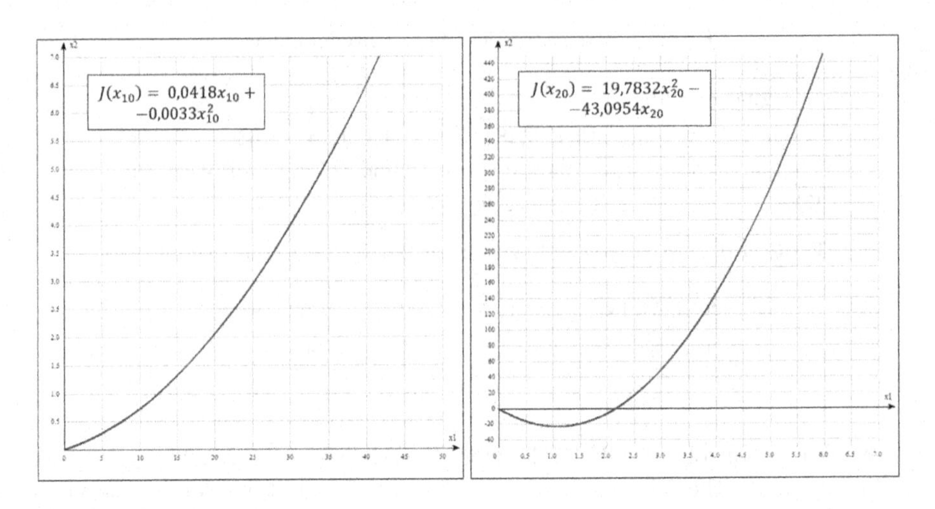


**Fig. 4.** A graph of the Gaelic language competitiveness function on the left, a graph of the English language competitiveness function on the right.

**Table 5.** Competitiveness functions of Gaelic and English languages.

| CF of Gaelic | $J(x_{10}) = 0,0418x_{10} + 0,0033x_{10}^2$ |
|---|---|
| CF of English | $J(x_{20}) = 19,7832x_{20}^2 - 43,0954x_{20}$ |

**Table 6.** Competitiveness functions of French, Dutch and other languages.

| CF of French | $J(x_{10}) = 20154,1098x_{10}^2 - 29,3216x_{10}$ |
|---|---|
| CF of Dutch | $J(x_{20}) = 14630,8569x_{20}^2 + 118,4622x_{20}$ |
| CF of other languaage | $J(x_{30}) = 3929,24x_{30}^2 + 20,8758x_{30}$ |

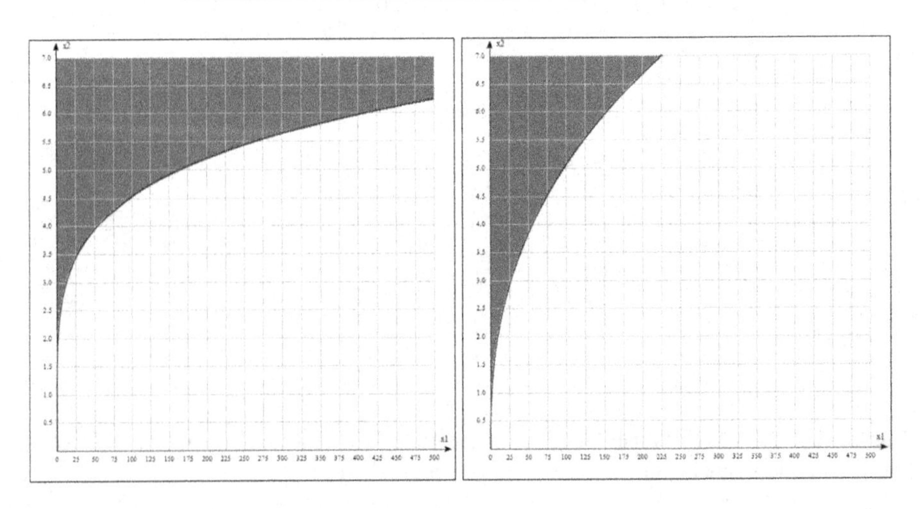

**Fig. 5.** CF comparison of Welsh and English-on the left, Gaelic and English-on the right. In the white area, the CF value of the first language is greater than the CF value of the second language. In the gray area, on the contrary, the CF value of the first language is greater than the CF value of the second. (Color figure online)

**Table 7.** The values of the competitiveness function as of 2001 and 1971.

| Language | $J(x_{i0})$ | Year |
|---|---|---|
| Welish | 0,0 | 2001 |
| English | 302306,3928 | 2001 |
| Gaelic | 0,0 | 1971 |
| English | 142593,762 | 1971 |

**Table 8.** The values of the competitiveness function as of 1947.

| Language | $J(x_{i0})$ | Year |
|---|---|---|
| French | 1857121,27 | 1947 |
| Dutch | 18235546,24 | 1947 |
| Other language | 11930362,28 | 1947 |

### 3.3 The Case of the Model with $k_i$, $k$ Coefficients Other Than Zero

In this case, the equation of the curves defining the separation boundaries of the basins of attraction is extremely difficult to express analytically, but using the numerical method of searching for the CF, the equation of the dividing boundary can be obtained with any accuracy. It is defined as the geometric locus of points in the phase space in which the CF of the first and second languages coincide. The phase space for (1), with $k_i$ and $k$ other than zero, is the locus of points with non-negative coordinate values. Model (1), in this case, has two unstable equilibrium states. The first of them is located at the origin and is an unstable node. The second is an unstable saddle type; its coordinates are determined by the following system of equations:

$$\begin{cases} x_i = \sqrt[\alpha_i - 1]{\dfrac{c(\sum_{j\neq i}^{N} s_j x_j^{\alpha_j - 1}) + k\sum_{j\neq i}^{N} x_j - k_i}{cs_i \sum_{j\neq i}^{N} x_j}}, \\ i = \overline{1, N}. \end{cases} \tag{3}$$

As in the first case, the phase space is divided into basins of attraction in such a way that the phase trajectories of these basins of attraction converge over time to one of the coordinate axes. For $N = 2$, the phase plane of system (1) is shown in Fig. 6. For $k_i$ and $k$ other than zero, selection processes are also carried out. The competitiveness functions and the values of their coefficients are presented in Tables 9, 10, 11.

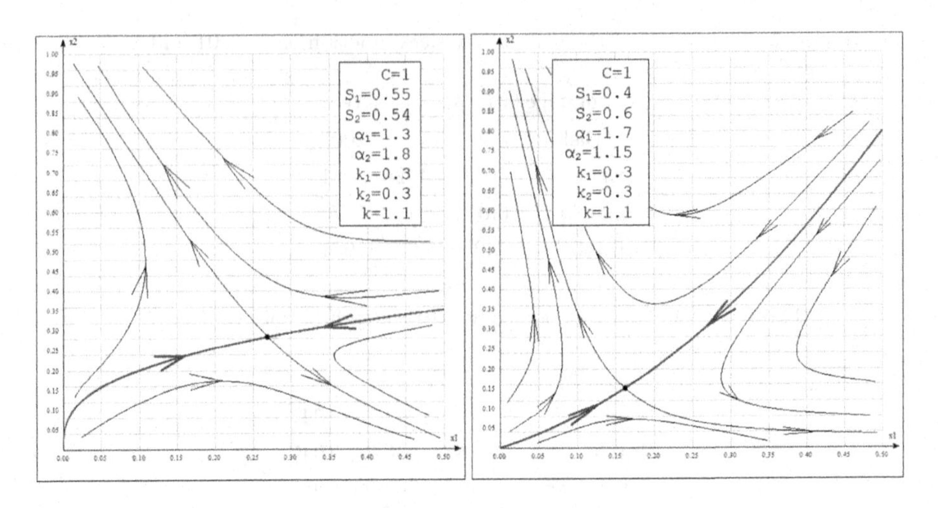


**Fig. 6.** Phase plane of system (1) for $N = 2$, with $k_i$ and $k$ other than zero.

**Table 9.** Coefficients of competitiveness functions.

| $\lambda_0$ | $\lambda_1$ | $\lambda_2$ | $\lambda_3$ | $\lambda_4$ | $\lambda_5$ | $\lambda_6$ |
|---|---|---|---|---|---|---|
| 6047,907 | -12173,6 | -3447,73 | 6865,4 | 112729,2 | -298,8 | -454935,3 |

| $\lambda_7$ | $\lambda_8$ | $\lambda_9$ | $\lambda_{10}$ | $\lambda_{11}$ | $\lambda_{12}$ | $\lambda_{13}$ | $\lambda_{14}$ |
|---|---|---|---|---|---|---|---|
| 460092,9 | 0,0 | 62,452 | 0,0 | 10,16 | 0,0 | 0,176 | 0,035 |

**Table 10.** Experimental values of language parameters.

| $s_1$ | $s_2$ | $\alpha_1$ | $\alpha_2$ | $k_1$ | $k_2$ | $k$ | $c$ |
|---|---|---|---|---|---|---|---|
| 0,55 | 0,54 | 1,3 | 1,8 | 0,3 | 0,3 | 1,1 | 1,0 |

**Table 11.** Competitiveness functions of First and Second languages.

| | |
|---|---|
| CF of first language | $J(x_{10}) = 651,0164x_{10}^2 - 55,6739x_{10}$ |
| CF of second language | $J(x_{20}) = 1241,6841x_{20}^2 - 217,9524x_{20}$ |

Graphs of competitiveness functions for two languages are shown in Fig. 7-a. By comparing the CF values for two languages, the boundary separating the basins of attraction of phase trajectories was determined in Fig. 7-b. The area in which the second language survives is marked in gray; the area in which the first language survives is marked in white.

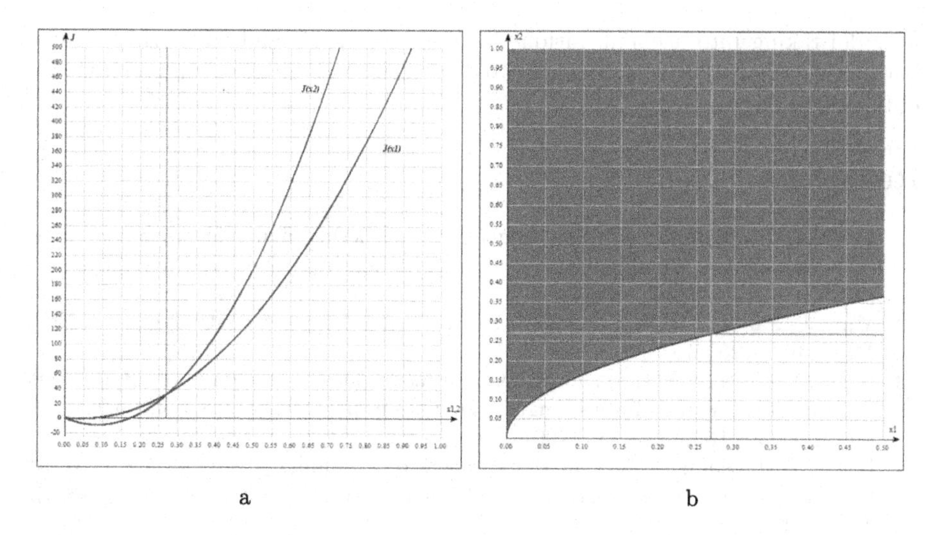


**Fig. 7.** a. - CF of first and second languages. b. - The boundary separating the basins of attraction of phase trajectories. In the white area the first language survives. In the gray area, the second language survives. (Color figure online)

## 4 Conclusion

The CF values allow us to conclude that in the future, under unchanged conditions, English will be dominant, and the number of Welsh and Gaelic languages in bilingual communities will decrease to zero over time. In a community of three competing languages, Dutch will be the dominant language. The forecast constructed as a result of modeling is in good agreement with the statistical data considered in the work. Comparing the CF values for different initial parameters is actually equivalent to determining the basins of attraction in the model. If the number of languages in the considered community is large enough (an example of such a community may be the Internet), then solving the research problem using classical methods becomes cumbersome and very complex. In this case, an algorithm based on maximizing the competitiveness function seems to be a convenient alternative to the classical approach. Using the constructed competitiveness function, it is possible to model language dynamics by numerically solving the minimax problem. The proposed methodology for constructing the competitiveness function is quite general and can be applied to a wide range of models of language dynamics. In this work, a modified Abrams-Strogatti model was constructed and studied to describe the competition of several languages. Languages are characterized and distinguished by the parameters of prestige, volatility and the initial number of native speakers. An analytical study based on the mathematical theory of selection was carried out. The search and construction of the competitiveness function was carried out. The result was applied to statistical data. The technique was tested on a model with an analytical solution. The adequacy and effectiveness of this method for constructing the CF on a model taking into account various language volatility and the variable number of native speakers was demonstrated. Forecasts on language dynamics were obtained.

## References


1. Castelly, X., Eguiluz, V., San Miguel, M.: Ordering dynamics with two non-excluding option: Bilingualism in language competition. New J. Phys. **1**(8), 308 (2006)
2. Mira, J., Paredes, B.: Interlinguistic similarity and language death dynamics. EPL **1**(69), 1031 (2005)
3. Baggs, I., Freedman, H.: A mathematical model for the dynamics of interactions between a unilingual and a bilingual population: persistence versus extinction. J. Math. Sociol. **16**(1), 51–75 (1990)
4. Baggs, I., Freedman, H.: Can the speakers of a dominated language survive as Unilinguals?: A mathematical model of Bilingualism. Math. Comput. Model. **18**(6), 9–18 (1993)
5. Wyburn, J., Hayward, J.: The future of Bilingualism: an application of the Baggs and freedman model. J. Math. Sociol. **32**(4), 267–284 (2008)
6. Diaz, M., Switkes, J.: A mathematical model of language preservation. Heliyon **7**(5), 2405–2425 (2021)
7. Abrams, D., Strogatz, S.: Modelling the dynamics of language death. Nature **1**(424), 900 (2003)
8. Birch, J.: Natural selection and the maximization of fitness. Biology **91**(3), 712–727 (2006)
9. Kuzenkov, O., Morozov, A., Kuzenkova, G.: Recognition of patterns of optimal diel vertical migration of zoo-plankton using neural networks. In: IJCNN 2019 - International Joint Conference on Neural Networks, Budapest, Hungary (2019)
10. Kuzenkov, O.: Fitness as a general technique for modeling the processes of transmission of non-innate information. Nizhny Novgorod State University, N. Novgorod (2019)
11. Kuzenkov, O., Ryabova, E., Krupoderova, K.: Mathematical models of selection processes, pp. 80–125. Nizhny Novgorod State University, N. Novgorod (2010). (in Russian)
12. Gorban, A.: Equilibrium bypass. Science, Novosibirsk (1984)
13. Gorban, A.: Selection Theorem for systems with inheritance. Math. Model. Nat. Phenom. (MMNP) **2**(4), 1–45 (2007)
14. Gorban, A.: Self-simplification in Darwin's systems. In: Gorban, A., Roose, D. (eds.) Coping with Complexity: Model Reduction and Data Analysis. Lecture Notes in Computational Science and Engineering, vol. 75, pp. 311–344. Springer, Berlin (2011). https://doi.org/10.1007/978-3-642-14941-2_17
15. Karev, G., Kareva, I.: Replicator equations and models of biological populations and communities. Math. Model. Nat. Phenom. (MMNP) **9**(1), 68–95 (2014)
16. Kuzenkov, O., Ryabova, E.: Variational principle for self-replicating systems. Math. Model. Nat. Phenom. (MMNP) **10**(2), 115–129 (2015)
17. Kuzenkov, O., Ryabova, E.: Limit possibilities of solution a hereditary control system. Diff. Eq. **51**(4), 500–511 (2015)
18. Kuzenkov, O., Morozov, A.: Towards the construction of a mathematically rigorous framework for the modelling of evolutionary fitness. Bull. Math. Biol. **81**(11), 4675–4700 (2019)

19. Mohri, M., Rostamizadeh, A., Talwalkar, A. Foundations of Machine Learning. The MIT, Cambridge (2018). ISBN:9780262039406
20. Kuzenkov, O.: Construction of the fitness function depending on a set of competing strategies based on the analysis of population dynamics. Izvestiya VUZ. Appl. Nonlinear Dyn. **30**(3), 276–298 (2022). https://doi.org/10.18500/0869-6632-2022-30-3-276-298. (in Russian)
21. Kuzenkov, O., Ryabova, E.: Optimal control of a hyperbolic system on a simplex. J. Comput. Syst. Sci. Int. **42**(2), 227–233 (2003)
22. Kuzenkov, O.: The investigation of the population dynamics control problems based on the generalized Kolmogorov model. J. Comput. Syst. Sci. Int. **48**(5), 839–846 (2009)
23. Kuzenkov, O., Ryabova, E.: Optimal control for a system on a unit simplex in infinite time. Autom. Remote. Control. **66**(10), 1594–1602 (2005). https://doi.org/10.1007/s10513-005-0193-z
24. Idy, B., Papa, I., Mahe, N., Aboubakary, D.: An Extension of two species Lotka-Volterra competition model. Biomath Commun. **8**(1), 1594–1602 (2021). https://doi.org/10.11145/bmc.2021.12.171
25. Medvedev, A., Kuzenkov, O.: Generalization of the Abrams-Strogatti model of language dynamics to the case of several languages. In: XXI International Scientific Conference on Differential, Mogilev, Belarus (2023)
26. Sutantawibul, C., Xiao, P., Richie, S., Fuentes-Rivero, D.: Revisit language modeling competition and extinction: a data-driven validation. J. Appl. Math. Phys. **6**(1), 1558–1570 (2018). https://doi.org/10.4236/jamp.2018.67132
27. Menghan, Z., Tao, G.: Principles of parametric estimation in modeling language competition. Proc. Nat. Acad. Sci. (PNAS) **110**(24), 194–212 (2013). https://doi.org/10.1073/pnas.1303108110
28. Leuvensteijn, J., van Tooren, M., Pijnenburg, W., van der Horst, M.: Language Faculty of Humanities. Amsterdam University Press, Amsterdam (1997)

# Parallel Implementation of a Computational Algorithm Based on the Explicit Iterative Scheme for Modeling Subsonic Reacting Gas Flows

E. E. Peskova(✉), O. S. Yazovtseva, E. Yu. Makarova, and N. A. Tingaeva

National Research Ogarev Mordovia State University, Bolshevistskaya Street, 68, Saransk, Russia
e.e.peskova@mail.ru

**Abstract.** The article presents a parallel implementation of a computational algorithm based on an explicitly iterative LI-M scheme constructed to calculate the dynamics of subsonic flows with radical chain chemical reactions, taking into account multicomponent diffusion and heat transfer, and also energy absorption during chemical transformations on the example of methane pyrolysis. The problem involves multi-scale processes, which leads to computational difficulties when using explicit difference schemes. The computational algorithm is partially simplified by using the principle of splitting by physical processes: kinetic equations are calculated in a separate block by the fifth-order three-stage Runge-Kutta method (Radau method); transfer processes are calculated using the Rusanov flow and the WENO scheme. However, even this approach does not provide an acceptable reduction in the estimated time for solving practical problems since the total calculated time step is determined by diffusion processes in the studied subsonic flows. The problem's solution is possible by using a scheme of local iterations to calculate dissipative terms. It is based on an explicit iterative process, the parameters of which are the roots of Chebyshev polynomials. The efficient use of parallel computing is possible since the scheme is explicit. The developed algorithm was programmatically implemented in C++ using MPI technology. The algorithm showed good performance and stability while using a sufficiently large time step. The parallel algorithm was analyzed in terms of efficiency for various spatial grids. The result of its work are profiles of substances' concentrations and temperatures along the reactor's length.

**Keywords:** mathematical modeling · MPI technology · subsonic flows · splitting by physical processes · Chebyshev explicit iterative scheme · methane conversion

The research was supported by the grant of the Russian Science Foundation No. 23-21-00202, https://rscf.ru/project/23-21-00202/.


D. Balandin et al. (Eds.): MMST 2023, CCIS 1914, pp. 112–121, 2024.
https://doi.org/10.1007/978-3-031-52470-7_9

## 1 Introduction

The use of parallel technologies is a modern tool of solving the problem of computing power's insufficiency for the implementation of computationally difficult algorithms. The use of one of two fundamentally different parallel programming paradigms (OpenMP or MPI) is determined by the physical and geometric features of the problem. Most practical problems are based on three-dimensional models, the computer implementation of which is complicated by multifactorial external and internal influences.

A fairly wide class of such problems are the problems of the chemical technology industry, for example, modeling chemical reactors for processing hydrocarbons [1]. The variety of its forms causes the complexity of their processing. Their mutual transformation, decay, isomerization and many other processes during the reactions occur that need to be taken into account. This requires a detailed description of all stages of chemical transformations, including radical mechanisms [2]. In terms of mathematical modeling, this fact greatly complicates both the model itself and the computational algorithm for it.

The decay and formation of radicals and substances in the reactor contribute to the active mixing of gases, as well as the occurrence of dissipative effects. This requires taking into account multicomponent diffusion, viscosity and thermal conductivity [3], which impose significant restrictions on the integration step in time in a developed computational algorithm.

The aim of this work is to develop and study a parallel computational algorithm using an explicitly iterative scheme for calculating dissipative terms, constructed to calculate the dynamics of subsonic flows with radical chain chemical reactions, taking into account multicomponent diffusion and heat transfer, as well as energy absorption during chemical transformations using the example of methane pyrolysis [4].

## 2 Mathematical Model

The article considers a cylindrical reactor. It makes possible to reduce a three-dimensional problem to a two-dimensional axisymmetric one, which significantly reduces the amount of calculations.

The reactor model for methane pyrolysis is based on a system of Navier-Stokes equations in the low Mach number approximation [6], modified by the peculiarities of the chemical process. Let's write it in vector form in a cylindrical coordinate system [5]:

$$\frac{\partial U}{\partial t} + \frac{\partial \left(F^{(1)}(U) - H^{(1)}(U)\right)}{\partial z} + \frac{1}{r}\frac{\partial \left(r\left(F^{(2)}(U) - H^{(2)}(U)\right)\right)}{\partial r} = W, \quad (1)$$

$$U = \begin{pmatrix} \rho Y_m \\ \rho u_z \\ \rho u_r \\ \rho h \end{pmatrix}, \quad F^{(1)}(U) = \begin{pmatrix} \rho u_z Y_m \\ \rho u_z^2 + \pi \\ \rho u_z u_r \\ \rho h u_z \end{pmatrix}, \quad F^{(2)}(U) = \begin{pmatrix} \rho u_r Y_m \\ \rho u_z u_r \\ \rho u_r^2 + \pi \\ \rho h u_r \end{pmatrix},$$

$$H^{(1)}(U) = \begin{pmatrix} J_{mz} \\ \tau_{zz} \\ \tau_{zr} \\ q_z \end{pmatrix}, \quad H^{(2)}(U) = \begin{pmatrix} J_{mr} \\ \tau_{rz} \\ \tau_{rr} \\ q_r \end{pmatrix}, \quad W = \begin{pmatrix} R_m \\ 0 \\ 0 \\ 0 \end{pmatrix}.$$

Here $m = \overline{1, M}$, $M$ — the number of components in the gas mixture, $\rho$ — the density of the mixture, $Y_m$ — the mass fraction of the $m$th component of the mixture, $u_z$, $u_r$ — projections of the velocity vector $\boldsymbol{u}$ on the $z$ and $r$ axes, $h$ — enthalpy of the mixture, $\pi = p - p_0$ — dynamic component of pressure, $p$ – pressure, $p_0$ – thermodynamic component of pressure, constant in the region, $R_m$ — the rate of formation or flow of the $m$th component of the mixture, $T$ — temperature.

The mixture enthalpy is calculated with respect to the multicomponence of the reaction mixture according to the law:

$$p_0 = \rho RT \sum_m \frac{Y_m}{M_{wm}}, \qquad h(T, Y_m) = \sum_m Y_m h_m(T).$$

$h_m(T)$ — enthalpy of the $m$th component of the mixture.

Diffusion in the multicomponent mixture is calculated by the formulas:

$$J_{mz} = \rho D_{m,mix} \frac{\partial Y_m}{\partial z}, \quad J_{mr} = \rho D_{m,mix} \frac{\partial Y_m}{\partial r},$$

where $\boldsymbol{J_m}$ — components of the diffusion flow vector, $D_{m,mix}$ — the average diffusion coefficient of the $m$ component over the mixture.

Axial and radial heat flows in vector form are calculated as

$$q_z = \lambda \frac{\partial T}{\partial z} + \sum_{m=1}^{M} h_m \rho D_{m,mix} \frac{\partial Y_m}{\partial z}, \quad q_r = \lambda \frac{\partial T}{\partial r} + \sum_{m=1}^{M} h_m \rho D_{m,mix} \frac{\partial Y_m}{\partial r},$$

where $\boldsymbol{q}$ — components of the heat flux vector, $\lambda$ — thermal conductivity of the mixture.

The stepwise record of the viscous stress tensor's vector $\overline{\overline{\tau}}$ is represented as

$$\tau_{zz} = \mu \left( 2\frac{\partial u_z}{\partial z} - \frac{2}{3} \left( \frac{\partial u_z}{\partial z} + \frac{1}{r} \frac{\partial r u_r}{\partial r} \right) \right), \quad \tau_{rr} = \mu \left( 2\frac{\partial u_r}{\partial r} - \frac{2}{3} \left( \frac{\partial u_z}{\partial z} + \frac{1}{r} \frac{\partial r u_r}{\partial r} \right) \right),$$

$$\tau_{zr} = \tau_{rz} = \mu \left( \frac{\partial u}{\partial r} + \frac{\partial v}{\partial z} \right).$$

where $\mu$ is the mixture's coefficient of dynamic viscosity.

A detailed description and expressions for finding the coefficients of multi-component mixtures is presented in [7].

It is necessary to fulfill the condition for the divergence of the velocity vector due to the physicochemical features of the process [6]:

$$\nabla \cdot \boldsymbol{v} = \frac{1}{\rho C_p T}\left(\nabla \cdot \lambda \nabla T + \sum_m \rho D_{m,mix} \nabla Y_m \nabla h_m + \alpha\theta\right) + \\ + \frac{1}{\rho}\sum_m \frac{M_w}{M_{wm}}\left(\nabla \cdot \rho D_{m,mix} \nabla Y_m\right) + \frac{1}{\rho}\sum_m \left(\frac{M_w}{M_{wm}} - \frac{h_m}{C_p T}\right) R_m, \quad (2)$$

where $C_p(Y_m, T)$ — the heat capacity of the mixture at constant pressure, $M_w$ — the molecular weight of the mixture.

# 3 Parallel Algorithm

## 3.1 Description of the Computational Algorithm

The above-mentioned features of the model lead to the need to develop a computational algorithm with a special structure. The rectangular grid uniform in each direction is used

$$\Omega_h = \left\{z_i = ih_z, r_j = jh_r; i = \overline{1, N_z}, j = \overline{1, N_r}; N_z h_z = L_z, N_r h_r = L_r\right\},$$

$L_{z(r)}$ — dimensions of the calculated area along the axes $\{z, r\}$. Initial data is set and thermodynamic and technological parameters are determined in each cell.

The abundance of gas-dynamic and chemical processes determine the use of the splitting by physical processes [8].

The calculation of changes in concentrations of substances and radicals during to chemical reactions is carried out in a separate plug-in module RADAU5 [9]. It is caused by the fact that the radical chain mechanism of methane pyrolysis implies sharp fluctuations in concentrations. The RADAU5 has proven itself well both for radical mechanisms [5] and for multistage heterogeneous kinetics [10].

After calculating the concentration changes, the convective terms are integrated without taking into account pressure using Rusanov flows (at the cell boundary) and the WENO scheme of the 5th order of accuracy.

The analysis of the flow showed a significant predominance of diffusion flows over convective ones. Dissipative processes lead to an unacceptably small step of integration in time due to rapid chemical reactions. The article proposes using of the Local Iterations Method [11,12] to calculate the dissipative terms: diffusion, viscosity, thermal conductivity, to avoid this problem. This method is based on an iterative process within a common time step, which parameters are the roots of Chebyshev polynomials. The calculation of the iterations's number is based on the scheme's stability condition. The sum of the values of the upper bounds of the discrete operators' spectra corresponding to approximations of dissipative diffusion processes, viscosity and thermal conductivity is calculated. The ordering of the Chebyshev polynomials' roots corresponds to preserve the scheme's stability. It is carried out in accordance with the algorithm presented in [13]. A sequence of iterative parameters is constructed based on the ordered roots of Chebyshev polynomials [11,12], used to find gas dynamic parameters taking into account the dissipation processes. The described scheme allows to maintain stability even when using a sufficiently large time integration step. The explicit scheme makes it easy to adapt it for parallel computing applications.

The final stage of integration is the correction of the velocity vector and the calculation of the dynamic component of the pressure from the Poisson equation.

The sequence of actions at each time iteration can be written as follows.

Step 1. Solution of chemical kinetics equations. Finding changes in the concentrations of substances during chemical reactions by solving a system obtained from cutting off the source terms of the primary problem.

Step 2. Solving the hyperbolic problem. Determination of changes in concentrations of substances, flow rate and energy due to heat and mass transfer from convective flows without using the dynamic component of pressure.

Step 3. Solving the parabolic problem. Calculation of the dissipative forces' contribution to changes in concentrations of substances, flow rate and energy.

Step 4. Solving an elliptic problem. Calculation of the pressure's dynamic component and correction of the velocity vector based on it.

### 3.2 Description of the Parallel Program

The decomposition of the original problem into a set of boundary value problems for the parallel implementation of the computational algorithm was carried out since the geometry of the axisymmetric flow assumes the division of the integration domain into structured cells. The integration domain was divided into disjoint subdomains. The boundary conditions of the initial problem are implemented at the outer boundaries of the integration domain, and information is exchanged between the subdomains at the boundaries of the subdomains (internal to the original problem).

The features of the constructed computational algorithm and the <<convenience>> geometric partitioning of the integration domain determine the use of the MPI standard with the connected MPICH library for interprocessor exchange.

The program has a modular structure. Each module corresponds to the steps of the computational algorithm, and its execution is implemented using a given number of processors. The block-scheme of the parallel program is shown in Fig. 1.

Steps in the diagram in Fig. 1 correspond to the steps from the description of the computational algorithm. Data exchange before Step 4 (calculation of the dynamic component of pressure) is necessary for calculation according to the data of the current iteration. After Step 4, updated pressure and velocity values are exchanged.

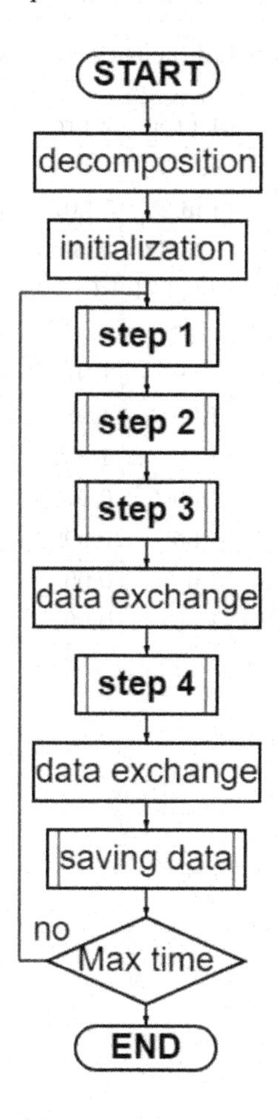


**Fig. 1.** The block-scheme of the parallel program.

### 3.3 Parallel Algorithm Efficiency Analysis

The effectiveness analysis of the algorithm's presented implementation is summarized in a table. Computational experiments were done for a different number of cells with a different number of processors involved for the same integration time. Calculations were made on the Intel Xeon CPU cluster of the National Research Mordovia State University with the number of nodes 12, each of which has 12 cores. Computational experiments were carried out on a single computing node (Table 1).

The comparative analysis is presented in the form of graphs in Fig. 2.

As can be seen from Fig. 2, the acceleration increases for grids of any dimension with an increase in the number of processors used. It approaches to the number of processors with an increase in the number of calculated cells. Unsatisfactory results are observed on a rough grid of 600 cells with a number of

**Table 1.** Efficiency Analysis

| Parameter | Cells | 1 proc | 2 proc | 4 proc | 8 proc | 12 proc |
|---|---|---|---|---|---|---|
| Time, sec | 600 cells | 40.25 | 20.66 | 10.6 | 7.07 | 5.01 |
| | 2400 cells | 146.24 | 74.07 | 37.73 | 19.25 | 13.06 |
| | 9600 cells | 523.32 | 264.39 | 135.15 | 67.27 | 45.54 |
| | 38400 cells | 2036.71 | 1027.73 | 519.37 | 258.46 | 174.79 |
| Speedup,% | 600 cells | 1.00 | 1.95 | 3.8 | 5.69 | 8.03 |
| | 2400 cells | 1.00 | 1.97 | 3.88 | 7.6 | 11.2 |
| | 9600 cells | 1.00 | 1.98 | 3.87 | 7.78 | 11.49 |
| | 38400 cells | 1.00 | 1.98 | 3.92 | 7.88 | 11.65 |
| Effectiveness,% | 600 cells | 1.00 | 0.97 | 0.95 | 0.71 | 0.67 |
| | 2400 cells | 1.00 | 0.99 | 0.97 | 0.95 | 0.93 |
| | 9600 cells | 1.00 | 0.99 | 0.97 | 0.97 | 0.96 |
| | 38400 cells | 1.00 | 0.99 | 0.98 | 0.99 | 0.97 |

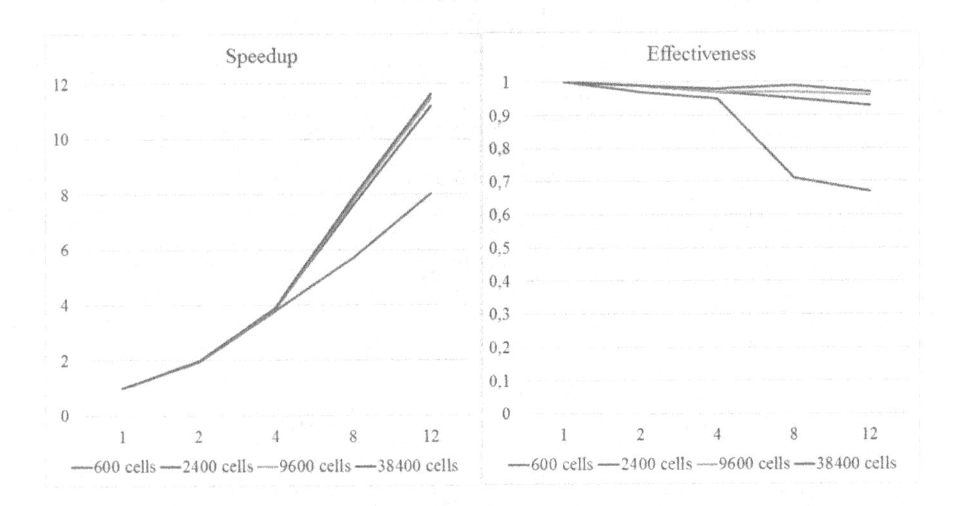


**Fig. 2.** Acceleration and efficiency of the parallel algorithm

processors greater than 4. Obviously, this is due to the commensurability of the calculation time and data exchange time at the boundaries of the areas. Efficiency is also maintained at a good level, except in the case of a small number of cells.

The authors cannot conduct computational experiments on a larger number of computing nodes due to the limitations of computing power. However, the obtained result gives reason to believe good parallelization efficiency with an increase in the processors' number, complicating the task by increasing the computational domain's size and supplementing physical processes.

## 4 Simulation Results

Computational experiments were done on a cylindrical reactor 280 mm long and 20 mm wide. At the initial moment of time, the area is filled with methane, the temperature is 700°C, the pressure is 101325 Pa. Methane flows in at a temperature of 700° along the entire border on the left side with a flow rate of 8 liter per hour. The outflow condition is set on the right, the outlet pressure is 101325 Pa. Wall temperature is 1100°C. The scheme from [4] with 15 components of the mixture is adopted to describe radical chain chemical reactions. It significantly increases the estimated time of the program.

The described initial and boundary conditions lead to strong heating of the mixture from the reactor walls (Fig. 3) and, as a consequence, methane conversion (Fig. 4).

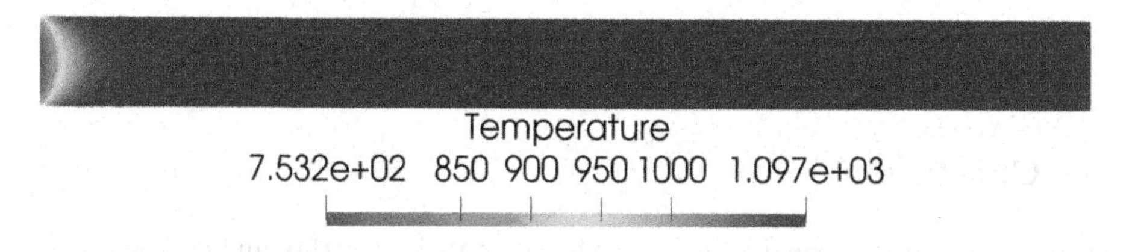


**Fig. 3.** Temperature distribution along the pipe

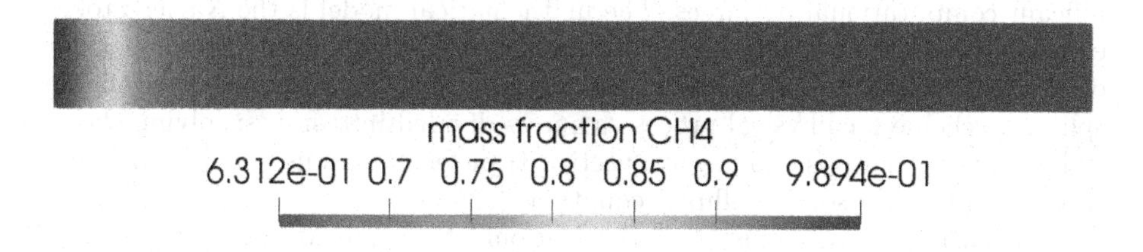


**Fig. 4.** Methane distribution along the pipe

The laminar flow is formed along the reactor's symmetry axis with an increase in velocity in the right part of the pipe (Fig. 5) due to the heating of the walls and active chemical transformations.

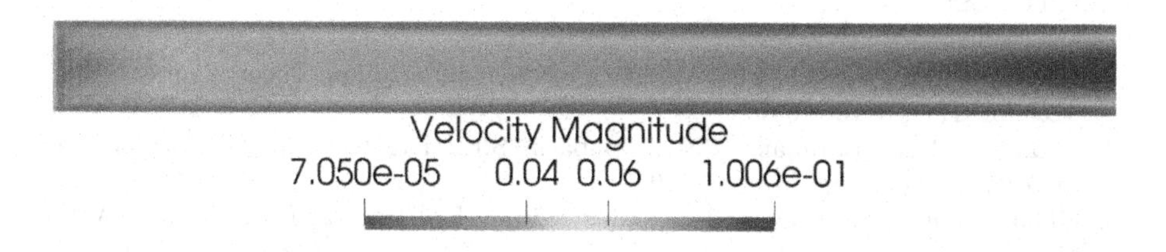


**Fig. 5.** Speed distribution along the pipe

A significant difference in temperature and the mixture density is observed in the left part of the pipe (Fig. 6) since methane flows at the left with a temperature significantly lower than the walls of the reactor.

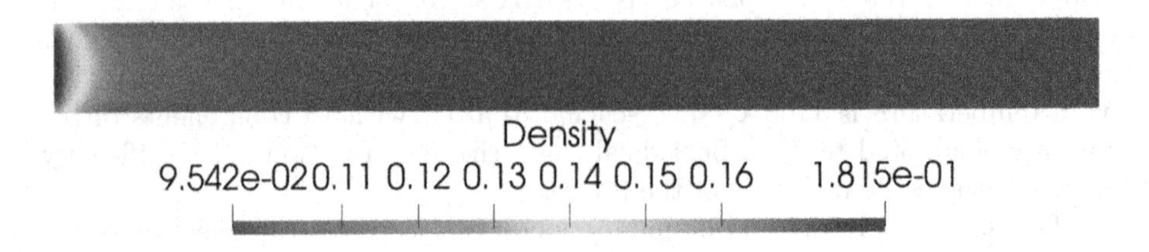


**Fig. 6.** Density distribution along the pipe

The results obtained well reflect the nature of the flow in the laboratory reactor.

## 5 Conclusion

The article presents the development of the parallel algorithm and its implementation in the program form with MPI technology for modeling subsonic reacting flows characterized by multiscale physico-chemical processes and requiring significant computational resources. The mathematical model is the Navier-Stokes equations in the low Mach number approximation, supplemented by equations of chemical kinetics. The numerical algorithm for this model is based on the splitting scheme for physical processes and is divided into stages: solving chemical kinetics equations, solving convective transfer equations, solving diffusion transfer equations, solving elliptic equations for pressure.

Calculations on the cluster at the National Research Mordovia State University showed good scalability of the developed program. Calculations of the methane conversion problem are done, the adequacy of the obtained results of the computational experiment is shown. The use of the developed program for multiparametric calculations for acceptable times can serve as a basis for the development of new chemical technologies for methane conversion.

## References

1. Boreskov, G.K., Slinko, M.G.: Modeling of chemical reactors. Theor. Found. Chem. Technol. **1**(1), 5–16 (1967)
2. Mukhina, T.N., Barabanov, N.L., Babash, S.E.: Pyrolysis of hydrocarbon raw materials. Chemistry, Moscow (1987)
3. Gubaydullin, I.M., Zhalnin, R.V., Masyagin, V.F., Peskova, E.E., Tishkin, V.F.: Simulation of propane pyrolysis in a flow-through chemical reactor under constant external heating. Math. Models Comput. Simul. **13**(3), 437–444 (2021)

4. Lashina, E.A., Snytnikov, V.N., Peskova, E.E.: Mathematical modeling of the dynamics of thermal conversion of methane-ethane mixtures in a wide temperature range. Chem. Sustain. Dev. **31**(3), 278–286 (2023)
5. Peskova, E.E. Numerical modeling of subsonic axisymmetric reacting gas flows. In: Journal of Physics: Conference Series, vol. 2057. https://doi.org/10.1088/1742-6596/2057/1/012071
6. Day, M.S., Bell, J.B.: Numerical simulation of laminar reacting flows with complex chemistry. Combust. Theor. Model. **4**(4), 535–556 (2000)
7. Stadnichenko, O.A., Snytnikov, V.N., Snytnikov, V.N., Masyuk, N.S.: Mathematical modeling of ethane pyrolysis in a flow reactor with allowance for laser radiation effects. Chem. Eng. Res. Des. **109**, 405–413 (2016)
8. Marchuk, G.I.: Splitting methods. Nauka, Moscow (1988)
9. Hairer, E., Wanner, G.: Solving Ordinary Differential Equations II. Stiff and Differential-Algebraic Problems. Springer-Verlag, Berlin (1996)
10. Yazovtseva, O.S., Gubaydullin, I.M., Peskova, E.E., Sukharev, L.A., Zagoruiko, A.N.: Computer simulation of coke sediments burning from the whole cylindrical catalyst grain. Mathematics **11**, 669 (2023). https://doi.org/10.3390/math11030669
11. Zhukov, V.T.: Explicit methods of numerical integration for parabolic equations. Math. Models Comput. Simul. **3**, 311–332 (2011). https://doi.org/10.1134/S2070048211030136
12. Zhukov, V.T., Feodoritova, O.B., Novikova, N.D., Duben, A.P.: Explicit-iterative scheme for the time integration of a system of Navier-stokes equations. Math. Models Comput. Simul. **12**, 958–968 (2020). https://doi.org/10.1134/S2070048220060174
13. Lebedev, V.I., Finogenov, S.A.: Utilization of ordered Chebyshev parameters in iterative methods. USSR Comput. Math. Math. Phys. **16**(4), 70–83 (1976)

# On the Theory of Fluctuations of Strongly Nonlinear (vibroimpact) Systems

L. A. Igumnov, V. S. Metrikin, and T. M. Mitryakova(✉)

National Research Lobachevsky State University of Nizhny Novgorod, Nizhny Novgorod, Russia
igumnov@mech.unn.ru, v.s.metrikin@mail.ru, tatiana.mitryakova@yandex.ru

**Abstract.** During the operation of a large group of technical devices and mechanisms with impact interactions, such as clock mechanisms, systems associated with vibro-driven piles and wells drilling for oil and gas production, pneumatic impact mechanisms, vibrating mills, cyclic automation mechanisms, impact interaction mechanisms due to the presence of gaps and others, there exist parameters, such as the restitution coefficient, adjusting gap, that do not remain constant but change randomly from impact to impact. However, most of the research results of strongly nonlinear systems contain a deterministic formulation. In this paper, we propose a numerical-analytical method for studying the dynamics of strongly nonlinear systems, taking into account fluctuations in system parameters. In this case, the mathematical apparatus of the point mappings method of Poincare surfaces is widely used. Random deviations of parameters can be both delta- and non-delta-correlated. The concepts of stochastic stability are introduced, which made it possible to determine not only the optimal values of system parameters, but also to find, to some extent, dangerous/safe areas of the stability regions boundaries of the systems motion modes.

**Keywords:** Strongly nonlinear systems · delta- and non-delta-correlated random deviations of parameters · Method of point mappings of Poincare surfaces · Stochastic stability

## 1 Introduction

A large number of experimental and theoretical works are devoted to the study of the dynamics of highly nonlinear (vibration-shock) systems. In this regard, we note that the interest of researchers in such essentially nonlinear systems still does not wane. This is due to the fact that applications of such systems are directly used in many industries and national economies. Along with obvious

Supported by the Ministry of Science and Higher Education of the Russian Federation, project no. FSWR-2023-0034, and by the Research and Education Mathematical Center <<Mathematics for Future Technologies>>.


D. Balandin et al. (Eds.): MMST 2023, CCIS 1914, pp. 122–129, 2024.
https://doi.org/10.1007/978-3-031-52470-7_10

practically important applications such as the construction of various civil facilities, medicine, military construction, etc., systems with impact interactions have found applications in capsule medicine, nanotechnology and other modern areas of science and technology. In the vast majority of problems, Newton's concept of the proportional relationship between the relative pre-impact and post-impact velocities of translationally moving bodies is used to describe direct impact interaction. It is accepted that the proportionality coefficient $R$, called the coefficient of restoration, depends on the properties of the impacting bodies and does not depend on the impact speed and is a constant value varying within $0 < R < 1$. It is known that in systems with collisions, movements with any finite number of impacts per period are fundamentally possible. Therefore, researchers consider only the movements that are practically the most important. Studies of more complex motion modes indicate a tendency toward a significant narrowing of the areas of existence of stable movements as the motion mode becomes more complex. In a number of cases, the range of feasibility of even two-shock oscillations turns out to be narrower than the stripes of natural dispersion of parameter values. In this paper, we study the dynamics of the vibration-impact mechanism, in the mathematical model of which it is assumed that the coefficient of recovery $R$ varies from blow to blow randomly with a zero average value and a dispersion different from zero. A numerical-analytical technology has been developed based on the point mapping method for finding statistical characteristics of post-impact velocity and impact time. The concept of stochastic stability was introduced and a numerical experiment was carried out to calculate the standard deviations of the speed and times of impacts.

## 2 Equations of Motion

Consider a vibroimpact mechanism consisting of a mass $M$, which is affected by an external periodic force $F \sin \omega t$ and a linear friction force $-h\dot{x}$. The mass itself oscillates along the stand attached to the base of the processed material. The physical scheme of the mechanism is presented in [7]. The dynamics of such a mechanism is of independent interest in terms of compaction of various materials (soil, sand, etc.). The research into dynamics of such systems is mainly carried out at constant parameters. Significantly less number of results on studying the dynamics of vibroimpact systems with randomly changing parameters can be observed. It is obvious that during the operation of vibro-impact compaction mechanisms for various materials, the restitution coefficient, when interacting with the medium, changes its initial value from impact to impact affecting the compaction process. In this regard, studying the dynamics of strongly nonlinear (vibro-impact) dynamic systems is a rather topical issue. In what follows, we will assume: 1) energy dissipation occurs through impacts and when friction forces are taken into account; 2) the impact is instantaneous with the restitution coefficient $R$, varying within $0 \leq R \leq 1$; 3) limiter offset is not taken into account. Under the assumptions made, the differential equation of the system in the interval between impacts ($x > 0$) can be written as:

$$M\ddot{x} + h\dot{x} = P + F\sin\omega t, \tag{1}$$

where $x$ — is the coordinate of the mass $M$, $P$ is a constant force acting on the impact mass $M$, and, generally speaking, $P \neq mg$, $h$ — is the friction coefficient, $F$, $\omega$ —are the amplitude and frequency of the driving force. At $x = 0$, $\dot{x} < 0$ an inelastic impact occurs with the restitution coefficient $R$, which is described using Newton's hypothesis $\dot{x}^+ = -R\dot{x}^-$, where $\dot{x}^-$, $\dot{x}^+$ are pre-impact and post-impact velocities of a mechanism. Introducing the dimensionless time $\tau = \omega t$ and the coordinate $\xi = \frac{M\omega^2 x}{F}$, Eq. (1) is rewritten as

$$\ddot{\xi} + \frac{h}{M\omega}\dot{\xi} = \frac{P}{F} + \sin\tau, \xi > 0, \tag{2}$$

$$\dot{\xi}^+ = -R\dot{\xi}^-, \xi = 0, \dot{\xi}^- < 0. \tag{3}$$

## 3 Point Mapping

Let us assume that during the operation of the mechanism, the restitution coefficient changes randomly from impact to impact, and the changes are so small that they do not take the system out of the vicinity of a stable periodic regime. Let in a deterministic system (the restitution coefficient is constant and equal to $R = R_0$) for some values of the parameter $R = R_0$ there is a stable periodic motion. The restitution coefficient for the $n$-th impact takes on the value: $R_n = R_0 + \Delta R_n$, where $\Delta R_n$ is a small random deviation from $R_0$ with zero mean value and non-zero variance. The phase space of the considered system (2), (3) in the coordinates $\xi, \dot{\xi}, \tau$ is three-dimensional. The plane $\xi = 0$ divides it into two subspaces: $\Phi_1(\xi \geq 0, \dot{\xi}, \tau)$ and $\Phi_2(\xi < 0, \dot{\xi}, \tau)$. The phase trajectory defined by (2), (3) is located in $\Phi_1$. It follows from the problem formulation that each time the image point falls on the $\xi = 0$ plane at the points $(\dot{\xi}_1, \tau_1), (\dot{\xi}_2, \tau_2), \ldots, (\dot{\xi}_n, \tau_n), \ldots$ with the restitution coefficient $R_1, R_2, \ldots, R_n, \ldots$. Therefore, it can be clearly concluded that the study of the dynamics of system (2), (3) can be carried out using a point mapping of the plane $\xi = 0$ into itself. The point mapping of the plane $\xi = 0$ is written as follows:

$$\begin{aligned}
\dot{\xi}_n &= \mu\Big(e^{-\mu(\tau_{n+1}-\tau_n)} - 1\Big)^{-1} \cdot \Big(e^{-\mu(\tau_{n+1}-\tau_n)} \cdot \Big(\frac{q}{\mu} - \frac{\nu}{\mu}\cos\tau_n + \nu\sin\tau_n\Big) - \frac{q}{\mu} - \\
&- q\tau_n + \frac{1}{\mu}\cos\tau_n + q\tau_{n+1} - \nu\sin\tau_{n+1} - \nu\mu\cos\tau_{n+1}\Big), \\
\dot{\xi}_{n+1} &= -R_{n+1}\Big(e^{-\mu(\tau_{n+1}-\tau_n)} \cdot \Big(\dot{\xi}_n - q + \nu\cos\tau_n - \nu\mu\sin\tau_n\Big) + q - \\
&- \nu\cos\tau_{n+1} + \nu\mu\sin\tau_{n+1}\Big).
\end{aligned} \tag{4}$$

Here $q = \frac{PM\omega}{Fh}$, $\mu^{-1} = \frac{M\omega}{h}$, $\nu = \frac{M^2\omega^2}{M^2\omega^2+h^2}$ It is directly seen from the point transformation that random sequences of points $\dot{\xi}_1, \tau_1; \dot{\xi}_2, \tau_2; \ldots; \dot{\xi}_n, \tau_n; \ldots$, determined from relations (4) with $R = R_1, R_2, \ldots, R_n, \ldots$, respectively, represent a Markov process to which the theory of Markov processes can be applied. Since, by assumption, the random deviations of the restitution coefficient $\Delta R_n$ for each $n$-th impact from $R_0$ are small, the set of points $\left\{\dot{\xi}_k, \tau_k\right\}$ will be located in the $\epsilon$ neighborhood of the stable periodic regime corresponding to the value of $R_0$. Therefore, it is possible to linearize point mapping (4) in the vicinity of this stable periodic motion, after which a system of equations in finite differences with respect to post-impact velocity deviations, impact time, and restitution coefficient is obtained in the form:

$$
\begin{aligned}
\Delta\dot{\xi}_n = \Delta\tau_{n+1} \cdot \Bigg( & \frac{\mu(q+\mu\nu)e^{-\mu T}}{e^{-\mu T}-1} - \frac{\mu e^{-\mu T}\cos\tau_0(1-\nu\mu^2-\nu e^{-\mu T})}{(e^{-\mu T}-1)^2} - q\mu - \nu\mu^2\sin\tau_0 + \\
& +\nu\mu\cos\tau_0\Bigg) + \Delta\tau_n \cdot \left( \frac{\mu e^{-\mu T}\cos\tau_0(\nu\mu^2-\nu+1)}{(e^{-\mu T}-1)^2} + \frac{\nu e^{-\mu T}(\sin\tau_0+\mu\cos\tau_0)}{e^{-\mu T}-1} - \frac{+\sin\tau_0}{e^{-\mu T}-1} \right), \\
\Delta\dot{\xi}_{n+1} = {} & \Delta R_{n+1} \cdot \left( (e^{-\mu T}-1)(q-\nu\cos\tau_0+\nu\mu\sin\tau_0) - e^{-\mu T}\dot{\xi}_0 \right) + \\
& +\Delta\tau_{n+1} \cdot \left( \mu R_0 e^{-\mu T}(\dot{\xi}_0 - q + \nu\cos\tau_0 - \nu\mu\sin\tau_0) + \nu\sin\tau_0 + \nu\mu\cos\tau_0 \right) + \\
& +\Delta\tau_n e^{-\mu T} \cdot \left( -R_0\mu(\dot{\xi}_0 - q + \nu\cos\tau_0 - \nu\mu\sin\tau_0) - \nu\sin\tau_0 - \nu\mu\cos\tau_0 \right) + \\
& +e^{-\mu T}\Delta\dot{\xi}_n,
\end{aligned} \tag{5}
$$

where $T$ denotes the period $2\pi n$. The system of two difference equations of the form (5) is reduced to a linear difference equation with the right side of the form:

$$
\Delta\tau_{n+2} + \alpha\Delta\tau_{n+1} + \beta\Delta\tau_n = \gamma\Delta R_{n+1}. \tag{6}
$$

Random changes in the restitution coefficient $\Delta R_{n+1}$ in Eq. (6) can be selected from the literature sources among the known dependencies.

The solution of the difference equation is sought as the sum of general solution $\Delta\tau_n^0 = C_1 z_1^n + C_2 z_2^n$ ($z_1, z_2$ — roots of the characteristic equation $z^2+\alpha z+\beta=0$) of a homogeneous equation and a particular solution $\Delta z_n^*$ of an inhomogeneous equation, which is found using the method of variation of arbitrary constants.

Since, by assumption, the periodic motion is stable, the roots of the characteristic equation of system (4) satisfy the condition $|z_1| < 1, |z_2| < 1$. With this in mind, the solution of the difference Eq. (6) will be written in the form:

$$
\Delta\tau_{n+1} = T \sum_{\nu=-\infty}^{n-1} \alpha_{n,\nu}\Delta R_{\nu+1}, \tag{7}
$$

where $\alpha_{n,\nu} = \frac{z_1^{n-\nu-1} - z_2^{n-\nu-1}}{z_1 - z_2}$.

For $\Delta\dot{\xi}_n$ we obtain:

$$\Delta \dot{\xi}_n = \frac{Tp}{1+R} \sum_{\nu=-\infty}^{n-1} \alpha_{n,\nu} \left\{ \left( 1 + \sqrt{\left(\frac{1+R}{p}\right)^2 - (1-R)^2(\pi n p)^2} \right) \Delta R_{\nu+1} - \right.$$

$$\left. -\Delta R_{\nu+2} \right\}, \tag{8}$$

where $p$ is the ratio of the weight of vibrostriker to the amplitude of driving force.

Assuming that random changes of $R$ are delta-correlated, from (7) and (8), we obtain expressions for standard deviations in the form:

$$\overline{\Delta \tau_n^2} = \frac{T^2(1+z_1 z_2}{(1+z_1 z_2)(1-Z_1^2)(1-z_2^2)} \overline{\Delta R^2}, \tag{9}$$

$$\overline{\Delta \dot{\xi}_n^2} = T^2 p^2 \left[ (1+m^2)(1+z_1 z_2) \right] \overline{\Delta R^2}, \tag{10}$$

where $m = 1 - \frac{1+R}{p} \cos \tau^*$.

Equalities (7), (8), (9), (9) show that neither $\overline{\Delta \tau_n^2}$, nor $\overline{\Delta \dot{\xi}_n^2}$ do not vanish for any values of the system parameters. This follows from the fact that $|z_i < 1|$, and $1 + z1z_2 = 1 + R^2$. However, it is possible to specify the values of the system parameters at which the variance $\overline{\Delta \dot{\xi}_n^2}$ equals zero, namely:

$$(1+R^2) \left\{ 2 + \left( \frac{1+R}{p} \cos \tau^* \right)^2 - 2 \frac{1+R}{p} \cos \tau^* \right\} - 2 \left\{ 1 - \frac{1+R}{p} \cos \tau^* \right\} \cdot$$

$$\cdot \left\{ 1 + R^2 - \frac{(1+R)^2}{p} \cos \tau^* \right\} = 0. \tag{11}$$

Hence we get the equation:

$$\cos \tau^* = \frac{\{2-(1-R)^2\} \left\{ 1 - \left( \frac{1-R}{1+R} \pi n p \right)^2 \right\}}{2p}. \tag{12}$$

For small $R << 1$, which corresponds to an almost absolutely inelastic impact, we obtain

$$\frac{1-(\pi n p)^2(1-2R)}{\sqrt{1-(\pi n p)^2}} = \frac{2R + 1 - (\pi n p)^2 (1-2R)}{2p}. \tag{13}$$

For $R = 0$ formula (13) gives $p = \left(4 + (\pi n)^2\right)^{-1/2}$. From the last equality we obtain that for $n = 1$ — $p = 0,276453$, for $n = 2$ — $p = 0,151653$, for $n = 3$ — $p = 0,10379$. For $R = 1$ the value of $p$ is 1. Hence it follows that the curve (12) lies near the upper stability limit.

## 4 Stochastic Stability

A dynamic system, the input of which is a random process $\xi(\lambda, t)$ and the output is a random process $y(\lambda, t)$, according to [2], will be called stochastically stable if for any $\epsilon > 0$ there exist $\delta > 0$ such that if $r(\xi) \leq \delta$ then $\rho(y) \leq \epsilon(\delta)$ and $\epsilon \to 0$ for $\delta \to 0$, where $r(\xi)$ and $\rho(y)$ are norms. The definition of stochastic stability depends on the way norms $r(\xi)$ and $\rho(y)$ are introduced.

It was shown above that the root mean square deviations of $\overline{\Delta\tau_n^2}$ and $\overline{\Delta\dot{\xi}_n^2}$ can be written as in the form of a product of some constant, which is the function of only the system parameters and the dispersion of the random deviation $\overline{\Delta R^2}$. Assume that $\overline{\Delta R} = 0$ and $\overline{\Delta R^2} \neq 0$. Then, using [2], we can assume that the system will be stochastically stable if there exists a constant $C$, that depends only on the parameters of the system, such that, for sufficiently large $n$, the inequalities hold

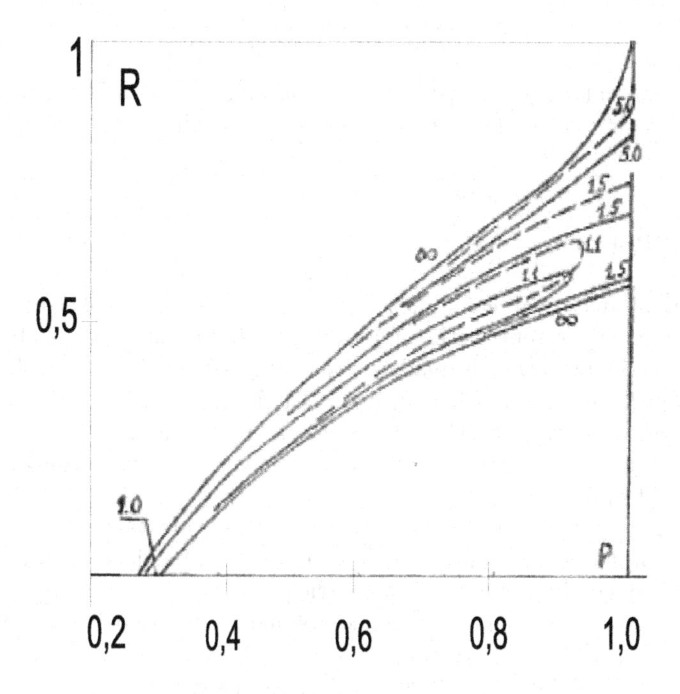


**Fig. 1.** Numerical experiment

$$\overline{\Delta\tau_n^2} + \overline{\Delta\dot{\xi}_n^2} \leq C\overline{\Delta R^2}. \tag{14}$$

Note that, since the changes of $R$ are small and parameter deviation occurs in the vicinity of such a value at which there is a stable periodic regime in the deterministic case, the value $\overline{\Delta R^2}$ is small, and therefore the expression of the right side of inequality (14) is also a small value (($C$ is a constant). Hence it follows that in inequality (14), due to the smallness of $\overline{C\Delta R^2}$ the root-mean-square values $\overline{\Delta\tau_n^2}$ and $\overline{\Delta\dot{\xi}_n^2}$ must also be small. In addition, it follows from (14) that if the deviations of the restitution coefficient tend to zero, then the root-mean-square ones also tend to zero if the system is stochastically stable.

## 5 Numerical Experiment

In this section, we present the results of the numerical experiment in order to identify quantitative characteristics of the effect of correlation functions with a change of $R$ on the value of the root-mean-square deviations.

In Fig. 1 dotted lines are level lines for delta-correlated changes of $R$, and solid lines are level lines corresponding to non-delta -correlated changes of $R$.

Inside the stability region there is a boundary curve

$$x = \frac{2(1+R)^2}{\sqrt{4(1+R)^2 + (1-R^2)^2 T^2}}, \quad (15)$$

meaning as follows. Correlation with a random change of $R$ when accounted for leads to a shift of the level lines to domain of smaller $p$ for $p \leq x$ and larger one $p$ for $p > x$.

## 6 Conclusions

1. A numerical-analytical technique for studying the dynamics of strongly nonlinear (vibro-impact) systems with randomly changing parameters has been developed. The main attention is paid to the study of the influence of the most significant parameters that change in time and depend on the behavior of the systems themselves in time. Thus, for example, the structure of the behavior of a vibro-impact system is studied when the velocity recovery coefficient changes from impact to impact.
2. A refined form of the stochastic stability of vibro-impact systems is given, with the help of which it is possible to determine, in particular, an analogue of dangerous and safe boundaries of stochastic stability.
3. The calculated data on the behavior of maximum velocities upon impact during material processing are given.
4. Exact ratios of solutions of difference equations depending on the choice of the form of random changes in the velocity recovery coefficient upon impact are given.

The work was supported by the RSF grant No. 22-19-00138.

## References

1. Bespalova, L.V.: On the theory of vibro-impact mechanism, no. 5. Izvestiya AN SSSR, OTN (1957)

2. Brusin, V.A., Tai, M.L.: Absolute stochastic stability. Radiophys. Quantum Electron. **10**(7) (1967)
3. Goldin, Y.M., Zaretsky, L.B.: Application of mathematical modeling for the study of shock-oscillatory systems with a random restitution coefficient, vol. 15. VNShK stroydormash, M. (1968)
4. Dimmentbert, M.F.: Amplitude-frequency characteristic of a system with randomly changing parameters, no. 2. Inzh. magazine MTT (1966)
5. Kobrinsky, A.A.: Forced motion of a vibroimpact system with a random restitution coefficient. Mashinovedenie (1968)
6. Kobrinsky, A.A.: Stochastic motion of a single-mass vibroimpact system, no. 1. Mashinovedenie (1968)
7. Metirikin, V.S., Nikiforova, I.V.: On the dynamics of systems with impact interactions with a non-analytical structure of the phase space. Autom. Remote Control 8 (2013)
8. Rytov, O.M.: On the theory of fluctuations in self-oscillating systems with piecewise linear characteristics. Izvestiya vysshikh uchebnykh zavedenii, Radiofizika **2**(1) (1959)
9. Sadekov, R.K.: To the question of fluctuations in piecewise linear self-oscillating systems. Izvestiya vysshikh uchebnykh zavedenii, Radiofizika **3**(5) (1960)

# A Network Evolution Model with Addition and Deletion of Nodes

Sergei Sidorov(✉), Sergei Mironov, and Timofei D. Emelianov

Saratov State University, Saratov 410012, Russia
sidorovsp@sgu.ru

**Abstract.** Most of complex network generation models study the mechanisms that drive the growth of systems. In this paper, we propose a network evolution model based on the simultaneous application of both node addition and deletion rules. The results show that the degree distribution in the networks generated based on this model follows a power law.

**Keywords:** Complex networks · growth network model · degree distribution

## 1 Introduction

The study of the rules and mechanisms of network generation is one of the most essential research topics in the theory of complex networks. The main problem that researchers face is to develop algorithms that simulate the evolution of networks so that the resulting synthetic graphs have some of the properties of real complex networks. As a property that it is desirable to reproduce in the process of generating a network, the power law for the degree distribution is often considered. Indeed, many real networks have the scale-free property, that is, the degree distribution in such networks follows a power law with a quite small exponent.

Most real networks grow by adding new nodes, just as social networks increase as new users join, World Wide Web grows when new pages or sites appear, citation networks are expanding through the emergence of new publications citing previously published ones. Therefore, the first network generation models were focused on modeling the process of network growth. These models include algorithms based on the preferential attachment mechanism. Over the past twenty years, many models have emerged that develop these ideas. However, almost all the works were devoted to the modeling of growing networks, while the study of contraction and shrinking mechanisms turned out to be practically beyond the area of interest of scientists. This can be explained by the following. Firstly, most real networks were growing, and the processes of reduction have become more pronounced and noticeable in many real systems only recently. Secondly,

The work was supported by the Russian Science Foundation, project 23-21-00148.


D. Balandin et al. (Eds.): MMST 2023, CCIS 1914, pp. 130–136, 2024.
https://doi.org/10.1007/978-3-031-52470-7_11

the modeling of shrinking processes has proved to be a very difficult task, since they are less intuitive and cannot be described by applying growth mechanisms triggered in the reverse direction of time.

At the same time, many systems and networks, undoubtedly, in addition to growth via the inclusion of new nodes or subsystems, show contraction due to the removal of its elements or links between them. Moreover, these processes (adding and removing network nodes) often occur simultaneously. However, in the scientific literature there is a clear lack of adequate models describing the evolution of such networks. One of such model was introduced in [5]. The model uses a node deletion as well as node duplication. The authors show that the generated graphs hold some important properties of protein interaction networks. Paper [8] presents a model in which one new node is added to the network and a random node is deleted with fixed probability. It was shown that the rate of node addition should be higher than the rate of deletion to insure the scale-free property of the evolving network. The work [7] utilizes the list of active nodes, deleting from which deactivates the node and excludes the node from the process of network growth in subsequent iterations.

The paper [1] proposes a model for growing network generation with both addition and removal of nodes. The model uses two independent steps at each iteration. With rate $0 < r < 1$, a node is attached to the graph and this node joins to a randomly selected node of the network. Then a randomly selected vertex is removed, and one of its neighbors obtains the links of its immediate descendants. Paper [4] examines a network growth mechanism which includes addition and removal of nodes. In particular, the authors of [4] study the effect of node removal on graph structure. They show that with the use of node removal rule, the degree distribution of a network transforms from scale-free to become exponential-type distributed.

Another mechanism for generating networks based on edge removal was proposed in [3], in which the authors investigated a model that involves the simultaneous use of the preferential attachment step and the step of removing a random edge from the network.

An efficient network generation model based on the removal of vertices by merging them was proposed recently in a paper [9] that shows that this mechanism is capable of simulating networks with the power-law degree distribution. The idea of merging blocks in networks is exploited in paper [10]. In paper [11] a model has been recently proposed for simulating network evolution with the state transition of network vertices between online and hidden based on the birth and death process.

In this article, we propose a new model for generating complex networks whose evolution involves the simultaneous use of growth and contraction mechanisms. Growth in the model will be provided by adding one new vertex at each iteration. The contraction will occur by merging two randomly chosen vertices. The result of the merging will be a vertex whose neighbors are the neighbors of these two vertices.

One of the problems that arises when analyzing complex networks is to coarse the graph through various methods, including dimension reduction, removing vertices or edges, so that the essential properties or structural features of the original system are preserved [2,6]. We would like to note that this study addresses a different problem related to the search for realistic mechanisms for the evolution of complex networks, and not to the development of graph coarsening techniques.

## 2 One-In and One-Out Model

### 2.1 The Model

Let us describe the model formally. Let at the initial time the network consists of $n$ vertices and $m$ edges. We assume that the vertices are numbered by integers $V_n := \{1, 2, \ldots, n\}$.

The evolution of the networks involves two steps at each iteration:

- (*Growth*). At each iteration $t = n+1, \ldots$ a new node $t$ is added to the network, i.e. at iteration $t$, the set of vertices $V_n := \{1, 2, \ldots, t\}$. This new vertex is connected by an edge to one of the already existing vertices in the network. The vertex to which a new vertex is attached is chosen randomly among all nodes with non-zero degree (the probability of choosing a vertex is $\frac{1}{n}$).
- (*Contraction*). In addition, at each iteration two vertices are randomly selected: a new vertex is formed, the identifier of which will be equal to the number of the first vertex, the second vertex is removed (i.e. takes degree 0), and all its neighbors are attached to the first vertex. If some neighbor of the second vertex was also a neighbor of the first, then the edge is not duplicated. Thus, the degree of the first vertex is increased by the degree of the second vertex, excluding the number of neighbors of the second node that are also neighbors of the first one.

Since the new vertex is joined with one edge, the total number of edges is increasing by 1 from iteration to iteration. Moreover, since at each iteration one new vertex is added to the network, but one vertex is removed as a result of the merging, the total number of nodes does not change from iteration to iteration. The edge between the merging vertices disappears if it exists.

Simultaneous use of the mechanism of growth and reduction of the network leads to the fact that the size of the network remain unchanged in the process of evolution, but its edge density is increasing. However, as a result of repeated application of the steps of the algorithm, the degree of each of the vertices may change, as well as the distribution of the degrees of the network. The subject of this paper is the analysis of the degree distribution evolution in networks generated by this model.

### 2.2 Degree Distribution Analysis

In this section, we will show that the limiting degree distribution in the networks generated by this model will converge to a power law, whatever the network's degree distribution is at the initial time.

Denote by $q_k(t)$ the fraction of network nodes that have degree $k$ at time $t$. Let $q_k = \lim_{t\to\infty} q_k(t)$ be the limit value of the probability that a randomly selected network node has degree $k$.

Note that the degree of a node can change

- if a new node chooses it (with probability $p_1 = \frac{1}{n}$),
- if this node is chosen as the second node when two randomly selected network nodes are merged (probability $p_2 = \frac{1}{n}$),
- if the node is chosen as the first node in the merge (probability $p_3 = \frac{1}{n}$).

Let us first study how the proportion of nodes with degree 1 changes at iteration $t + 1$. First, a node can have degree 1 at iteration $t$ and none of the possible situations leading to a change in its degree has occurred. The probability of such an outcome is $1 - p_1 - p_2 - p_3$. Secondly, at each iteration a new node is added with degree 1 (with one node removed at the step of merging nodes, so the number of nodes in the network remains unchanged and equal to $n$). Therefore, the fraction of nodes with degree 1 increases by $\frac{1}{n}$ at each iteration. Thus, we get the equation

$$q_1(t+1) = q_1(t)(1 - p_1 - p_2 - p_3) + \frac{1}{n},$$

whence for $t \to \infty$ we get $q_1 = \frac{1}{3}$.

Further, the proportion of nodes that have degree 2 can change at iteration $t + 1$ in two cases: first, if the node had degree 1 at iteration $t$ and a new node has joined it; second, if a node had degree 1, it was chosen as the first node in the merging step, and the degree of the second node turned out to be 1. In addition, a node can have degree 2 at iteration $t$ and none of the possible situations leading to a change in its degree did not occur (its probability is equal to $1 - p_1 - p_2 - p_3$). Thus we have

$$q_2(t+1) = q_2(t)(1 - p_1 - p_2 - p_3) + q_1(t)p_1 + p_2 q_1^2(t),$$

whence for $t \to \infty$ we get $q_2 = \frac{4}{27}$, taking into account $q_1 = \frac{1}{3}$.

In general, the share of vertices that have degree $k$ may change at iteration $t + 1$ in two situations. First, if the vertex with degree $k - 1$ at iteration $t$ joins a new node. Second, if a vertex with degree $i$ is chosen as the first node in the merging step, while the degree of the second vertex is equal to $k - i$. Also, the share of nodes with degree $k$ at iteration $t$ may do not change if none of the possible situations has occurred, i.e. with the probability $1 - p_1 - p_2 - p_3$). Thus we have

$$q_k(t+1) = q_k(t)(1 - p_1 - p_2 - p_3) + q_{k-1}(t)p_1 + p_2 \sum_{i=1}^{k-1} q_i(t) q_{k-i}(t),$$

and we obtain the limit value as follows

$$q_k = \frac{1}{3}\left(q_{k-1} + \sum_{i=1}^{k-1} q_i q_{k-i}\right). \tag{1}$$

It can be shown that the sequence $(q_k)_{k\geq 1}$ obtained from (1) follows the power law with exponent $-\frac{3}{2}$ (see Fig. 1 (a)).

### 2.3 Empirical Results

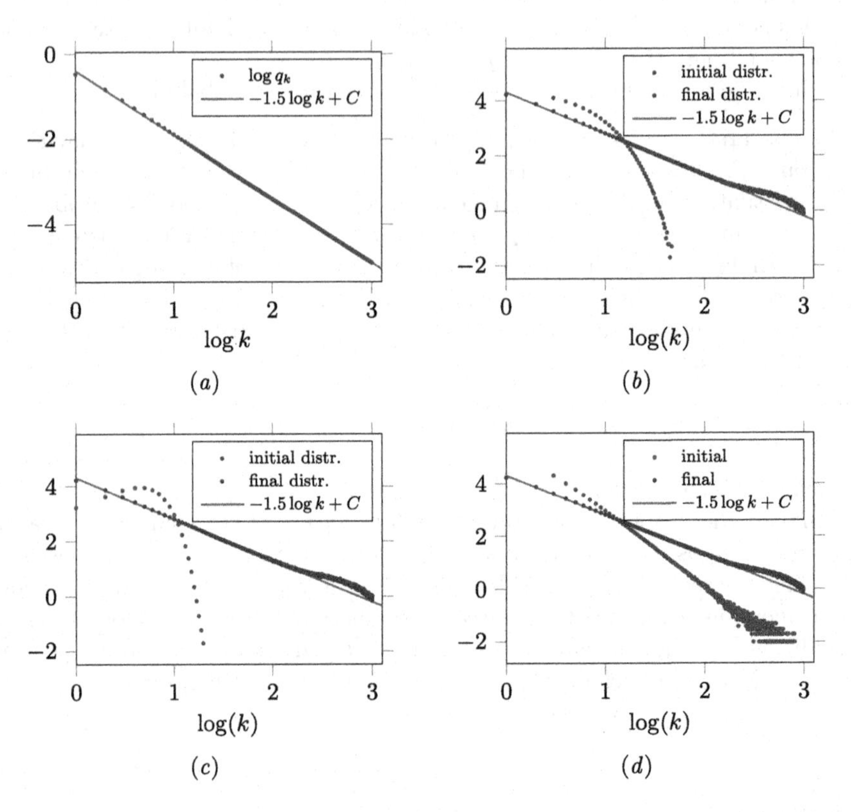


**Fig. 1.** (*a*) Probability distribution according to Eq. (1); (*b*) The degree distribution in the graph in the first series of experiments, averaged over 100 simulations. The initial graph consisted of $n = 50,000$ vertices and was built as a growing network; (*c*) The degree distribution obtained in the second series of experiments. The initial network is the Erdős-Rényi random graph with $n = 50000$ vertices constructed with the parameter $p = 0.0001$; (*d*) The average degree distribution in the graphs obtained in the third series of experiments. The initial networks are the Barabási–Albert graphs with $n = 50000$ vertices, constructed with the parameter $m = 3$

Three series of experiments were carried out in which the proposed model was applied for three different types of initial graphs.

In each experiment, 1M iterations were applied to the initial graph of $n =$ 50k vertices, at each of which two random vertices were selected with equal probability from the entire set of vertices. Neighboring vertices of the first of them, which are not neighbors of the second, were connected by an edge with the second vertex. After that, all incident edges of the first vertex were removed from the graph, and instead of them, an edge was drawn to a random vertex of the graph. The choice of a vertex for joining was carried out equiprobably. For the initial graph and for the graph obtained as a result of applying the model, the distribution of vertex degrees was calculated. In each series, 100 independent experiments were carried out and the degree distribution was averaged.

In the first series of experiments, we took the initial graph obtained as the result a growing network generation process as follows: the growth begins from a complete graph of $m$ vertices, after which, at each growth step, $m$ vertices were chosen equiprobably in the graph, to which an edge was drawn from the new vertex. In Fig. 1 (b) the average distribution of degrees of the initial graph (constructed in this way with parameter $m = 3$) is represented by blue color in the log-log scale. The purple color in Fig. 1 (b) shows the degree distribution obtained after 1M iterations of the proposed model applied to the initial graph, averaged over 100 independent simulations. Figure 1 (b) shows that degrees of graphs obtained by means of simulations are the power-law distributed with an exponent $\gamma \approx 1.5$.

In the second series of experiments, the initial graphs were constructed according to the Erdős-Rényi model with the parameter $p = 0.0001$. The degree distribution of the graph degrees averaged over 100 simulations is shown in Fig. 1 (c). The blue colored line shows the average degree distribution of the original graphs, the purple colored line shows the average degree distribution of graphs after 1M iterations of the model. As in the previous case, the log-log plot of the degree distribution are concentrated around the same straight line.

In the third series of experiments, the graph constructed according to the Barabási-Albert model with the parameter $m = 3$ acted as the initial graph. The results of the experiments are shown in Fig. 1 (d). The blue colored plot shows the average degree distribution of the initial graph, the purple colored plot shows the average degree distribution of the graphs after 1M iterations of the model.

## 3 Conclusion

In this paper, we propose a novel model for network generation that takes into account both the growth and reduction of the network during evolution. Our model involves the addition of a new vertex at each iteration, randomly attaching it to an existing node in the network. Additionally, the network experiences node loss through the merging of two vertices.

Analytical findings indicate that by applying an infinite number of iterations to an initial graph, the resulting degree distribution follows a power law with an exponent of $-\frac{3}{2}$. To validate these results, we conducted three sets of experiments. Regardless of the initial graph's degree distribution, our experiments consistently revealed that the degree distribution of generated networks follows a power law with the aforementioned exponent.

## References

1. Ben-Naim, E., Krapivsky, P.L.: Addition-deletion networks. J. Phys. A: Math. Theor. **40**(30), 8607 (2007). https://doi.org/10.1088/1751-8113/40/30/001
2. Chen, J., Saad, Y., Zhang, Z.: Graph coarsening: from scientific computing to machine learning. SeMA J.: Bull. Span. Soc. Appl. Math. **79**, 187–223 (2022). https://doi.org/10.1007/s40324-021-00282-x
3. Deijfen, M., Lindholm, M.: Growing networks with preferential addition and deletion of edges. Phys. A **388**(19), 4297–4303 (2009). https://doi.org/10.1016/j.physa.2009.06.032
4. Deng, K., Zhao, H., Li, D.: Effect of node deleting on network structure. Physica A: Stat. Mech. Appl. **379**(2), 714–726 (2007). https://doi.org/10.1016/j.physa.2007.02.039, https://www.sciencedirect.com/science/article/pii/S0378437107001823
5. Farid, N., Christensen, K.: Evolving networks through deletion and duplication. New J. Phys. **8**(9), 212 (2006). https://doi.org/10.1088/1367-2630/8/9/212, https://app.dimensions.ai/details/publication/pub.1032738127
6. Huang, Z., Zhang, S., Xi, C., Liu, T., Zhou, M.: Scaling up graph neural networks via graph coarsening. In: Proceedings of the 27th ACM SIGKDD Conference on Knowledge Discovery & Data Mining, KDD '21, pp. 675–684. Association for Computing Machinery, New York (2021). https://doi.org/10.1145/3447548.3467256
7. Klemm, K., Eguíluz, V.M.: Growing scale-free networks with small-world behavior. Phys. Rev. E **65**, 057102 (2002). https://doi.org/10.1103/PhysRevE.65.057102
8. Moore, C., Ghoshal, G., Newman, M.E.J.: Exact solutions for models of evolving networks with addition and deletion of nodes. Phys. Rev. E **74**, 036121 (2006). https://doi.org/10.1103/PhysRevE.74.036121
9. Naglić, L., Šubelj, L.: War pact model of shrinking networks. PLoS ONE **14**(10), 1–14 (2019). https://doi.org/10.1371/journal.pone.0223480
10. Takemoto, K., Oosawa, C.: Evolving networks by merging cliques. Phys. Rev. E **72**, 046116 (2005). https://doi.org/10.1103/PhysRevE.72.046116
11. Zeng, Z., Feng, M., Kurths, J.: Temporal network modeling with online and hidden vertices based on the birth and death process. Appl. Math. Model. **122**, 151–166 (2023). https://doi.org/10.1016/j.apm.2023.05.034, https://www.sciencedirect.com/science/article/pii/S0307904X23002433

# Quantitative Study on the Friendship Paradox in Networks with Power-Law Degree Distribution

Alexey Grigoriev, Sergei Mironov, and Sergei Sidorov(✉)

Saratov State University, Saratov 410012, Russia
sidorovsp@sgu.ru

**Abstract.** For a node of a complex network, the friendship index is obtained as the average degree of neighbors divided by its own degree. This index is broadly used in social network analysis. It is known that for the vast majority of real complex networks, their degree distributions follow a power-law with some exponent $\gamma$. In this paper, we establish a quantitative relationship between the parameter $\gamma$ and the proportion of network nodes for which the values of their friendship index is greater than 1. We will explore scale-free networks with degree-degree neutral mixing and compare our findings with the empirical behavior for several real networks.

**Keywords:** Social network analysis · Complex networks · Friendship index · Friendship paradox · Configuration model

## 1 Introduction

The friendship index is a significant measure used to determine the popularity of a node in comparison to its friends on a social network. Popularity is typically determined by the number of contacts a node has, known as its degree. The friendship index is calculated by dividing the average degree of a node's friends by its own degree [11]. By definition, the value of node's friendship index more than one means that, on average, its neighbors (friends) are more popular than the node itself. Conversely, a value less than one shows that the node is more popular than its neighbors. The friendship index is broadly applied in various cases of social network analysis [1,2,5–8,10].

Numerous studies on real social networks have consistently shown that the majority of nodes has value of friendship index more than one. In other words, in real networks, on average, the friends are most likely to be more popular than you. In sociological sciences this phenomenon is called the friendship paradox, which has been extensively explored in various research papers, including the study by [11], which examined the aggregated friendship index value across entire networks, both real and synthetic. Other characteristics, associated with node popularity among its neighbors, are studied in [4,5,8,10,11,13,14].

The work was supported by the Russian Science Foundation, project 23-21-00148.


D. Balandin et al. (Eds.): MMST 2023, CCIS 1914, pp. 137–144, 2024.
https://doi.org/10.1007/978-3-031-52470-7_12

Expanding the definition of friendship index for a node, we get that it is defined as the ratio of the sum of the degrees of its neighbors to the square of its own degree. In many cases, we can consider the numerator of this ratio as the sum of random variables distributed according to the cumulative degree distribution function of the network. Real complex networks often exhibit a power-law degree distribution with an exponent $\gamma$. In this article, we establish a quantitative relationship between the parameter $\gamma$ and the proportion of network nodes whose friendship index values exceed 1. We investigate scale-free networks with degree-degree neutral mixing. Additionally, we compare our findings to the empirical distribution of friendship index in several real networks.

This study aims to evaluate the friendship paradox in both real and synthetic networks. We focus on how characteristics of degree distributions for scale-free networks impact the proportion of nodes with a friendship index greater than 1.

## 2 The Quantitative Characterization of the Friendship Paradox

### 2.1 Notations

Denote as $G_n = (V_n, E_n)$ a graph with $n$ vertices, where $V_n = \{1, 2, \ldots, n\}$ are its (integer) vertices and $E_n$ is the set of (undirected) edges. Let $d_i$ denote the degree of vertex $i$ and let $D_n = \{d_1, d_2, \ldots, d_n\}$ be the degree sequence of graph $G_n$. Then the total degree is obtained as follows: $\sum_{i=1}^{n} d_i = 2|E_n|$.

Denote as $(i, j)$ the directed (half-edge) edge between nodes $i$ and $j$, then the undirected edge $e_{ij}$ would be the sum of two half-edges $(i, j)$ and $(j, i)$. Let $|E_{ij}|$ be the sum of all half-edges between $i$ and $j$. $|E_{ij}|$ can be greater than 1 if $G_n$ is multi-graph. In case of undirected graph $G_n$, $|E_{ij}| = |E_{ji}|$. In such graphs self-loops of vertex $i$ should be counted twice. Following the notations, the degree of node $i$ is defined as $d_i = \sum_{j=1}^{n} |E_{ij}|$.

Let $\xi$ be a positive integer-valued random variable with the probability density function $f_\xi(k) = p(\xi = k)$ and the cumulative distribution function $F_\xi(k) = \sum_{j \leq k} f_\xi(k)$.

We assume that the degree sequences $D_n$ are obtained as $n$ independent and identically distributed samples of random variable $\xi$. If $\sum_{i=1}^{n} d_i$ turns out to be odd then without the loss of generality we set $d_1 = d_1 + 1$, since the addition of 1 to the total degree of graph $G_n$ does not affect its asymptotic properties when $n$ is tending to infinity.

Denote $V_n(k) = \{i : d_i = k\} \subset V_n$ the set of vertices of graph $G_n$ having the degree $k$. Denote $\beta_i = \frac{\sum_{j \in E_{ij}} d_j}{d_i^2}$ the friendship index of node $i$.

### 2.2 Configuration Model

The configuration model (CM) [3] is a graph generation model, which produces graphs with a given size $n$ and degree sequence $D_n$. This particular model is important for analyzing degree-degree correlations in complex networks since it creates graphs $G_n$ with a specific degree distribution $D_n$. By assigning stubs to each node and randomly pairing them, one could obtain a multi-graph with the desired degree sequence.

Given $\xi$ is a positive integer-valued random variable with probability density function $f$, the degree sequence $D_n = (d_1, d_2, \ldots, d_n)$ can be obtained as $n$ independent and identically distributed samples of random variable $\xi$. One of known traits of CM is that it produces random multi-graphs with the empirical degree distribution $f_n$ that converges to $f$ as $n$ approaches infinity.

### 2.3 The Main Result

First we assume that the network degree distribution of underlying random variable $\xi$ follows the power law with parameter $\gamma$, i.e. $f(k) = p(\xi = k) = c(\gamma)k^{-\gamma-1}$, $k = 1, 2, \ldots$, where $c(\gamma) = \frac{1}{\zeta(\gamma+1)}$, where $\zeta(s) := \sum_{k \geq 1} k^{-s}$ is the Riemann zeta function.

**Theorem 1.** *Let $m \in \mathbb{N}$ is integer. Now we assume that the network degree distribution of underlying random variable $\xi$ follows the power law with parameter $\gamma$, while its values begin with $m$, i.e. $f(k) = p(\xi = k) = c_m(\gamma)k^{-\gamma-1}$, $k = m, m+1, \ldots$. Then the constant $c_m(\gamma)$ can be found from the equality*

$$c_m(\gamma) \sum_{k=m}^{\infty} k^{-\gamma-1} = 1.$$

*i.e. $c_m(\gamma) = \frac{1}{\zeta(\gamma+1) - \sum_{k=1}^{m-1} k^{-\gamma-1}}$, where $\zeta(s) := \sum_{k \geq 1} k^{-s}$ is the Riemann zeta function. Denote $\kappa_n$ the proportion of nodes which friendship index is more than 1 in multi-graph $G_n$ and let $\kappa := \lim_{n \to \infty} \kappa_n$. Then*

$$1 - \sum_{s=m}^{\infty} \frac{1}{(\zeta(\gamma+1) - \sum_{k=1}^{m-1} k^{-\gamma-1})^{s+1}} s^{-\gamma-1} \left( \sum_{j=m}^{s(s-1)+1} j^{-\gamma-1} \right)^{s} \leq \kappa \leq$$

$$1 - \sum_{s=m}^{\infty} \frac{1}{(\zeta(\gamma+1) - \sum_{k=1}^{m-1} k^{-\gamma-1})^{s+1}} s^{-\gamma-1} \left( \sum_{j=m}^{s} j^{-\gamma-1} \right)^{s}. \quad (1)$$

*Proof.* To find the total share of nodes that have the friendship index value greater than 1, we will estimate the proportion of those nodes that have the friendship index less than or equal to 1.

The fraction of nodes with degree 1 is equal to $f_1 = p(\xi = 1) = \frac{1}{\zeta(\gamma+1)}$. If a node has degree 1, then its friendship index is equal to the degree of its only neighbor. Obviously, the value of the friendship index will be less than or equal to 1 only if the neighbor has degree 1. Given the assumption of neutral mixing in the configuration model, we find that the proportion of nodes of degree 1 whose friendship index is less than or equal to 1 is $a(1) := f_1^2 = \frac{1}{\zeta^2(\gamma+1)}$.

The share of nodes with degree 2 is equal to $f_2 = p(\xi = 2) = \frac{2^{-\gamma-1}}{\zeta(\gamma+1)}$. If a node has degree 2, then its friendship index is equal to the average of the degrees of its two neighbors divided by 2. Obviously, the value of the friendship index will be less than or equal to 1 only if its two neighbors have degrees 1 and 1, 1 and 2, 2 and 1, 2 and 2, 1 and 3, 3 and 1. Using the neutral mixing assumption in the configuration model, we get that the share nodes of degree 2 whose friendship index is less than or equal to 1 is equal to

$$a(2) := f_2((f_1 + f_2)^2 + 2f_1f_3) = \frac{2^{-\gamma-1}((1+2^{-\gamma-1})^2 + 2 \cdot 3^{-\gamma-1})}{\zeta^3(\gamma+1)}.$$

We have $f_2(f_1 + f_2)^2 \leq a(2) \leq f_2(f_1 + f_2 + f_3)^2$.

In general case, the fraction of nodes with degree $s$ is equal to $f_s = p(\xi = s) = \frac{s^{-\gamma-1}}{\zeta(\gamma+1)}$. If a node has degree $s$, then its friendship index is equal to the average of the degrees of its $s$ neighbors divided by $s$. Obviously, the value of the friendship index will be less than or equal to 1 only if its $s$ neighbors have all possible combinations of degrees such that its average value less than $s$. The neutral mixing assumption implies that the share nodes of degree $s$ whose friendship index is less than or equal to 1 is equal to $a(s) := f_s(f_1+\ldots+f_s)^s + sf_1^{s-1}f_{s+1}\ldots$. Then we get $f_s(f_1 + \ldots + f_s)^s \leq a(s) \leq f_s(f_1 + \ldots + f_{2s-1})^s$. We have

$$f_s(f_1 + \ldots + f_s)^s = \frac{s^{-\gamma-1}(1 + 2^{-\gamma-1} + \ldots + s^{-\gamma-1})^s}{\zeta^{s+1}(\gamma+1)}.$$

Adding the shares $a(s)$ for all $s \geq 1$, and subtracting this sum from 1, we get the estimate (1) in the case of $m = 1$.

Repeating the steps in proof for the case of $m > 1$, we may estimate the proportion of those nodes that have the friendship index less than or equal to 1. □

Figure 1 shows the dependence of $\kappa$ on $\gamma$ obtained by Eq. (1) for the networks generated by the configuration model with neutral mixing. The share reaches its maximum value of 1 when $\gamma$ is equal to 1, and then the share slowly decreases with increasing $\gamma$. It is noticeable that the friendship paradox is clearly manifested for scale-free networks in which the exponent $\gamma$ in the power-law distribution is between 1 and 2.

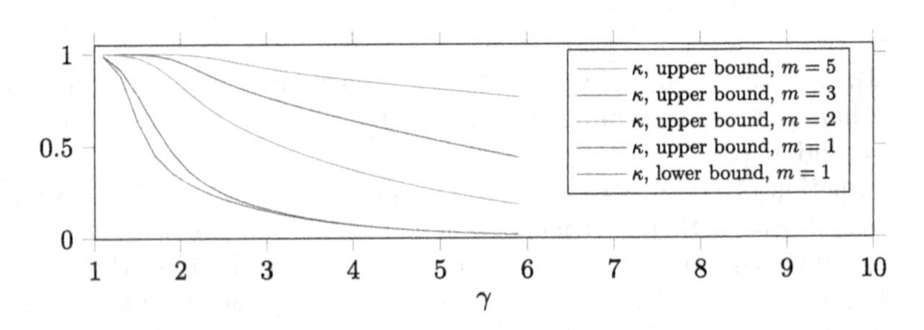


**Fig. 1.** The dependence of $\kappa$, the share of nodes which the friendship index is more than 1, on $\gamma$, the exponent of the power law distribution

### 2.4 Numerical Experiments on Simulated Networks with a Finite Size

We capture the behavior of $\Psi_n(k)$-distributions for networks generated by the CM as can be seen in Fig. 3. The degree sequences $D_n$ are generated using the Pareto-distributed random variable $\xi$ with parameter $\gamma > 1$, $p(\xi = k) \sim \gamma m^\gamma k^{-\gamma-1}$, $p(\xi > t) = m^\gamma t^{-\gamma}$ for $t \geq 1$. For experiments, the scaling factor $m$ equal to 5 is set when generating random variables distributed according to the Pareto law.

Next, a number of networks of size $n = 300,000$ are generated using the degree sequences $D_n$ obtained for different $\gamma$. Then the values of $\kappa$ are calculated. The characteristics of simulated networks are presented in Table 1.

We observe the dependence of $\kappa_n$ (i.e. the share of network nodes for which the friendship index is greater than 1) on the parameter $\gamma$. Since this share also depends on the size of the network, we plotted this dependence for various $n$. For different $\gamma$ in range 1 to 5, we create graphs according to the configuration model and find the corresponding value of $\kappa_n$ in it. The plots obtained are shown in Fig. 2, for different $m = 1, 2, 3, 5, 8, 10$.

**Table 1.** Simulated network statistics

| Model | $\lvert V\rvert$ | $\lvert E\rvert$ | vertices with FP | $\kappa$ |
|---|---|---|---|---|
| Barabási–Albert | 300006 | 1500015 | 278853 | 0.929 |
| CM with $\gamma = 1.5$ | 300006 | 2204918 | 298981 | 0.997 |
| CM with $\gamma = 2.0$ | 300006 | 1493771 | 275775 | 0.919 |
| CM with $\gamma = 2.5$ | 300006 | 1246494 | 252113 | 0.840 |

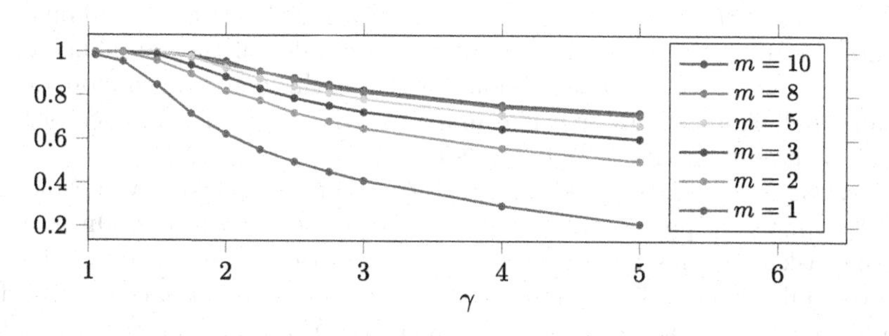


**Fig. 2.** The proportion of vertices in the graph for which the friendship paradox is satisfied, depending on the scaling factor $m$ of values distributed according to the Pareto law with the parameter $\gamma$

We denote the shares of nodes with degrees for which their friendship index is higher than 1, as $\kappa_n(k)$. Next, we find $\kappa(k)$ for all $k = 1, 2, \ldots$ in four simulated networks. The dependence of $\kappa_n(k)$ on $\log k$ can be seen in Fig. 3.

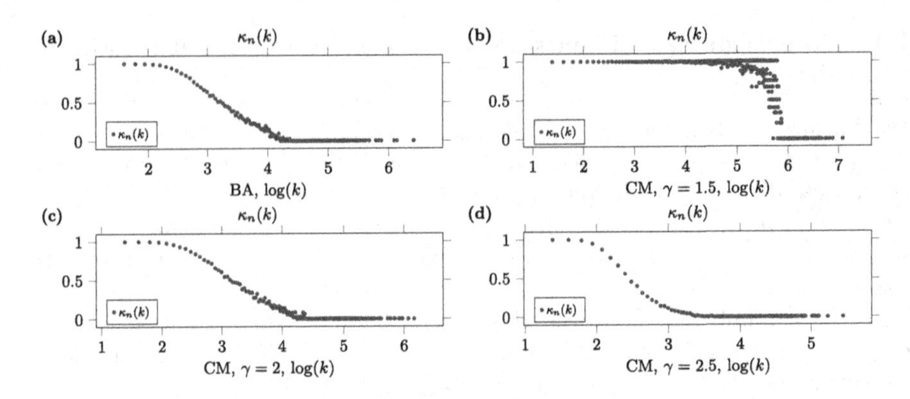


**Fig. 3.** The share of vertices for which the friendship paradox holds among vertices with the same degree $k$ in the model-simulated graph.

# 3 The Distribution of Friendship Index in Real Networks

Now let us illustrate the distributions of the friendship index found in real networks In this section, we examine the distribution of the friendship index in four real networks:

- *Amazon recommendation network.* Amazon, a well known global e-commerce platform, allows one to purchase and rate products from different merchants. Network data consists of user recommendations of products displayed on the website. Thus a bipartite network is constructed [12].
- *Yahoo email network* Yahoo dataset represents a social network where users, which had email conversations, are connected. This dataset, focused on users of the email service, was collected in [12].
- *Enron mail network* is an another example of social network based on Enron employee interactions [9]. Two employees are linked if they exchange corporate email messages. Unlike previous networks, this one is much more densely connected. Additionally, it's one more trait is a notable presence of nodes with high and extremely high degrees.
- *SuperUser network* is based on data of user interactions on the SuperUser platform, where people hold conversions on IT-related topics. On it people either ask questions or answer them. Nodes (users) are connected if either of the following conditions are satisfied: a user answers a question asked by another user and a user adds a comment on a topic or answer by another user [12].

Table 2 shows the main features of observed networks: $|V|$ is the number of nodes, $|E|$ is the number of links, $\gamma$ is the exponent of the power law degree distribution, $\kappa$ is the share of network nodes for which their friendship index is more than 1.

Similarly to plots for simulated networks in Fig. 3, we display the dependence of $\kappa_n$ on the degree $n$ for four real networks. The results can be seen in

**Table 2.** Real network statistics

| *Real network* | $\|V\|$ | $\|E\|$ | $\gamma$ | $\kappa$ | Other traits |
|---|---|---|---|---|---|
| *Amazon* products | 2.100.000 | 5.800.000 | 2.96 | 0.9662 | Bipartite network |
| *Enron* mails | 87.100 | 1.100.000 | 1.09 | 0.8425 | Close community |
| *Yahoo* messages | 100.000 | 3.200.000 | 1.31 | 0.9743 | – |
| *SuperUser* reactions | 194.000 | 1.400.000 | 2.38 | 0.9798 | – |

Fig. 4. The results for most networks are fairly similar to distributions for simulated networks. Moreover, the shapes of distributions are generally alike despite networks being different in terms of size, density, and origin of data.

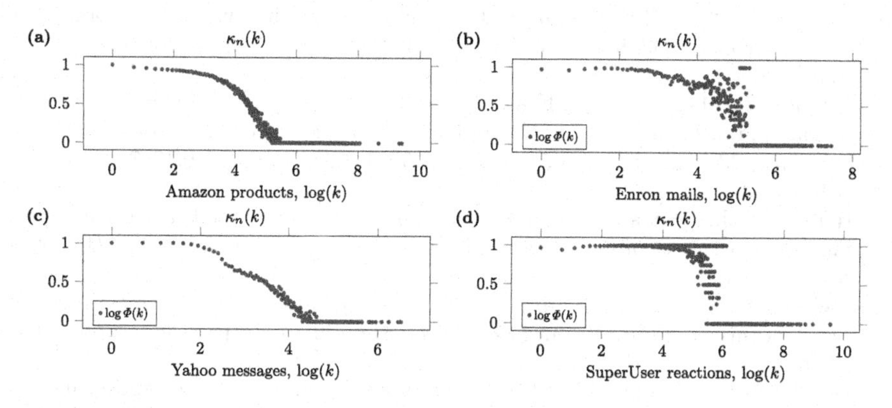


**Fig. 4.** The share of nodes $\kappa_n(k)$ for which the friendship paradox holds among all nodes with the same degree $k$ in four real networks.

## 4 Conclusion

In the social sciences, the so-called friendship paradox is well known, which is associated with the disparity in the popularity of users at the local level of a social network. The explanation of this very common phenomenon in previous work was limited to either empirical studies of social networks or particular varieties of synthetic networks. In this paper, we show that this paradox is inherent in all networks whose degree distribution follows a power law. Note that most real networks have this property, or in other words, many networks are scale-free. We study the dependence of the share of network nodes, for which the friendship index is greater than one, on the parameters of the power-law distribution of node degrees. The results show that the smaller the power law exponent, the greater the number of network nodes will present the friendship paradox. Moreover, the greater the minimum degree of the vertices of the network, the stronger this paradox will manifest itself in this graph. The findings of the paper make it possible to obtain a quantitative assessment of the friendship paradox in complex social networks, based only on the known value of the node degree distribution exponent, and without making calculating the friendship indices for all nodes in the graph, which for real networks can be a very computationally intensive task.

## References

1. Alipourfard, N., Nettasinghe, B., Abeliuk, A., Krishnamurthy, V., Lerman, K.: Friendship paradox biases perceptions in directed networks. Nat. Commun. **11**(1), 707 (2020). https://doi.org/10.1038/s41467-020-14394-x
2. Bollen, J., Gonçalves, B., van de Leemput, I., Ruan, G.: The happiness paradox: your friends are happier than you. EPJ Data Sci. **6**(1), 1–17 (2017). https://doi.org/10.1140/epjds/s13688-017-0100-1
3. Chen, N., Olvera-Cravioto, M.: Directed random graphs with given degree distributions. Stochast. Syst. **3**(1), 147–186 (2013). https://doi.org/10.1214/12-SSY076
4. Eom, Y.H., Jo, H.H.: Generalized friendship paradox in complex networks: The case of scientific collaboration. Sci. Rep. **4**, 1–6 (2014). https://doi.org/10.1038/srep04603
5. Fotouhi, B., Momeni, N., Rabbat, M.G.: Generalized friendship paradox: an analytical approach. In: Aiello, L.M., McFarland, D. (eds.) SocInfo 2014. LNCS, vol. 8852, pp. 339–352. Springer, Cham (2015). https://doi.org/10.1007/978-3-319-15168-7_43
6. Higham, D.J.: Centrality-friendship paradoxes: when our friends are more important than us. J. Complex Netw. **7**(4), 515–528 (2018). https://doi.org/10.1093/comnet/cny029
7. Jackson, M.O.: The friendship paradox and systematic biases in perceptions and social norms. J. Polit. Econ. **127**(2), 777–818 (2019). https://doi.org/10.1086/701031
8. Lee, E., Lee, S., Eom, Y.H., Holme, P., Jo, H.H.: Impact of perception models on friendship paradox and opinion formation. Phys. Rev. E **99**(5), 052302 (2019). https://doi.org/10.1103/PhysRevE.99.052302
9. Leskovec, J., Lang, K., Dasgupta, A., Mahoney, M.: Community structure in large networks: natural cluster sizes and the absence of large well-defined clusters. Internet Math. **6**, 29–123 (2008). https://doi.org/10.1080/15427951.2009.10129177
10. Momeni, N., Rabbat, M.: Qualities and inequalities in online social networks through the lens of the generalized friendship paradox. PLoS ONE **11**(2), e0143633 (2016). https://doi.org/10.1371/journal.pone.0143633
11. Pal, S., Yu, F., Novick, Y., Bar-Noy, A.: A study on the friendship paradox – quantitative analysis and relationship with assortative mixing. Appl. Net. Sci. **4**(1), 71 (2019). https://doi.org/10.1007/s41109-019-0190-8
12. Paranjape, A., Benson, A.R., Leskovec, J.: Motifs in temporal networks. In: In Proceedings of the Tenth ACM International Conference on Web Search and Data Mining. ACM (2017). https://doi.org/10.1145/3018661.3018731
13. Sidorov, S.P., Mironov, S.V., Grigoriev, A.A.: Friendship paradox in growth networks: analytical and empirical analysis. Appl. Net. Sci. **6**, 35 (2021). https://doi.org/10.1007/s41109-021-00391-6
14. Sidorov, S., Mironov, S., Malinskii, I., Kadomtsev, D.: Local degree asymmetry for preferential attachment model. In: Benito, R.M., Cherifi, C., Cherifi, H., Moro, E., Rocha, L.M., Sales-Pardo, M. (eds.) COMPLEX NETWORKS 2020 2020. SCI, vol. 944, pp. 450–461. Springer, Cham (2021). https://doi.org/10.1007/978-3-030-65351-4_36

# Computation in Optimization and Optimal Control

# An Algorithm for Finding the Global Extremum of a Partially Defined Function

Marina Usova(✉) and Konstantin Barkalov

Lobachevsky State University of Nizhny Novgorod, Nizhny Novgorod, Russian Federation
{usova,konstantin.barkalov}@itmm.unn.ru

**Abstract.** The paper discusses the problem of finding the global minimum of a function that may be partially defined in the search domain. The objective function may not be defined within some search regions due to the nature of the optimized object and of the simulation method (for example, numerical instability of the simulation method used). In some cases, these regions are known; but in most cases, the information about them is missing. The paper gives a description of a global search algorithm for solving such class of problems. Numerical experiments confirming the efficiency of the proposed algorithm were carried out.

**Keywords:** Lipschitz global optimization · Multiextremal functions · Partially defined functions · Dimensionality reduction

## 1 Introduction

Global optimization deals with the development of methods for solving problems of finding the global (absolute) minimum of multidimensional multiextremal functions. The possibility of constructing a reliable estimate of the global extremum in such problems is principally based on the available information on the problem to be solved. This information makes it possible to relate the unknown values of the objective function to the known values at the points of the performed search trials.

Often such information is represented in the form of suggestion that the objective function $\phi(y)$ satisfies the Lipschitz condition with a priori unknown constant $L$. This suggestion can be interpreted (in the relation to the applied problems) as the reflection of a limited power generating the changes in the simulated system. In this case, the objective function is defined in the form of a "*black box*", may be a non-differentiable one, and each computation of its value at some point within a feasible domain may require considerable computational resources.

This work was supported by the Ministry of Science and Higher Education of the Russian Federation, project no. FSWR-2023-0034, and by the Research and Education Mathematical Center "Mathematics for Future Technologies".


D. Balandin et al. (Eds.): MMST 2023, CCIS 1914, pp. 147–161, 2024.
https://doi.org/10.1007/978-3-031-52470-7_13

A number of efficient deterministic algorithms for solving the Lipschitz global optimization problems have been developed [7,9,12–14]. The comparisons conducted have shown the deterministic algorithms outperform (in various criteria) the widespread nature-inspired algorithms [11,15].

The present work continues the development of one of the efficient deterministic algorithms for solving the Lipschitz global optimization problems, namely the information-statistical global search algorithm [16,20]. This algorithm suggests the solving of a multidimensional problem by means of its reduction to an equivalent one-dimensional optimization problem using the Peano-Hilbert space-filling curves.

Note that the use of complex mathematical models is natural in the optimization of complex real world objects that, as a consequence, essentially increases the computation costs for searching the optimum. In last decades, the experts in the fields of optimization and parallel computations have proposed many methods for reducing the computation costs and speed-up of the algorithms related to solving the appearing optimization problems [10,17] as well as to the numerical analysis of the original models [4,5].

However, a principally new problem in applications, namely the numerical instability of the models investigated in some (unknown a priori) subregions of the parameters' variations recently became relevant. In such subregions, one cannot perform the numerical modeling and compute the objective function values correctly. This phenomenon can be interpreted either as the presence of some hidden constraints [19], or as unknown discontinuous regions [1], or as a partial computability of the objective function in the search domain [3,18,21]. In such a statement, the optimization problem becomes more complex since the region of feasible parameters' combinations is not defined in advance.

This paper presents the results in a new research direction related to the use of imputation[1] (restoring) of the objective function values in the course of the global search algorithm operation when solving the problem with partially defined objective function. The approach proposed is based on an adaption of the computation rules of the global search algorithm. The efficiency of the approach proposed has been demonstrated numerically.

## 2 Problem Statement

In the general form, the global optimization problem can be formulated as follows:

$$\phi^* = \phi(y^*) = \min_{y \in D} \phi(y), D = \left\{y \in R^N : a_i \leq y_i \leq b_i,\ 1 \leq i \leq N\right\}, \tag{1}$$

where $y = (y_1, y_2, ..., y_N)$ is the vector of the varied parameters, $D$ is an $N$-dimensional hypercube, and $N$ is the dimension of the problem being solved.

[1] The term "imputation" in the paper is used following the established terminology in the field of machine learning related to the restoring of the missing data.

Using the Peano-Hilbert curves unambiguously mapping the interval $[0,1]$ onto the $N$-dimensional unit hypercube

$$D = \{y \in R^N : -2^{-1} \leq y_i \leq 2^{-1}, 1 \leq i \leq N\} = \{y(x) : 0 \leq x \leq 1\},$$

we can reduce the initial problem (1) to a one-dimensional problem

$$f^*(x) = \phi(y(x^*)) = \min_{x \in [0,1]} \{\phi(y(x))\}, \tag{2}$$

that allows applying efficient one-dimensional optimization algorithms for its solving.

We make the following assumptions on the objective function.

First, the objective function may be multiextremal, non-differentiable, and, moreover, defined in the form of a "black box" (i.e. in the form of some subroutine, which entry the argument is supplied to, and the output is respective function value).

Second, each computing of the function at some point of the feasible region may require considerable computation resources.

Third, the objective function satisfies the Lipschitz condition

$$|\phi(y') - \phi(y'')| \leq L\|y' - y''\|,\ y', y'' \in D, \tag{3}$$

where $L$ is the Lipschitz constant. The dimensionality reduction scheme using Peano curves is known to juxtapose the multidimensional problem with a Lipschitz objective function (1) to problem (2) with the one-dimensional objective function satisfying Hölder condition

$$|f(x') - f(x'')| \leq K\rho(x', x''),\ x', x'' \in [0,1], \tag{4}$$

where $\rho(x', x'') = |x' - x''|^{1/N}$ and $N$ is the dimensionality of the initial multidimensional problem and coefficient $K$ is related to the Lipschitz constant $L$ by the relation $K \leq 2L\sqrt{N+3}$ [20].

Various versions of algorithms for solving the problems of such a class and respective theories of convergence are presented in [16,20].

Fourth, the objective function may be undefined in some regions of search domain $D$ (in particular cases, at one or several points).

From our experience in solving the applied optimization problems [2,8], we will assume the total volume of the non-computability regions to be a small fraction of the whole volume of the search domain $D$ (about 10–20%)[2].

The last assumption makes the application of the information-statistical global search algorithm [20] or other Lipschitz optimization methods [12,14] impossible and requires developing a modification for solving the problems with the objective functions, which are not defined everywhere.

[2] When testing the algorithm, we have considered also more complex problems, in which the volume of the non-computability region achieved 50–80%. The problems of this kind were successfully solved by the proposed algorithm as well but a large number of trials was required to do it since the method attempted to explore the whole non-computability region. In this connection, in the rules of constructing the test problems for the dimensionality $N = 5$ were modified so that the non-computability region were located only in the corners of the search domain.

# 3 Method for Solving the Problems with Partially Defined Objective Function

## 3.1 Global Search Algorithm

Let us give more detailed description of the **global search algorithm** (GSA) for solving the problem (2).

At every iteration of the global search a *trial* is executed. We will refer to computing the objective function value $\phi(y(x))$ from (2) as a trial. The first trial will be executed in the middle point $x^1 \in (0, 1)$. The choice of the point $x^{k+1}, k \geq 1$ of the next $(k+1)^{\text{th}}$ trial is performed on the base of the following rules.

*Rule 1.* Renumber (by the lower indices) the points $x^i, 0 \leq i \leq k$ of the previous trials in the ascending order by their coordinates, i.e.

$$0 = x_0 < x_1 < ... < x_i < ... < x_k = 1 \tag{5}$$

and juxtapose them with the values $z_i = f(x_i), 0 < i < k$, computed at these points and to the indices $v_i = v(x_i)$ defined by the rule

$$v_i = v(x_i) = \begin{cases} -1 & \text{if } x_i \text{ is a boundary point,} \\ 1 & \text{if } x_i \text{ is an internal point.} \end{cases} \tag{6}$$

The points $x_0 = 0$ and $x_k = 1$ are introduced additionally (the values $z_0$ and $z_k$ are undefined) for convenience of the following notations.

*Rule 2.* Compute the estimate

$$\mu = \max \left\{ \frac{|z_i - z_{i-1}|}{\Delta_i}, 2 \leq i \leq k \right\}, \Delta_i = (x_i - x_{i-1})^{1/N}, \tag{7}$$

of the unknown Hölder constant $K$ from (4) for the reduced function $f(x)$, where $r > 1$ is a parameter of the algorithm; if $\mu = 0$ accept $\mu = 1$.

*Rule 3.* Determine the best current value of the objective function

$$z^* = \min \{f(x_i) : 1 \leq i \leq k\}. \tag{8}$$

*Rule 4.* For each interval $(x_{i-1}, x_i), 1 \leq i \leq k$, compute the value $R(i)$ called a *characteristic* of the interval according to the formula

$$R(x_i) = \begin{cases} \Delta_i + \frac{(z_i - z_{i-1})^2}{(r\mu)^2 \Delta_i} - 2\frac{z_i + z_{i-1} - 2z^*}{r\mu}, & v(x_i) = v(x_{i-1}) = 1, \\ 2\Delta_i - 4\frac{(z_i - z^*)}{r\mu}, & v(x_{i-1}) = -1, v(x_i) = 1, \\ 2\Delta_i - 4\frac{(z_{i-1} - z^*)}{r\mu}, & v(x_{i-1}) = 1, v(x_i) = -1. \end{cases} \tag{9}$$

*Rule 5.* Select the interval $(x_{t-1}, x_t)$ with the largest value of characteristic $R(i)$:

$$R(t) = \max\{R(i) : 1 \le i \le k\}.$$

*Rule 6.* Perform the next trial in the middle point of the interval $(x_{t-1}, x_t)$, if the indices of its end points are not the same, i.e.

$$x^{k+1} = \frac{x_t + x_{t-1}}{2}, v(x_{t-1}) \neq v(x_t). \tag{10}$$

Otherwise, perform the next trial in the point

$$x^{k+1} = \frac{x_t + x_{t-1}}{2} - \text{sign}(z_t - z_{t-1}) \frac{1}{2r} \left[ \frac{|z_t - z_{t-1}|}{\mu} \right]^N, v(x_{t-1}) = v(x_t). \tag{11}$$

*Stop Criterion.* The search is terminated, if the length of the interval $(x_{t-1}, x_t)$ containing current point $x^{k+1}$ is less than or equal to the predefined accuracy $\varepsilon$, i.e. $\Delta_t \le \varepsilon$, where $t$ is from Rule 5 and $\varepsilon > 0$ is the algorithm parameter.

### 3.2 Approaches to the Imputation of the Objective Function Values

The optimization problem with not everywhere defined objective function described above is a partial analog of the long-standing missing data problem in Data Science. The origin of this problem may be, for example, the technical problems or assembling the data set from several sources with different sets of parameters.

It is worth noting that some machine learning algorithms can take into account and even restore the missing values in the data. For example, LightGBM has a regime of ignoring the missing values while XGBoost restores the data by means of the decrease in the loss function when learning. Nevertheless, the majority of models cannot process the missing values and require the valid data without NaN or "missing" values. It makes this problem to be serious enough.

The experts in Data Science have proposed various approaches to the imputation (restoring) of the missing data:

- Deleting of all cases, which have missing values of some variables (at the worst, the algorithm loses the access to the useful information contained in the on-missing observations' values);
- Replacing any missing value with the average of the available values of the variable (doesn't work with the qualitative variables, doesn't account for the correlation of the data);
- Replacing by the most frequent value or by a constant (differs from the use of the median only by the ability to work with the qualitative parameters);
- Replacing the data using the method of $k$-nearest neighbors (unlike the former, accounts for the correlation between the parameters, sensitive to the fadeouts, computation costly);

- Multiple imputation by chained equations (MICE) – imputation of each value is performed not one but many times (allows understanding how reliable or unreliable is the value suggested);
- Imputation of the data using a deep learning (restores the missing values by means of training a neural network on the points, for which all parameters are available, computation costly for big data).

The absence of a universal algorithm and the necessity to find the most suitable methods for each particular problem are the common drawbacks of the approaches listed above.

For the imputation of the objective function values, one can consider, in general, a similar set of approaches, however, taking into account the Lipschitz condition and using the information-statistical approach. It allows to propose a good algorithm for this problem class. The points, at which the non-computability was found, are interpreted as the ones having the zero index, and (6) is replaced by an extended formula

$$v_i = v(x_i) = \begin{cases} -1, & \text{if } x_i \text{ is boundary point,} \\ 0, & \text{if } x_i \text{ is non-computable point,} \\ 1, & \text{if } x_i \text{ is internal point.} \end{cases} \tag{12}$$

Let us describe various possible approaches to the imputation of the objective function values.

**The First Approach.** To ignore the absence of the objective function value at the point. The characteristic of an interval with even a single non-computable point is equal to the length of the interval, i.e.

$$R(i) = \Delta_i, \text{ if } v(x_i) = 0 \text{ or } v(x_{i-1}) = 0. \tag{13}$$

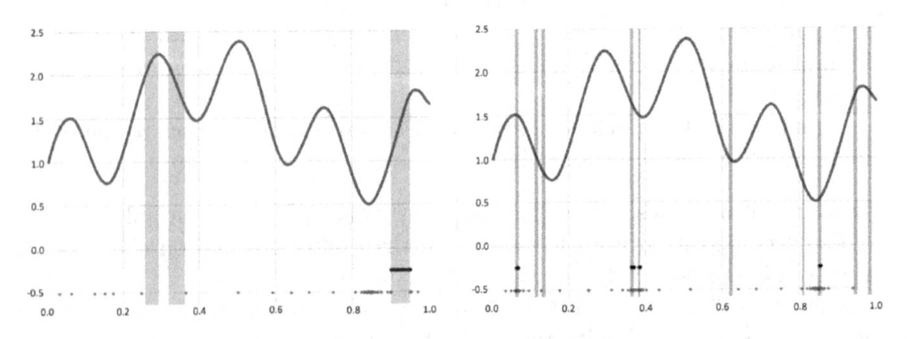


**Fig. 1.** Examples of the search trials distribution in the non-computability regions for the first approach to the values imputation

A large number of excess trials after getting into the non-computability region due to a large value of the characteristic obtained according to (13) is an essential drawback of this approach.

**The Second Approach.** To ignore the absence of the objective function value at the point. The characteristic of the interval with two non-computable point or with one non-computable point and one boundary point is equal to the interval length multiplied by a coefficient, i.e.

$$R(i) = (1 - 1/r)^N \Delta_i, \tag{14}$$

if $v(x_i) = v(x_{i-1}) = 0$ or $v(x_i) = 0$ and $v(x_{i-1}) = -1$ or $v(x_i) = -1$ and $v(x_{i-1}) = 0$. The characteristic of the interval with one non-computable point and one computable point is calculated according to the rules of operation with the boundary interval

$$R(i) = 2\Delta_i - 4\frac{(z_i - z^*)}{r\mu}, \text{ if } v(x_{i-1}) = -1 \text{ or } v(x_{i-1}) = 0, v(x_i) = 1, \tag{15}$$

$$R(i) = 2\Delta_i - 4\frac{(z_{i-1} - z^*)}{r\mu}, \text{ if } v(x_{i-1}) = 1, v(x_i) = -1 \text{ or } v(x_i) = 0. \tag{16}$$

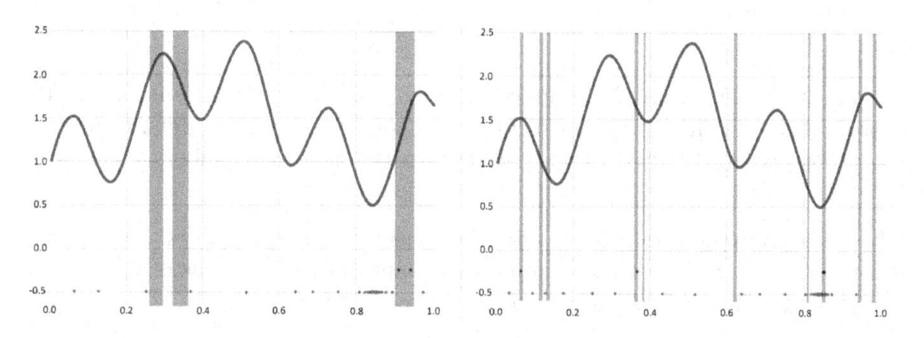


**Fig. 2.** Examples of the search trials distribution in the non-computability regions for the second approach to the values imputation

This approach partly resolves the problems of the first approach outlined above. However, it doesn't solve the problem of excess trials and requires a considerable increasing of the parameter $r$ for more complex test problem. The examples for the one-dimensional problems using the first two approaches are presented in Fig. 1 and Fig. 2, where the trials falling into the region with the undefined values are marked by the black points and the computable points of the search domain are marked by the green ones.

**The Third Approach.** To restore the objective function values at the points of the intervals with two non-computable points as the averages of the neighbor points and to apply corresponding computation rule from formula (9) assuming the points to be computable ones. If the neighbor points are non-computable as well, use formula (14). In the case of an interval with one non-computable point, apply formulae (15) and (16) (Fig. 3).

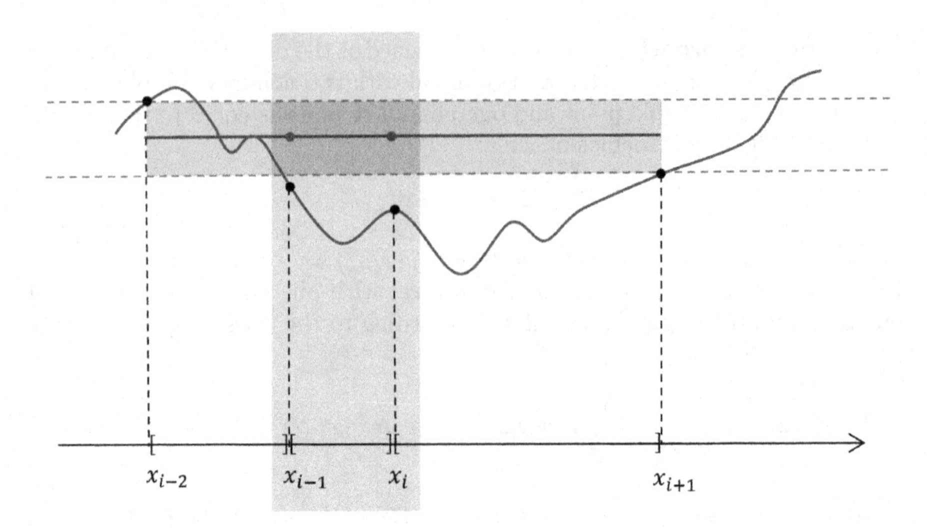


**Fig. 3.** Imputation of the non-computable values by the averages of the neighbor points. Approach No. 3

Essential drawbacks of this approach are not utilizing the information on the coordinates aw well as seldom executing the computation rule added because of searching the substituting values only among the nearest neighbors.

**The Fourth Approach.** To restore the objective function values at the points of the intervals with two non-computable points as the averages of the nearest computable points taking into account the coordinates, i.e.

$$\tilde{z}(x) = z' + \mathrm{sign}(z'' - z')\frac{|z'' - z'| \cdot \rho(x, x')}{\rho(x', x'')}, \tag{17}$$

where $z'$ and $z''$ are the objective function values in the computable points $x'$ and $x''$ nearest to the interval $(x_{i-1}, x_i)$ and $\rho(x_1, x_2) = |x_1 - x_2|^{1/N}$. The rest of the approach is identical to the third one. Formula (14) in this case will be applied when the non-computable region is located only near the boundary point (Figs. 4 and 5).

**The Fifth Approach.** To utilize the search information and the suggestions on the function form most completely, one can restore the objective function values in the points taking onto account the Hölder condition.

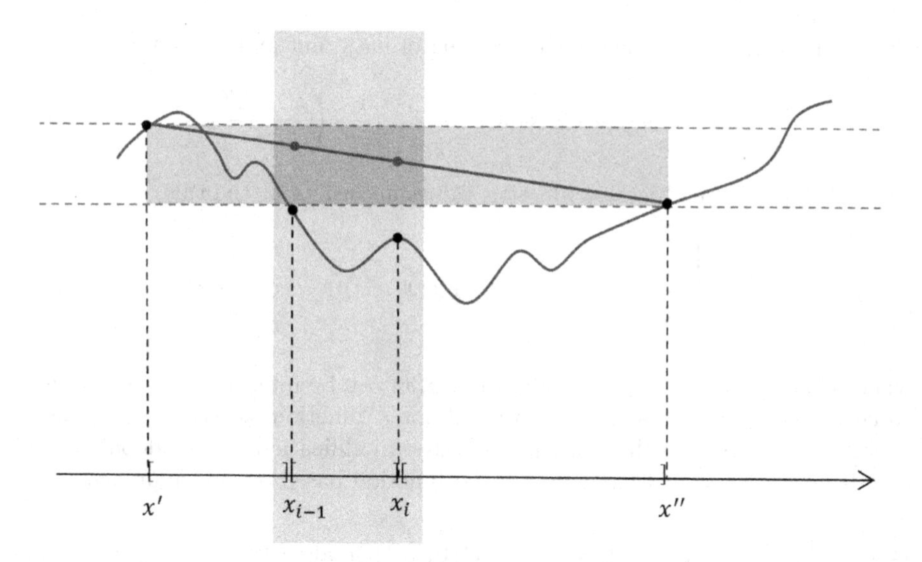


**Fig. 4.** Imputation of non-computable values by averages of the neighboring points. Approach No. 4

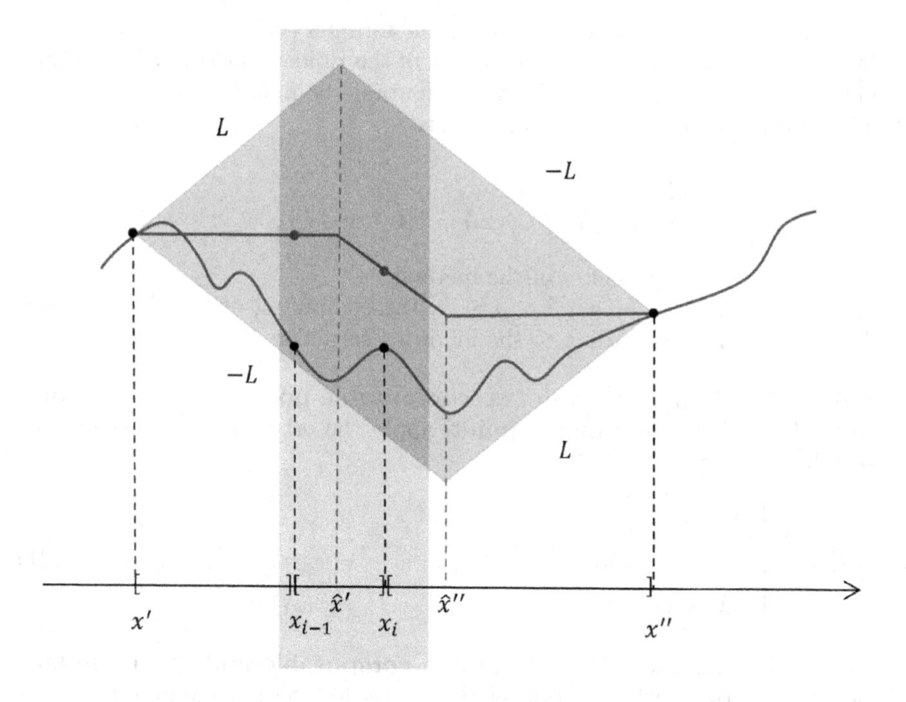


**Fig. 5.** Imputation of non-computable values by averages of the minorant and the majorant. Approach No. 5

*Step 1.* Find the end points of the crossing of majorant and minorant

$$\hat{x}', \hat{x}'' = \frac{x' + x''}{2} \pm \text{sign}(z'' - z') \cdot \frac{1}{2r} \cdot \left(\frac{z'' - z'}{\mu}\right)^N, \tag{18}$$

*Step 2.* Restore the objective function values according to the rule

$$\tilde{z}(x) = \begin{cases} z', & x \leq \hat{x}', \\ \frac{z'+z''}{2} - \frac{r\mu}{2}(\rho(x, x') + \rho(x, x'')), & \hat{x}' < x < \hat{x}'', \\ z'', & x \geq \hat{x}'', \end{cases} \tag{19}$$

where $\rho(x_1, x_2) = |x_1 - x_2|^{1/N}$. The value $\tilde{z}(x)$ can be interpreted as an average value of the minorant and the majorant for a function satisfying the Hölder condition. Let us give the description of the modification of the global search algorithm in the case of partially defined function based on this approach.

**Description of the Modified Algorithm.** The algorithm remains the same with an introduction of the following minor additions in the computing formulae in the case of falling into the non-computable search regions.

1. The index $v_i = v(x_i)$ is determined by the formula (12).
2. When computing current lower estimate of the Hölder constant $K$ in formula (7), use only the intervals with two computable boundary points.
3. The value $z_v^*$ depends on the index value

$$z_v^* = \begin{cases} -\varepsilon_r, & v = 0, \\ \min\{f(x_i) : v(x_i) = 1, 0 < i < k\}, & v = 1, \end{cases} \tag{20}$$

   where $\varepsilon_r > 0$ is a parameter of the method.
4. For each interval $(x_{i-1}, x_i), 1 \leq i \leq k$, the computing of the characteristic $R(i)$ is performed according to the following formulae.

   *Case 1.* For the intervals with **two computable points** or with **one computable and one boundary point**, apply the original rules of the global search algorithm

$$R(i) = \begin{cases} \Delta_i + \frac{(z_i - z_{i-1})^2}{(r\mu)^2 \Delta_i} - 2\frac{z_i + z_{i-1} - 2z_1^*}{r\mu}, & v(x_i) = v(x_{i-1}) = 1, \\ 2\Delta_i - 4\frac{(z_i - z_1^*)}{r\mu}, & v(x_{i-1}) = -1, v(x_i) = 1, \\ 2\Delta_i - 4\frac{(z_{i-1} - z_1^*)}{r\mu}, & v(x_{i-1}) = 1, v(x_i) = -1. \end{cases} \tag{21}$$

   *Case 2.* For the intervals with **one non-computable and one boundary point**, perform the imputation of the values for the non-computable point based on the neighbor computable point and apply the formula

$$R(i) = \begin{cases} 2\Delta_i - 4\frac{(z_{i+1} - z_0^*)}{r\mu}, & v(x_{i-1}) = -1, v(x_i) = 0, \\ 2\Delta_i - 4\frac{(z_{i-2} - z_0^*)}{r\mu}, & v(x_{i-1}) = 0, v(x_i) = -1. \end{cases} \tag{22}$$

*Case 3.* For the intervals with **one non-computable and one computable point**, perform the imputation of the value for the non-computable point based on the neighbor computable point and apply the formula

$$R(i) = \begin{cases} \Delta_i + \frac{(z_i - z_{i-2})^2}{(r\mu)^2 \Delta_i} - 2\frac{z_i + z_{i-2} - 2z_0^*}{r\mu}, & v(x_{i-1}) = 0, v(x_i) = 1, \\ \Delta_i + \frac{(z_{i+1} - z_{i-1})^2}{(r\mu)^2 \Delta_i} - 2\frac{z_{i+1} + z_{i-1} - 2z_0^*}{r\mu}, & v(x_{i-1}) = 1, v(x_i) = 0. \end{cases} \quad (23)$$

*Case 4.* For the intervals with **two non-computable points**, perform the imputation of the values for the non-computable points based on the rule (19) and apply the formula

$$R(i) = \Delta_i + \frac{(\tilde{z}_i - \tilde{z}_{i-1})^2}{(r\mu)^2 \Delta_i} - 2\frac{\tilde{z}_i + \tilde{z}_{i-1} - 2z_1^*}{r\mu}, \; v(x_{i-1}) = v(x_i) = 0. \quad (24)$$

*Case 5.* In the case when any imputation is impossible, use the formula

$$R(i) = \Delta_i \cdot \left(1 - \frac{1}{r}\right)^N + z_0^*. \quad (25)$$

5. If several intervals have equal characteristics, prefer the interval with computable points that has the lowest number.
6. If the best interval appears to be the one with two non-computable points, perform the next trial in the middle point of the interval.

## 4 Results of Numerical Experiments

The numerical experiments were carried out using a computer with Intel® Core™ i7-10750H CPU @ 2.60 GHz using the developed software package implementing the modification described in Sect. 3. To generate a series of problems, GKLS generator described in [6] was used. It allows generating the multiextremal optimization problems with the properties known in advance. These problems were appended with the non-computable regions of various kinds.

Each series of experiments consisted of 100 GKLS problems with everywhere computable objective function (original), 100 problems with the non-computabilities at the boundaries of the search domains (round), and 100 problems with non-computability regions defined randomly (random).

In the first series of experiments, the correctness of the modification operation was checked and the features of executing the trials in the non-computability regions were revealed. The problems of the dimensionality $N = 2$ were used in the experiment. The reliability parameter was set to $r = 5.5$, the accuracy was $\varepsilon = 0.01$. The results of experiments are presented in Table 1.

Figure 6 presents the examples of solving several problems with the modified GSA. The trials inside the computable regions are marked by the blue points, the global minima of the problems are marked by the red points, and the points

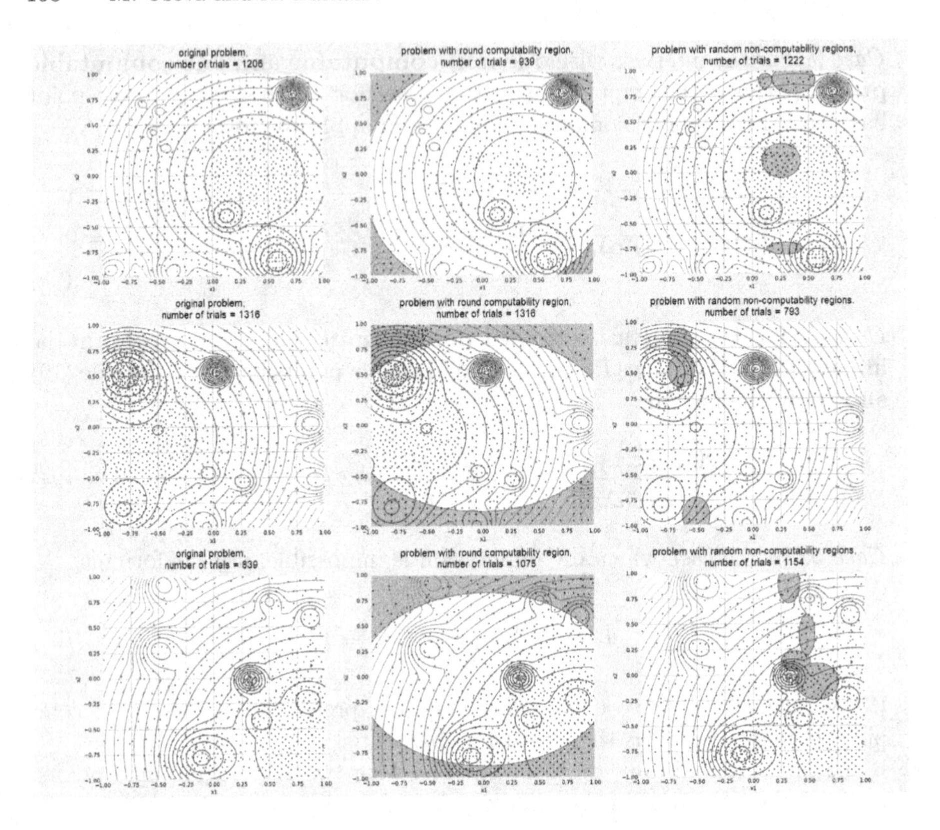


**Fig. 6.** Examples of problems No. 19, 25, and 35 from the first series of experiments

of trials performed inside the non-computability regions are marked by the black points. Also, the non-computability regions are filled with grey for clarity.

Since the modified algorithm works with the original GKLS problem series according to the rules of base algorithm, one can compare it with the GSA. Thus, the general behavior of GSA was preserved for the modification, and the same regions of trials' accumulation near the local minima were observed that can be explained by the realization of the computation rules (21) in most cases. One can see in Table 1 that the number of trials executed inside the non-computable regions is small. Total number of trials for the original problems and for the problems with partially defined objective function are comparable, the method doesn't perform excess trials in the non-promising regions due to the computing formulae utilizing the information on the nearest computable points.

The use of rule (25) forms a grid of trials inside the non-computability regions that is visible clearly for problem No. 35. In this case, the trial points are denser at the boundary of the non-computable region close to potential minimum when attempting to approach this one from different sides due to the computational rules (22)–(24) that was observed in all problems presented as the examples. 100% of the problems were solved. However, for the series with the boundary

**Table 1.** Comparison of the original and modified GSA

| | original | round | random |
|---|---|---|---|
| Average number of trials | 781.7 | 854.9 | 839.0 |
| Average number of trials in non-computability regions | – | 185.8 | 42.3 |
| Number of solved problems | 100/100 | 100/100 | 100/100 |

non-computability region the value of parameter $r = 6.0$ was required to solve one problem, and for the series with the random non-computability regions the same value of $r$ was required to solve two problems.

**Table 2.** Problem series of various dimensions

| N | Problem type | Average number of trials | Average number of trials in non-computability regions | Solved problems |
|---|---|---|---|---|
| 3 | original | 12251.1 | – | 100/100 |
| | round | 12201.8 | 4763.6 | 100/100 |
| | random | 12210.7 | 63.4 | 100/100 |
| 4 | original | 25144.5 | – | 100/100 |
| | round | 29840.3 | 15004.4 | 100/100 |
| | random | 25287.5 | 26.4 | 100/100 |
| 5 | original | 56725.9 | – | 100/100 |
| | round | 47840.1 | 24109.2 | 100/100 |
| | random | 57874.8 | 82.6 | 100/100 |

In the second series of experiments, the problem series of various dimensions were solved. In the experiments with the problems of the dimensions $N = 3, 4, 5$, the reliability parameter was set to $r = 4.5$, the precision $\varepsilon = 0.02$. The global minimum was considered to be found if the algorithm generates a trial point inside the $\varepsilon$-vicinity of the global minimum. The results of the experiments are presented in Table 2.

The results of the experiments show that the method resolves successfully the hits into the non-computability regions randomly scattered inside the search domain whereas in the case of large boundary regions it tries the execute a complete exploration without executing too many excess trials.

## 5 Conclusions

In the present work, the problem of finding the global minimum of a computation-costly "black box" function was considered. As compared to the traditional formulation used in the global optimization, the considered problem has an important difference: the objective function may be undefined in some regions of the search domain. The appearing of such a formulation is caused by a frequent problem of numerical instability of the models investigated in the applied tasks.

The main problem is the impossibility to identify the non-computability regions in advance in the most cases. In fact, the region of feasible combinations of parameters is unknown a priori. This phenomenon can be interpreted as partial computability of the objective function.

The problem investigated in the present work echoes to problem of missing data existing in Data Science for a long time. As a solution, the experts in the field have proposed various approaches to the imputation (restoring) of the missing values. The applicability of already existing approaches appended with the different variants to account for the Lipschitz condition was investigated. As a result, efficient algorithm for solving the problems of this class has been proposed. This algorithm is a modification of the deterministic method for solving the Lipschitz global optimization problems, namely information-statistical global search algorithm. As part of the modification, the computational scheme of the algorithm was adapted with the introduction of the data imputation rules.

The numerical experiments were carried out using series of multiextremal GKLS problems of the dimensions $N = 2, 3, 4, 5$. The results have demonstrated the reliability and efficiency of the proposed approach. The modified algorithm handles successfully the problems, in which the non-computability regions are randomly scattered within the search domain. In the case of large boundary non-computability regions, the method solves the problems efficiently as well performing an exploration of the non-computability regions.

## References


1. Audet, C., Batailly, A., Kojtych, S.: Escaping unknown discontinuous regions in blackbox optimization. SIAM J. Optim. **32**(3), 1843–1870 (2022). https://doi.org/10.1137/21m1420915
2. Barkalov, K.A., et al.: On solving the problem of finding kinetic parameters of catalytic isomerization of the pentane-hexane fraction using a parallel global search algorithm. Mathematics **10**(19), 3665 (2022). https://doi.org/10.3390/math10193665
3. Candelieri, A.: Sequential model based optimization of partially defined functions under unknown constraints. J. Glob. Optim. **79**(2), 281–303 (2019). https://doi.org/10.1007/s10898-019-00860-4
4. Dongarra, J.J.: The evolution of mathematical software. Commun. ACM **65**(12), 66–72 (2022). https://doi.org/10.1145/3554977
5. Duwe, K., et al.: State of the art and future trends in data reduction for high-performance computing. Supercomput. Front. Innov. **7**(1) (2020). https://doi.org/10.14529/jsfi200101
6. Gaviano, M., Kvasov, D.E., Lera, D., Sergeyev, Y.D.: Software for generation of classes of test functions with known local and global minima for global optimization. ACM Trans. Math. Softw. **29**(4), 469–480 (2003)

7. Grishagin, V.A., Israfilov, R.A.: Global search acceleration in the nested optimization scheme. In: AIP Conference Proceedings, vol. 1738, p. 400010 (2016). https://doi.org/10.1063/1.4952198
8. Gubaydullin, I.M., Enikeeva, L.V., Barkalov, K.A., Lebedev, I.G., Silenko, D.G.: Kinetic modeling of isobutane alkylation with mixed c4 olefins and sulfuric acid as a catalyst using the asynchronous global optimization algorithm. Commun. Comput. Inf. Sci. **1618**, 293–306 (2022). https://doi.org/10.1007/978-3-031-11623-0_20
9. Jones, D., Martins, J.: The direct algorithm: 25 years later. J. Glob. Optim. **79**(3), 521–566 (2021). https://doi.org/10.1007/s10898-020-00952-6
10. Kvasov, D.E., Sergeyev, Y.D.: Lipschitz global optimization methods in control problems. Autom. Remote Control **74**(9), 1435–1448 (2013). https://doi.org/10.1134/s0005117913090014
11. Liberti, L., Kucherenko, S.: Comparison of deterministic and stochastic approaches to global optimization. Int. Trans. Oper. Res. **12**, 263–285 (2005)
12. Paulavičius, R., Žilinskas, J.: Simplicial Global Optimization. Springer, New York (2014). https://doi.org/10.1007/978-1-4614-9093-7
13. Paulavičius, R., Sergeyev, Y.D., Kvasov, D.E., Žilinskas, J.: Globally-biased BIRECT algorithm with local accelerators for expensive global optimization. Expert Syst. Appl. **144**, 113052 (2020). https://doi.org/10.1016/j.eswa.2019.113052
14. Sergeyev, Y.D., Kvasov, D.E.: Deterministic Global Optimization: An Introduction to the Diagonal Approach. Springer, New York (2017). https://doi.org/10.1007/978-1-4939-7199-2
15. Sergeyev, Y.D., Kvasov, D.E., Mukhametzhanov, M.S.: On the efficiency of nature-inspired metaheuristics in expensive global optimization with limited budget. Sci. Rep. **8**(1), 435 (2018)
16. Sergeyev, Y.D., Strongin, R.G., Lera, D.: Introduction to Global Optimization Exploiting Space-Filling Curves. Springer Briefs in Optimization, Springer, New York (2013). https://doi.org/10.1007/978-1-4614-8042-6
17. Sergeyev, Y.D., Candelieri, A., Kvasov, D.E., Perego, R.: Safe global optimization of expensive noisy black-box functions in the $\delta$-Lipschitz framework. Soft Comput. **24**(23), 17715–17735 (2020). https://doi.org/10.1007/s00500-020-05030-3
18. Sergeyev, Y.D., Pugliese, P., Famularo, D.: Index information algorithm with local tuning for solving multidimensional global optimization problems with multiextremal constraints. Math. Program. **96**(3), 489–512 (2003). https://doi.org/10.1007/s10107-003-0372-z
19. Stripinis, L., Paulavičius, R.: A new DIRECT-GLh algorithm for global optimization with hidden constraints. Optim. Lett. **15**(6), 1865–1884 (2021). https://doi.org/10.1007/s11590-021-01726-z
20. Strongin, R.G., Sergeyev, Y.D.: Global Optimization with Non-Convex Constraints. Sequential and Parallel Algorithms. Kluwer Academic Publishers, Dordrecht (2000)
21. Strongin, R.G., Barkalov, K.A., Bevzuk, S.A.: Global optimization method with dual Lipschitz constant estimates for problems with non-convex constraints. Soft Comput. **24**(16), 11853–11865 (2020). https://doi.org/10.1007/s00500-020-05078-1

# Evolutionary Global Optimization Survival of the Fittest Algorithm

D. Perov(✉) and O. Kuzenkov

Lobachevsky State University, Gagarin Av. 23, Nizhny Novgorod 603950, Russia
unn@unn.ru
http://www.unn.ru

**Abstract.** This research presents the improved evolutionary Survival of the Fittest algorithm for global optimization. The algorithm uses a measure of probability concentration near the best solution to control the search process. It is strictly established in the paper that the sequence of points generated by the algorithm approaches the global optimum with the probability of unity. The improvements focus on enhancing the crossover operation, generalizing convergence conditions, and increasing overall algorithm efficiency in finite-dimensional spaces. Through testing with classical multidimensional and randomly generated functions, this method demonstrates superior performance compared to other evolutionary algorithms. Additionally, comparative analysis is conducted to select optimal hyperparameters for the method. The improved Survival of the Fittest algorithm offers a promising approach to efficiently solving complex optimization challenges across various domains.

**Keywords:** Global optimization · Survival of the Fittest algorithm · Differential Evolution · Evolutionary strategies · Convergence proof

## 1 Introduction

Evolutionary computation plays a critical role in solving global optimization problems [1–3]. These methods use concepts of biological evolution such as mutation, crossover, and selection. It is widely employed in biology, engineering, artificial intelligence, and other fields where global optimization problems exist [4–11]. However, despite advancements in developing new modifications of evolutionary algorithms to improve optimization efficiency [12–14], crucial theoretical concerns, such as convergence, remain [15,16].

The Survival of the Fittest algorithm (SoFa) is the stochastic global optimization method [17,18]. It relies on the concept of biological evolution, known as "Survival of the fittest." In practical terms, this method utilizes the concentration of probability measures around the global maximum. In these previous works [17,18], it was proven that the algorithm generates a sequence of points that converges to the global solution with a probability of unity. Initially, the algorithm was described for Hilbert spaces [17]. Subsequently, the improved


D. Balandin et al. (Eds.): MMST 2023, CCIS 1914, pp. 162–175, 2024.
https://doi.org/10.1007/978-3-031-52470-7_14

and more efficient simplification was presented specifically tailored for finite-dimensional spaces [18]

The objective of this work is to modify the SoFa to enhance the convergence performance and provide novel convergence proof of the modification. By simplifying the integration and working on the versatility of the SoFa mutation and crossover operator, we demonstrated the efficacy of the improved method through experimental evaluations on classical multidimensional and randomly generated functions. Additionally, comparative analysis is conducted to select optimal hyperparameters for the algorithm.

## 2 Improved Survival of the Fittest Algorithm

Let's formulate an optimization problem. We have a continuous positive objective function $J(x_1, x_2, \ldots, x_D)$, referred to as a fitness function, defined in a $D$-dimensional cube:

$$\Pi = \{X = [x_1, x_2, \ldots, x_D] : x_{min} \leq x_j \leq x_{max}, j = \overline{1, D}\}, \tag{1}$$

where $x_{min}$, $x_{max}$ are constant bounds. The main goal of optimization is to find the unique global maximum point $X^* = [x_1^*, x_2^*, \ldots, x_D^*]$ of the function $J(X)$ within the bounds of $\Pi$.

Now, let's delve into the Survival of the Fittest algorithm (SoFa). This is a stochastic population-based optimization method that is used to solve complex global optimization problems [5,17,18]. The algorithm has the following modified steps:

1) *Initialization*
   Create the initial random population $P_g = [X_{1,g}, X_{2,g}, \ldots, X_{NP,g}]$ of $NP$ candidate solutions (points in the search space $\Pi$), where $g = 1$ is the iteration index. Each component of every point is assigned a random value drawn from a uniform distribution within the bounds: $x_{i,j,1} = rand(x_{min}, x_{max})$, $j = \overline{1, D}$, $i = \overline{1, NP}$.
2) *Mutation*
   During each iteration $g$ in the population of $P_g$, each point is assigned the probability of being the reference point $X_{r,g} = [x_{r,1,g}, x_{r,2,g}, \ldots, x_{r,D,g}]$, $r \in \{1, 2, \ldots, NP\}$:

$$p_{i,g} = \frac{(\frac{J(X_{i,g})}{J_g^*})^{\omega_g}}{(\frac{J(X_{1,g})}{J_g^*})^{\omega_g} + \ldots + (\frac{J(X_{NP,g})}{J_g^*})^{\omega_g}}. \tag{2}$$

   Here $\omega_g$ represents a sequence that regulates the probability of choosing the point with the maximum fitness value $J_g^*$ in the population $P_g$ as the reference point $X_{r,g}$. For each mutant point $\widetilde{X}_{i,g+1} = [\widetilde{x}_{i,1,g+1}, \ldots, \widetilde{x}_{i,D,g+1}]$ the reference point $X_{r,g}$ is selected, and the components of the mutant point are selected from the segment $[x_{min}, x_{max}]$ using the Cauchy probability density function:

$$\frac{A_{i,r,j,g}\upsilon_{i,g}}{1+\upsilon_{i,g}^{2}(\widetilde{x}_{i,j,g+1}-x_{r,j,g})^{2}}. \tag{3}$$

The sequence $\upsilon_{i,g}$ regulates the convergence rate to the reference point, and $A_{i,r,j,g}$ serves as the normalization constant for the probability density over the segment $[x_{min}, x_{max}]$:

$$A_{i,r,j,g} = (arctan(\upsilon_{i,g}(x_{max} - x_{r,j,g})) - arctan(\upsilon_{i,g}(x_{min} - x_{r,j,g})))^{-1}. \tag{4}$$

In other words, the mutation can be formulated as stated below:

$$\widetilde{x}_{i,j,g+1} = x_{r,j,g} + \frac{1}{\upsilon_{i,g}} tan(\frac{rand(0,1) - \frac{1}{2}}{A_{i,r,j,g}}). \tag{5}$$

3) *Crossover*
With the crossover probability $CR_{i,j,g}$, each component of the mutant point $\widetilde{X}_{i,g+1} = [\widetilde{x}_{i,1,g}, \widetilde{x}_{i,2,g}, \ldots, \widetilde{x}_{i,D,g}]$ replaces the component of the current point $X_{i,g} = [x_{i,1,g}, x_{i,2,g}, \ldots, x_{i,D,g}]$ in population $P_g$ to create the test point $\bar{X}_{i,g+1} = [\bar{x}_{i,1,g}, \bar{x}_{i,2,g}, \ldots, \bar{x}_{i,D,g}]$ as follows:

$$\bar{x}_{i,j,g+1} = \begin{cases} \widetilde{x}_{i,j,g+1}, & \text{if } rand(0,1) \leq CR_{i,j,g} \text{ or } j = j_r \\ x_{i,j,g}, & \text{otherwise.} \end{cases} \tag{6}$$

where the randomly selected index $j_r \in \{1, 2, \ldots, D\}$ ensures that the test point includes at least one component from the mutant point.

4) *Selection*
Evaluate the fitness value $J(\bar{X}_{i,g+1})$ of the new test point. Then, if the fitness $J(\bar{X}_{i,g+1})$ of the new test point is superior to the fitness $J(X_{i,g})$ of the current point, the test point will replace the current point in the next population:

$$X_{i,g+1} = \begin{cases} \bar{X}_{i,g+1}, & \text{if } J(X_{i,g}) \leq J(\bar{X}_{i,g+1}) \\ X_{i,g}, & \text{otherwise.} \end{cases} \tag{7}$$

5) *Termination criteria*
Repeat steps $2-4$ for all points in the population until the index of iterations $g$ reaches the predefined maximum number of iterations $G$ ($g = \overline{1, G}$). It is imperative to note that the objective function is evaluated a total of $NP$ times during each iteration.

The main feature of the improved SoFa is the flexibility in parameter selection, which is an important practical advantage that allows the algorithm to be adapted to a wide range of optimization problems. The new SoFa has three adjustable parameters: $\omega_g$, $\upsilon_{i,g}$, and $CR_{i,j,g}$. It will be proven below that this algorithm will generate test points that converge to the global maximum with the probability of unity for a continuous positive function $J$ in a $D$-dimensional cube $\Pi$, if the following conditions on these parameters are met.

**Condition 1.** *The sequence $\upsilon_{i,g}$, which regulates the convergence rate to the reference point, approaches infinity as $g$ tends to infinity, and $\upsilon_{i,g}^{D}$ approaches infinity slower than $g$ tends to infinity:*

$$\upsilon_{i,g} \xrightarrow{g \to \infty} \infty,$$

$$\frac{\upsilon_{i,g}^{D}}{g} \xrightarrow{g \to \infty} 0.$$

**Condition 2.** *The crossover probability $CR_{i,j,g}$, which regulates the mutation rate, approaches unity as $g$ tends to infinity:*

$$CR_{i,j,g} \xrightarrow{g \to \infty} 1.$$

**Condition 3.** *The sequence $\omega_g$, which regulates the probability of choosing the point with the maximum fitness value in the population as the reference point, approaches infinity as $g$ tends to infinity:*

$$\omega_g \xrightarrow{g \to \infty} \infty.$$

These conditions can be easily met by setting the appropriate parameters. For example, in our work we used $\upsilon_{i,g} = \ln^4(g+e)$, $CR = \frac{g}{G}$, and $\omega = ln^2(g+e)$.

## 3 Results

### 3.1 Convergence Proof for Improved SoFa

In this section, we demonstrate the convergence prove of the improved SoFa within a $D$-dimensional cube $\Pi$ for a continuous positive function $J$. The proof involves three theorems.

**Theorem 1.** *If Conditions 1 and 2 are met, then the sequence of test points $\bar{X}_{i,g+1}$, $\forall i \in \{1, 2, \ldots, NP\}$ becomes an everywhere dense sequence in the cube $\Pi$ as $g$ approaches infinity. In other words, for any point $X \in \Pi$ and any positive number $\sigma$, the probability of finding the point $\bar{X}_{i,g+1}$ in the vicinity $O_\sigma(X)$ approaches unity as $g \to \infty$.*

*Proof.* The cubic vicinity, denoted as $O_\sigma(X)$, is proposed for the point $X = [x_1, \ldots, x_D]$, where $0 < \sigma < \frac{1}{2}$. The point $\bar{X} = [\bar{x}_1, \ldots, \bar{x}_D]$ is within $O_\sigma(X)$, if $|\bar{x}_j - x_j| \leq \sigma$, $j = \overline{1, D}$. The sequence $\bar{X}_{i,g+1}$, $g = 1, 2, ...$, is everywhere dense in the cube $\Pi$ if: $\forall X \in \Pi, \forall \sigma > 0, \exists g > 0, \exists i \in \{1, 2, \ldots, NP\}$: $\bar{X}_{i,g+1} \in O_\sigma(X)$.

The test point $\bar{X}_{i,g+1} = [\bar{x}_{i,1,g}, \ldots, \bar{x}_{i,D,g}]$ is generated by selecting a reference point $X_{r,g} = [x_{r,1,g}, \ldots, x_{r,D,g}]$ randomly from the population $P_g$. The coordinate of the test point $\bar{x}_{i,j,g+1}$ with the crossover probability $CR_{i,j,g}$ has the following probability density:

$$\frac{A_{i,r,j,g}\upsilon_{i,g}}{1 + \upsilon_{i,g}^2(\widetilde{x}_{i,j,g+1} - x_{r,j,g})^2}, \tag{8}$$

where $\widetilde{x}_{i,j,g+1}$ is component of mutation point $\widetilde{X}_{i,g+1} = [\widetilde{x}_{i,1,g+1}, \ldots, \widetilde{x}_{i,D,g+1}]$, and $A_{i,r,j,g}$ represents the constant used to normalize probability density in the segment $[x_{min}, x_{max}]$:

$$A_{i,r,j,g} = (arctan(\upsilon_{i,g}(x_{max} - x_{r,j,g})) - arctan(\upsilon_{i,g}(x_{min} - x_{r,j,g})))^{-1}. \quad (9)$$

The subsequent inequality holds:

$$A_{i,r,j,g} \geq \frac{1}{\pi}. \quad (10)$$

The probability that all components of the test point $\bar{x}_{i,j,g+1}$, $j = \overline{1,D}$ will receive the mutation is determined by $\prod_{j=1}^{D} CR_{i,j,g}$. With the probability $\prod_{j=1}^{D} CR_{i,j,g}$ the probability $P(\forall j \in \{1, 2 \ldots, D\}, |\bar{x}_{i,j,g+1} - x_j| \leq \sigma)$ that the test point $\bar{X}_{i,g+1}$ will be included in $O_\sigma(X)$ at the generation $g$ is established as:

$$P(\forall j \in \{1, 2 \ldots, D\}, |\bar{x}_{i,j,g+1} - x_j| \leq \sigma)$$
$$= \prod_{j=1}^{D} \frac{\int_{x_j-\sigma}^{x_j+\sigma} (\frac{A_{i,r,j,g}\upsilon_{i,g}}{1 + \upsilon_{i,g}^2(\widetilde{x}_{i,j,g+1} - x_{r,j,g})^2}) d\widetilde{x}_{i,j,g+1}}{\int_{x_{min}}^{x_{max}} (\frac{A_{i,r,j,g}\upsilon_{i,g}}{1 + \upsilon_{i,g}^2(\widetilde{x}_{i,j,g+1} - x_{r,j,g})^2}) d\widetilde{x}_{i,j,g+1}}. \quad (11)$$

In this case, we can estimate this probability:

$$\prod_{j=1}^{D} \int_{x_j-\sigma}^{x_j+\sigma} (\frac{A_{i,r,j,g}\upsilon_{i,g}}{1 + \upsilon_{i,g}^2(\widetilde{x}_{i,j,g+1} - x_{r,j,g})^2}) d\widetilde{x}_{i,j,g+1}$$
$$\geq \prod_{j=1}^{D} (\frac{A_{i,r,j,g}\upsilon_{i,g}}{1 + \upsilon_{i,g}^2(x_{max} - x_{min})^2} \int_{x_j-\sigma}^{x_j+\sigma} d\widetilde{x}_{i,j,g+1})$$
$$= \prod_{j=1}^{D} \frac{2A_{i,r,j,g}\sigma\upsilon_{i,g}}{1 + \upsilon_{i,g}^2(x_{max} - x_{min})^2}$$
$$> \prod_{j=1}^{D} \frac{2A_{i,r,j,g}\sigma}{\upsilon_{i,g}(1 + (x_{max} - x_{min})^2)}$$
$$\geq \prod_{j=1}^{D} \frac{2\sigma}{\pi\upsilon_{i,g}(1 + (x_{max} - x_{min})^2)} = \frac{c}{\upsilon_{i,g}^D}, \quad (12)$$

where the constant $c = \prod_{j=1}^{D} \frac{2\sigma}{\pi(1 + (x_{max} - x_{min})^2)}$ does not depend on the index of the iteration.

With the probability $\prod_{j=1}^{D} CR_{i,j,g}$ the probability of not falling into the vicinity $O_\sigma(X)$ can be estimated:

$$P(\exists j \in \{1, 2, \ldots, D\} : |\bar{x}_{i,j,g+1} - x_j| > \sigma) \leq 1 - \frac{c}{\upsilon_{i,g}^D}. \quad (13)$$

If Conditions 1 and 2 are met, then for $g$ iterations, the probability of the point $\bar{X}_{i,g+1}$ not being chosen from the vicinity $O_\sigma(X)$ approaches zero as $g \to \infty$:

$$(1 - \frac{c}{v_{i,g}^D})^g \xrightarrow{g\to\infty} 0. \tag{14}$$

As a result, the produced test points $\bar{X}_{i,g}$ become the everywhere dense sequence in $\Pi$ with the probability of unity.

**Theorem 2.** *Let Condition 3 be met, $J(X)$ is a continuous positive function defined within $\Pi$, and $X^* = [x_1^*, \ldots, x_D^*]$ representing the unique point of its global maximum, then, for any positive value $\sigma$, the probability of selecting a reference point $X_{r,g}$, $r \in \overline{1, NP}$ from the vicinity $O_\sigma(X^*)$ approaches unity as $g \to \infty$.*

*Proof.* The following definitions are introduced: $J_0 = \sup_{X\in\Pi\setminus O_\sigma(X^*)} J(X)$; $\bar{I}_g$ represents the set of indexes corresponding to points $X_{i,g}$, $i = \overline{1, NP}$, not belonging to $O_\sigma(X^*)$; $I_g$ represents the set of indexes corresponding to points $X_{i,g}$, $i = \overline{1, NP}$, belonging to $O_\sigma(X^*)$.

According to the Theorem 1, as $g$ approaches infinity, $\bar{X}_{i,g+1}$ forms an everywhere dense sequence within $\Pi$, with the probability of unity. There invariably exists a point $X_{m,n} \in O_\sigma(X^*)$, $m \in \{1, 2, \ldots, NP\}$ such that $J(X_{m,n}) > J_0$, since $J$ is a continuous function. Then the probability of selecting the point $X_{r,g}$ as a reference point, with index $r$ from the set $\bar{I}_g$ for $g > n$, can be estimated as:

$$\begin{aligned} &\frac{\sum_{i\in\bar{I}_g} (\frac{J(X_{r,g})}{J_g^*})^{\omega_g}}{\sum_{i=1}^{NP} (\frac{J(X_{i,g})}{J_g^*})^{\omega_g}} = \frac{\sum_{i\in\bar{I}_g} J^{\omega_g}(X_{r,g})}{\sum_{i=1}^{NP} J^{\omega_g}(X_{i,g})} \\ &\le \frac{\sum_{i\in\bar{I}_g} J_0^{\omega_g} J^{-\omega_g}(X_{m,n})}{\sum_{i=1}^{NP} J^{\omega_g}(X_{i,g}) J^{-\omega_g}(X_{m,n})} < (\frac{J_0}{J(X_{m,n})})^{\omega_g} NP \xrightarrow{g\to\infty} 0. \end{aligned} \tag{15}$$

Observably, as $g$ tends to infinity, this probability approaches zero. Consequently, the probability of selecting point $X_{r,g}$ as a reference point with the index $r$ from the set $I_g$ approaches unity.

**Theorem 3.** *Let Conditions 1, 2, and 3 be met, and the continuous positive function $J(X)$ defined within $\Pi$ has the unique global maximum, which is achieved at $X^* = [x_1^*, \ldots, x_D^*]$, then for any positive value $\sigma$, the probability of choosing new test points $\bar{X}_{i,g+1}$ from the vicinity $O_\sigma(X^*)$ approaches unity as $g \to \infty$. In other words, the sequence $\bar{X}_{i,g+1}$ converges in probability to the global maximum point:*

$$\forall\sigma > 0, \lim_{g\to\infty} P(||\bar{X}_{i,g+1} - X^*|| > \sigma) = 0. \tag{16}$$

*Proof.* Taking into account that $O_\sigma(X^*) = \{X = [x_1, \ldots, x_D] : |x_j - x_j^*| \le \sigma, j = \overline{1, D}\}$, we assess the probability of selecting test points $\bar{X}_{i,g+1}$ from the vicinity

$O_\sigma(X^*)$, under the condition that $X_{r,g} \in O_\sigma(X^*)$ is taken as the reference point with the index $r$ from the set $I_g$:

$$\mathbb{P}(\forall j, |\bar{x}_{i,j,g} - x_j^*| \le \sigma) = \prod_{j=1}^{D} \frac{\int_{x_j^*-\sigma}^{x_j^*+\sigma} (\frac{A_{i,r,j,g} \upsilon_{i,g}}{1+\upsilon_{i,g}^2 (\widetilde{x}_{i,j,g+1} - x_{r,j,g})^2}) d\widetilde{x}_{i,j,g+1}}{\int_{x_{min}}^{x_{max}} (\frac{A_{i,r,j,g} \upsilon_{i,g}}{1+\upsilon_{i,g}^2 (\widetilde{x}_{i,j,g+1} - x_{r,j,g})^2}) d\widetilde{x}_{i,j,g+1}}$$

$$\ge \prod_{j=1}^{D} F(x_j^* + \sigma - x_{r,j,g}) - \prod_{j=1}^{D} F(x_j^* - \sigma - x_{r,j,g}), \tag{17}$$

here

$$F(x) = \frac{1}{\pi} arctan(x \upsilon_{i,g}) + \frac{1}{2}. \tag{18}$$

Considering that $\upsilon_{i,g} \xrightarrow{g \to \infty} \infty$, $x_j^* + \sigma - x_{r,j,g} > 0$ and $x_j^* - \sigma - x_{r,j,g} < 0$, we have, respectively:

$$F(x_j^* + \sigma - x_{r,j,g}) \xrightarrow{g \to \infty} 1, \tag{19}$$

$$F(x_j^* - \sigma - x_{r,j,g}) \xrightarrow{g \to \infty} 0. \tag{20}$$

As the result:

$$P(\forall j, |\bar{x}_{i,j,g} - x_j^*| \le \sigma)$$
$$\ge \prod_{j=1}^{D} F(x_j^* + \sigma - x_{r,j,g}) - \prod_{j=1}^{D} F(x_j^* - \sigma - x_{r,j,g}) \xrightarrow{g \to \infty} 1. \tag{21}$$

That fact serves as the final piece of evidence confirming the convergence of the improved algorithm.

It is worth noting that if the point $X^*$ of the global maximum of the function $J$ is not unique, then the algorithm will generate test points $\bar{X}_{i,g+1}$ that will converge to a shared vicinity of the global maximums points.

### 3.2 Comparison of the Improved SoFa Efficiency for Various Parameters

All approaches and algorithms are compared on the test functions with the same stopping criteria (maximum number of fitness evaluations). The test functions have the following modification to satisfy the condition of a positive function:

$$J(k, x_1, x_2, \ldots, x_D) = \frac{1}{1 + e^{\frac{1}{D} J_k(x_1, x_2, \ldots, x_D)}}. \tag{22}$$

Here we compare the efficiency of the improved SoFa algorithm depending on the parameter $\upsilon_{i,g}$. For the test, there are several ways to select it:

$$\upsilon_{i,g} = ln^{2k}(g + e),\ k = 1,\ 2,\ 3; \tag{23}$$

$$v_{i,g} = \begin{cases} \ln^4(g+e), & \text{if } p_{i,g} > \dfrac{1}{NP} \\ g, & \text{otherwise;} \end{cases} \tag{24}$$

$$v_{i,g} = g^k,\ k = 1,\ 2. \tag{25}$$

It is imperative to note that the last approach (25) violates the convergence of the algorithm. The other parameters are fixed as follows: $\omega = ln^2(g+e)$, $CR = \frac{g}{G}$, and $NP = 10D$. The choice of these parameters is based on numerous tests, convergence conditions, and previously performed works [5,17,18].

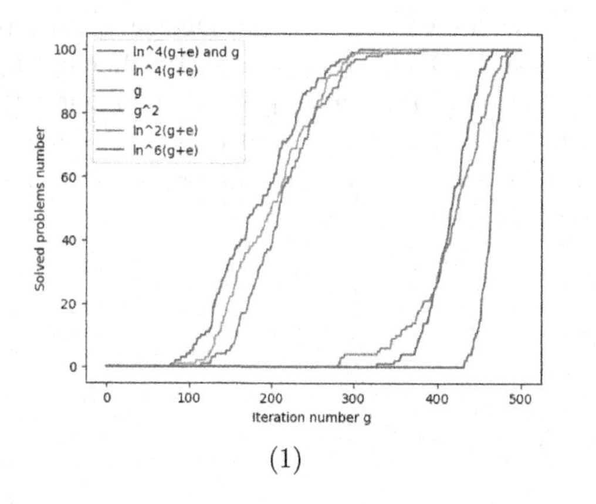


(1)

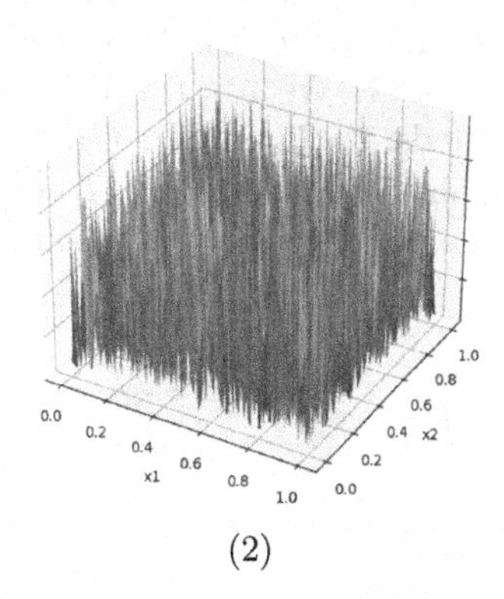


(2)

**Fig. 1.** Comparison of the average (over 10 runs) number of solved problems by SoFa with various parameter $v_{i,g}$ on 100 randomly generated two-dimensional modified Grishagin functions (1); Graphical representation of a randomly generated two-dimensional modified Grishagin function (2).

For the test, 100 randomly generated two-dimensional $(D = 2)$ modified Grishagin functions [19] are used with the following $J_k$, $k = \overline{1,100}$:

$$\begin{aligned} J_k(x_1,x_2) = \{(&\sum_{i=1}^{7}\sum_{j=1}^{7} A_{k,i,j}a_{k,i,j}(x_1,x_2) + B_{k,i,j}b_{k,i,j}(x_1,x_2))^2 \\ +(&\sum_{i=1}^{7}\sum_{j=1}^{7} C_{k,i,j}a_{k,i,j}(x_1,x_2) - D_{k,i,j}b_{k,i,j}(x_1,x_2))^2\}^{\frac{1}{2}}, \end{aligned} \tag{26}$$

$$\begin{aligned} a_{k,i,j}(x_1,x_2) &= \sin(\pi i x_1)\sin(\pi j x_2), \\ b_{k,i,j}(x_1,x_2) &= \cos(\pi i x_1)\cos(\pi j x_2), \\ x_{min} &= 0, x_{max} = 1, \\ A_{k,i,j}, B_{k,i,j}, C_{k,i,j}, D_{k,i,j} &= rand(-1,1). \end{aligned} \tag{27}$$

The problem of finding the maximum value for the function $J(k, x_1, x_2)$ is solved if the following inequality holds:

$$|J(k, x_1, x_2) - 0.5| \leq 10^{-4}. \qquad (28)$$

The average (over 10 runs) dependencies of the solved problem number on the iteration number $g$ are shown in Fig. 1. It can be seen that the combining approach (24) of choosing the parameter $\upsilon_{i,g}$ shows greater efficiency in finding the maximum on 100 modified two-dimensional Grishagin functions than others.

### 3.3 Comparison of the Improved SoFa with Other Methods

The SoFa algorithm is compared to previous version [18] and DE using various mutation strategies [2,18]: DE, DE best 1, DE current-to-best 1, DE current-to-*p*best 1. We use the following parameters for the improved SoFa and DE methods: $NP = 10D$, $CR = \frac{g}{G}$, $\omega = ln^2(g + e)$, to select $\upsilon_{i,g}$ we use the combined approach (24), $F = rand(0, 1)$, and $p = \frac{NP}{10}$. For the previous version of the SoFa we used parameters from work [18].

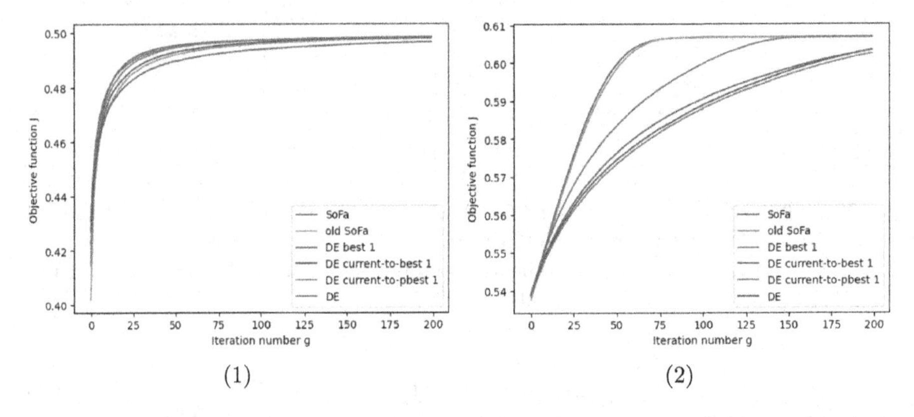


**Fig. 2.** Comparison of the average (over 10 runs) convergence rate to the maximum on all 100 randomly generated functions: (1) two-dimensional modified Grishagin functions; (2) ten-dimensional modified Exponential functions.

We tested methods on 100 random modified two-dimensional Grishagin functions with $J_k$, $k = \overline{1, 100}$, and on 100 random modified ten-dimensional Exponential functions (inspired by Gaussian functions [20], you can see graphical representations in Fig. 3) with the following $J_k$, $k = \overline{101, 200}$:

$$J_k(x_1, x_2, \ldots, x_D) = \sum_{j=1}^{D} \frac{1}{2} cos(2\pi|x_j - y_{j,k}|)e^{-\frac{1}{5}|x_j - y_{j,k}|}. \qquad (29)$$

where $y_{j,k} = rand(x_{min}, x_{max})$, $x_{max} = 50$, $x_{min} = -50$, $D = 10$.

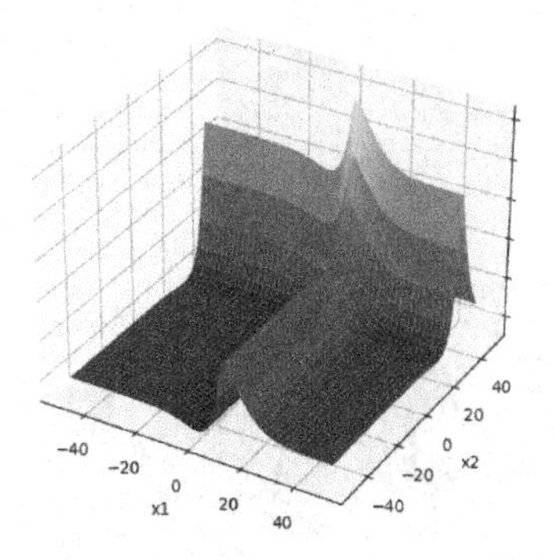


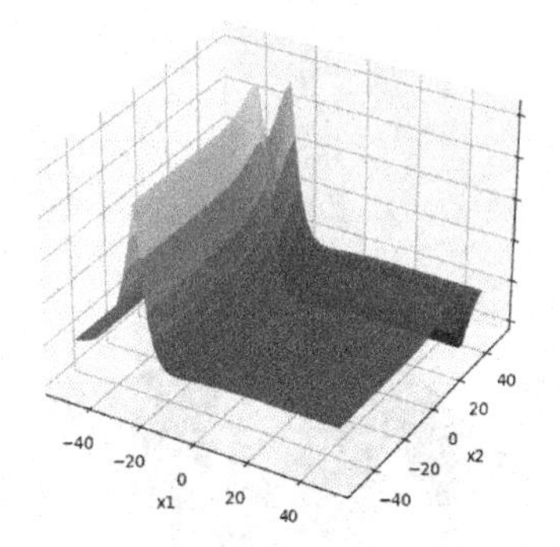


**Fig. 3.** Graphical representations of a randomly generated two-dimensional modified Exponential functions.

The average (over 10 runs) convergence rates to the maximum on all 100 randomly generated two-dimensional modified Grishagin and ten-dimensional modified Exponential functions are shown in Fig. 2. It can be seen that the improved SoFa shows the better convergence rate than other algorithms.

Additionally, the algorithms are tested on 4 modified classical multidimensional functions ($D = 10$) with the following $J_k$, $k = \overline{201, 204}$:

**Rastrigin function**

$$J_{201}(x_1, x_2, \ldots, x_D) = \sum_{i=1}^{D}(x_i^2 - 10cos(2\pi x_i) + 10). \tag{30}$$

The optimum is achieved at $x_i^* = 0$, $i = \overline{1, D}$,
within the constraints $x_{min} = -5.12$, $x_{max} = 5.12$.
**Weierstrass function**

$$J_{202}(x_1, x_2, \ldots, x_D) = \sum_{i=1}^{D}(\sum_{j=0}^{j_{max}}[a^j cos(2\pi b^j(x_i + 0.5))]) - D\sum_{j=0}^{j_{max}} a^j cos(\pi b^j). \tag{31}$$

Here $a = 0.5$, $b = 3$, $j_{max} = 20$. The optimum is achieved at $x_i^* = 0$, $i = \overline{1, D}$,
within the constraints $x_{min} = -0,5$, $x_{max} = 0,5$.
**Wavy function**

$$J_{203}(x_1, x_2, \ldots, x_D) = 100 - \frac{100}{D}\sum_{i=1}^{D} cos(10x_i)e^{-\frac{x_i^2}{2}}. \tag{32}$$

The optimum is achieved at $x_i^* = 0$, $i = \overline{1, D}$,
within the constraints $x_{min} = -3.14$, $x_{max} = 3.14$.

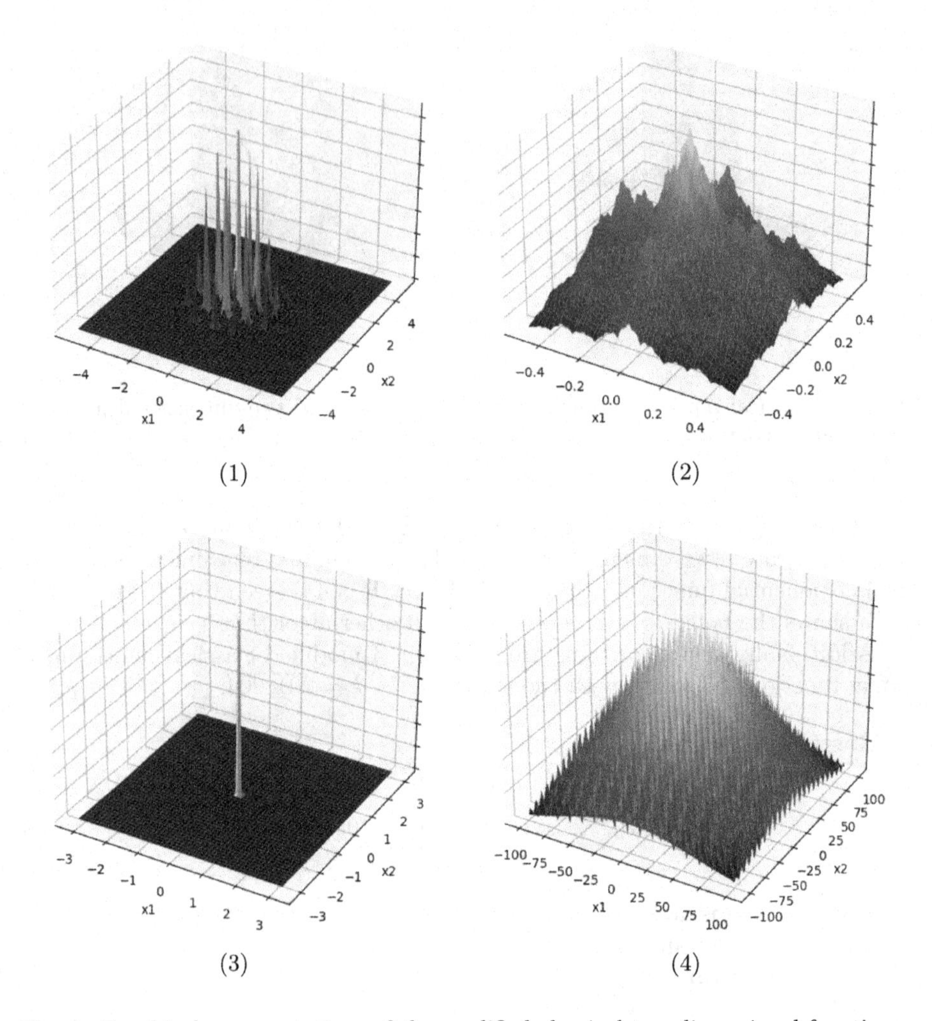


**Fig. 4.** Graphical representations of the modified classical two-dimensional functions: (1) Rastrigin function; (2) Weierstrass function; (3) Wavy function; (4) Griewank function.

**Griewank function**

$$J_{204}(x_1, x_2, \ldots, x_D) = \sum_{i=1}^{D} \frac{x_i^2}{4000} - 10 \prod_{i=1}^{D} cos(\frac{x_i}{\sqrt{i}}) + 10. \tag{33}$$

The optimum is achieved at $x_i^* = 0$, $i = \overline{1, D}$,
within the constraints $x_{min} = -100$, $x_{max} = 100$.

Graphs of the modified classical two-dimensional functions can be seen in Fig. 4. The average (over 10 runs) convergence rates to the maximum on the modified classical ten-dimensional functions are shown in Fig. 5. There also shown the

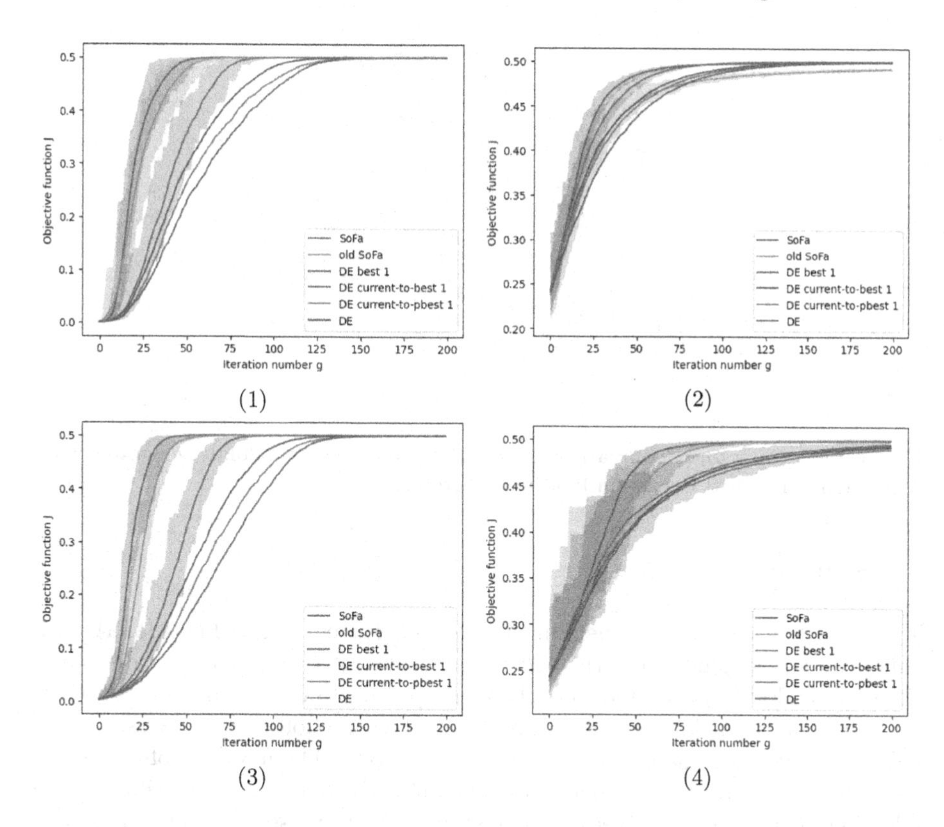


**Fig. 5.** Comparison of the average (over 10 runs) convergence rates to the maximum with lower and upper bounds on the modified classical ten-dimensional functions: (1) Rastrigin function; (2) Weierstrass function; (3) Wavy function; (4) Griewank function.

upper and lower bounds of the convergence rates for the SoFa, old SoFa, and DE best 1. It can be seen that on all modified ten-dimensional classical functions the improved SoFa shows better convergence to the maximum.

We also compared the improved SoFa with other simple evolutionary strategies that use the Cauchy (CES) and Gaussian (GES) distributions [21] with parameter $NP = 100$. We tested the performance on 100 randomly generated two-dimensional Grishagin functions over 20 runs. To evaluate the performance, we used operational zones from the works [22,23] and the condition described earlier (28). In Fig. 6 shown the average number of solved problems, as well as the best and worst cases. It can be seen that the improved SoFa shows the more stable behavior and outperforms the other two evolutionary algorithms.

Overall, the results demonstrate that the improved SoFa exhibits a superior convergence rate and outperforms other considered methods. The algorithm shows fast convergence to the vicinity of the solution. Furthermore, it is superior to its previous version.

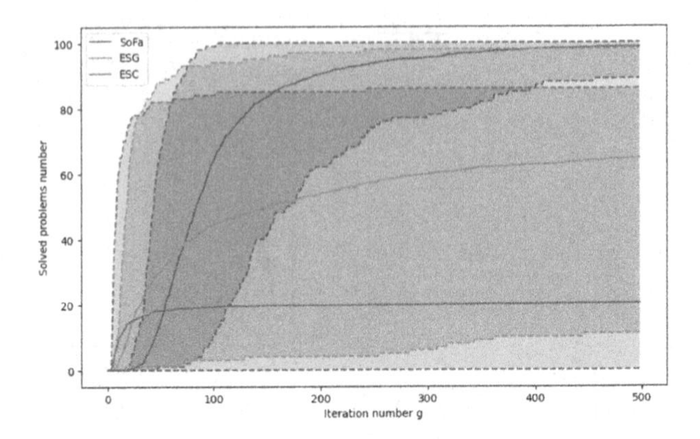


**Fig. 6.** Operation zones comparison (over 20 runs) for 100 randomly generated two-dimensional modified Grishagin functions – problems.

## 4 Summary

The improved version of the Survival of the Fittest algorithm (SoFa) with enhanced convergence performance was presented. The experimental evaluations of randomly generated and classical optimization functions demonstrate that the improved SoFa outperforms its previous version and some variants of DE on tested functions in terms of convergence rate. The novel proof of convergence further increases the algorithm reliability as well as its versatility in the selection of parameters. Overall, the enhanced SoFa presents a novel approach to global optimization problems.

## References

1. Del Ser, J., et al.: Bio-inspired computation: where we stand and what's next. Swarm Evol. Comput. **48**, 220–250 (2019). https://doi.org/10.1016/j.swevo.2019.04.008
2. Storn, R., Price, K.: Differential evolution - a simple and efficient heuristic for global optimization over continuous spaces. J. Glob. Optim. **11**(4), 341–359 (1997). https://doi.org/10.1023/A:1008202821328
3. Vikhar, P.A.: Evolutionary algorithms: a critical review and its future prospects. In: 2016 International Conference on Global Trends in Signal Processing, In-formation Computing and Communication (ICGTSPICC), pp. 261–265 (2016). https://doi.org/10.1109/ICGTSPICC.2016.7955308
4. Deng, W., Shang, S., Cai, X., Zhao, H., Song, Y., Xu, J.: An improved differential evolution algorithm and its application in optimization problem. Soft Comput. **25**(7), 5277–5298 (2021). https://doi.org/10.1007/s00500-020-05527-x
5. Kuzenkov, O., Perov, D.: Construction of optimal feedback for zooplankton diel vertical. Migration **1739**, 139–152 (2022). https://doi.org/10.1007/978-3-031-22990-9_10
6. Slowik, A., Kwasnicka, H.: Evolutionary algorithms and their applications to engineering problems. Neural Comput. Appl. **32**(16), 12363–12379 (2020). https://doi.org/10.1007/s00521-020-04832-8

7. Salimans, T., Ho, J., Chen, X., Sidor, S., Sutskever, I.: Evolution Strategies as a Scalable Alternative to Reinforcement Learning (2017). https://doi.org/10.48550/ARXIV.1703.03864
8. Rasool, A., Jiang, Q., Wang, Y., Huang, X., Qu, Q., Dai, J.: Evolutionary approach to construct robust codes for DNA-based data storage. Front. Genet. **14** (2023). https://doi.org/10.3389/fgene.2023.1158337
9. Mishra, V., Kane, L.: A survey of designing convolutional neural network using evolutionary algorithms. Artif. Intell. Rev. **56**(6), 5095–5132 (2023). https://doi.org/10.1007/s10462-022-10303-4
10. Sharma, V., Tripathi, A.K.: A systematic review of meta-heuristic algorithms in IoT based application. Array **14**, 100164 (2022). https://doi.org/10.1016/j.array.2022.100164
11. Ismayilov, G., Topcuoglu, H.R.: Neural network based multi-objective evolutionary algorithm for dynamic workow scheduling in cloud computing. Futur. Gener. Comput. Syst. **102**, 307–322 (2020). https://doi.org/10.1016/j.future.2019.08.012
12. Brest, J., Maučec, M.S., Bošković, B.: The 100-digit challenge: algorithm jde100. In: 2019 IEEE Congress on Evolutionary Computation (CEC), pp. 19–26(2019)
13. Viktorin, A., Senkerik, R., Pluhacek, M., Kadavy, T., Zamuda, A.: Dish algorithm solving the CEC 2019 100-digit challenge. In: 2019 IEEE Congress on Evolutionary Computation (CEC), pp. 1–6 (2019)
14. Kumar, A., Misra, R.K., Singh, D., Das, S.: Testing a multi-operator based differential evolution algorithm on the 100-digit challenge for single objective numerical optimization. In: 2019 IEEE Congress on Evolutionary Computation (CEC), pp. 34–40 (2019)
15. Hu, Z., Xiong, S., Su, Q., Zhang, X.: Sufficient conditions for global convergence of differential evolution algorithm. J. Appl. Math. **2013**, 193196 (2013). https://doi.org/10.1155/2013/193196
16. He, J., Yu, X.: Conditions for the convergence of evolutionary algorithms. J. Syst. Architect. **47**(7), 601–612 (2001). https://doi.org/10.1016/S1383-7621(01)00018-2
17. Morozov, A.Y., Kuzenkov, O.A., Sandhu, S.K.: Global optimisation in Hilbert spaces using the Survival of the Fittest algorithm. Commun. Nonlinear Sci. Numer. Simul. **103**, 106007 (2021). https://doi.org/10.1016/j.cnsns.2021.106007
18. Kuzenkov, O., Perov, D.: Global Optimization Method Based on the Survival of the Fittest algorithm, pp. 187–201 (2022). https://doi.org/10.1007/978-3-031-24145-1_16
19. Gergel, V., Grishagin, V., Isra lov, R.: Local tuning in nested scheme of global optimization. Procedia Comput. Sci. **51**, 865–874 (2015). https://doi.org/10.1016/j.procs.2015.05.216
20. Gallagher, M.R., Yuan, B.: A general-purpose tunable landscape generator. IEEE Trans. Evol. Comput. **10**, 590–603 (2006)
21. Hansen, N., Ostermeier, A.: Completely derandomized self-adaptation in evolution strategies. Evol. Comput. **9**(2), 159–195 (2001)
22. Sergeyev, Y.D., Kvasov, D.E., Mukhametzhanov, M.S.: Operational zones for comparing metaheuristic and deterministic one-dimensional global optimization algorithms. Math. Comput. Simul. **141**, 96–109 (2017). https://doi.org/10.1016/j.matcom.2016.05.006
23. Sergeyev, Y.D., Kvasov, D.E., Mukhametzhanov, M.S.: On the efficiency of nature-inspired metaheuristics in expensive global optimization with limited budget. Sci. Rep. **8**(1), 453 (2018). https://doi.org/10.1038/s41598-017-18940-4

# Path Tracking Control of a Spherical Robot with Pendulum-Like Driver

Dmitry V. Balandin, Ruslan S. Biryukov(✉), and Alexander M. Tuzikov

Lobachevsky State University of Nizhni Novgorod, Nizhny Novgorod, Russia
{dmitriy.balandin,ruslan.biryukov}@itmm.unn.ru

**Abstract.** We consider the path tracking problem for a rolling spherical robot on an arbitrary uneven surface. A full mathematical model of motion as well as its reduced version without slipping and twisting are derived. A state-feedback control laws that provide robot motion along a given trajectory are synthesized. The effectiveness of the proposed control laws is demonstrated by results of computer simulation.

**Keywords:** spherical robot · mathematical model · path tracking control

## 1 Introduction

During last decade, works aimed at design of mobile robots, including spherical robots, applied in different fields of human activity, have been conducted. A spherical mobile robot has a ball-shaped outer shell to include all its mechanism, control devices and energy sources in it. As a consequence, it has several advantages over wheeled and legged robots including no chance of rolling over due to their symmetrical structure, the capacity of rapid maneuvers and movement because of omnidirectional driving, no problem with obstacles, inner component protection from the external environment, such as liquid, dust, and gas, and a high energy efficiency due to single contact point with the ground. And the most important, it can resume stability in a quick manner even if a collision happened.

Typically, the drive system of the spherical robot can be realized using different technical constructions, such as a system of rotating flywheels, a set of moving masses, located on orthogonally related axes, as well as a pendulum, connected with the spherical shell by a joint, and moved by a system of motors [1–6]. In fact, each type of driver is based on shifting the sphere's center of gravity, that is why spherical robots are referred to the class of systems with internal

D.V. Balandin is supported by the Ministry of Science and Higher Education of the Russian Federation, (Agreement 075-10-2021-093, Project RAI-RND-2126 Artificial intelligence methods and algorithms development for solving problems of dynamic manipulation in robotic applications). R.S. Biryukov and A.M. Tuzikov are supported by the Ministry of Science and Higher Education of the Russian Federation, project no. FSWR-2023-0034, and by the Research and Education Mathematical Center Mathematics for Future Technologies.


D. Balandin et al. (Eds.): MMST 2023, CCIS 1914, pp. 176–189, 2024.
https://doi.org/10.1007/978-3-031-52470-7_15

moving masses [7]. Such system have a number of specific features. First, the motion of a spherical robot is performed by friction forces applied to the spherical shell at the point of its contact with the surface on which it moves. Second, a spherical robot is constantly rotating, but remains in functionality.

In this paper, a model of a spherical robot is considered, which is two thin-walled concentric spherical shells. The movable element of such a design that drives the robot is the inner shell rotating relative to the outer spherical shell of the robot. The basic principles of motion of such a system can be borrowed from the theory and practice of spherical motors [8–10]. This rotation can be carried out, for example, by electromagnets distributed over the spherical surface of the robot. As further analysis shows, the control of such a robot is much easier to implement than a robot with a pendulum suspended in the center of a spherical shell discussed in [1].

The rest of the paper is organized as follows. Section 2 describes the mathematical model of the spherical robot rolling on an arbitrary uneven surface. Section 3 deals with the path-tracking controller design, which is based on linearization near references trajectory with some additional assumptions. A linear-model path-tracking controller depends on the deviations of the robot from a reference path. To suppress disturbances and minimize deviations, we use generalized $H_2$ controller. Basic information about the generalized $H_2$ control and multi-objective control problem are introduced in Sect. 4. Section 5 presents simulation results to illustrate the effectiveness and validity of the proposed controllers. Finally, Sect. 5 presents concluding remarks.

## 2 Derivation of Equations of Motion

A schematic representation of the spherobot rolling on a uneven surface $\Sigma$ is given in Fig. 1. It consists of two rigid thin-walled uniform concentric spherical shells. The inner shell has the radius $r$, the mass $m$ and the moments of inertia $j$, similar parameters for the outer shell are equal to $R > r$, $M$ and $J$, respectively. The center of the robot coinciding with its center of mass. The inner shell can rotate freely inside the outer one. Suppose that the outer shell has a single point of contact with the surface $\Sigma$.

The motion of the spherobot is controlled by rotating the inner shell. More precisely, a change in the angular acceleration of the inner shell leads to the robot changes its motion direction and velocity.

We derive the dynamic model of the spherical robot using the Newton's laws of motion. Let $Oxyz$ be a Cartesian frame of reference fixed in space so that the axis $Oz$ is directed vertically upward. In such a frame, the balance equations of linear and angular momentum written for the shells have the form:

$$M\ddot{\boldsymbol{\xi}} = \boldsymbol{F} + \boldsymbol{F}_{\text{fric}} + N\boldsymbol{n} - Mg\boldsymbol{k}, \tag{1a}$$

$$J\dot{\boldsymbol{\Omega}} = \boldsymbol{P} - R\,[\boldsymbol{n} \times \boldsymbol{F}_{\text{fric}}], \tag{1b}$$

$$m\ddot{\boldsymbol{\xi}} = -\boldsymbol{F} - mg\boldsymbol{k}, \tag{1c}$$

$$j\dot{\boldsymbol{\omega}} = -r\boldsymbol{P}/R, \tag{1d}$$

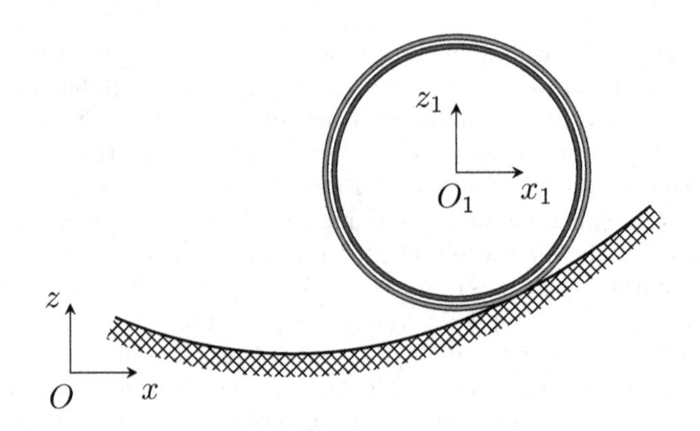


**Fig. 1.** Side view of the spherobot rolling on a uneven surface.

where $\boldsymbol{\xi} = (x, y, z)$ is the position of the center of the robot, $\boldsymbol{\Omega} = (\Omega_x, \Omega_y, \Omega_z)$ and $\boldsymbol{\omega} = (\omega_x, \omega_y, \omega_z)$ are the angular velocities of the outer and inner shells, respectively, $\boldsymbol{F}$ is the reaction force, $\boldsymbol{P}$ is the moment of force of the outer shell on the inner shell, calculated relative to the geometric center of the shells, $N$ is the normal force, $\boldsymbol{F}_{\text{fric}}$ is the friction force, $\boldsymbol{n}$ is the unit normal of the surface $\Sigma$ at the point of contact, $g$ is the gravitational acceleration, and $\boldsymbol{k} = (0, 0, 1)$ is the unit vector along the $z$-axis.

Eliminating the reaction force $\boldsymbol{P}$ from the Eq. (1), we obtain the final system

$$(M + m)\ddot{\boldsymbol{\xi}} = \boldsymbol{F}_{\text{fric}} + N\boldsymbol{n} - (M + m)g\boldsymbol{k}, \tag{2a}$$

$$rJ\dot{\boldsymbol{\Omega}} = -jR\dot{\boldsymbol{\omega}} - rR\,[\boldsymbol{n} \times \boldsymbol{F}_{\text{fric}}]. \tag{2b}$$

We assume that the normal force $N$ is strictly positive and the robot remains in permanent contact with the surface $\Sigma$, so the scalar product $\langle \dot{\boldsymbol{\xi}}, \boldsymbol{n} \rangle$ is zero. To write down the expression for the normal force, we calculate the scalar product of the Eq. (2a) and the vector $\boldsymbol{n}$, and take into account that $\langle \ddot{\boldsymbol{\xi}}, \boldsymbol{n} \rangle = -\langle \dot{\boldsymbol{\xi}}, \dot{\boldsymbol{n}} \rangle$, then

$$N = g(M + m)\,\langle \boldsymbol{n}, \boldsymbol{k} \rangle - (M + m)\langle \dot{\boldsymbol{\xi}}, \dot{\boldsymbol{n}} \rangle. \tag{3}$$

In addition, we have to consider supplementary conditions about contact between the robot and the surface, according to the rolling or slipping. In the case of slipping, the friction force $\boldsymbol{F}_{\text{fric}}$ is modelled by the Coulomb friction law,

$$\boldsymbol{F}_{\text{fric}} = -\mu N \frac{\boldsymbol{v}_c}{|\boldsymbol{v}_c|}, \qquad \boldsymbol{v}_c = \dot{\boldsymbol{\xi}} - R\,[\boldsymbol{\Omega} \times \boldsymbol{n}], \tag{4}$$

where $\boldsymbol{v}_c$ is the velocity at the contact point and $\mu$ denotes the friction coefficient. In the case of rolling, the kinematic (nonholonomic) constraint $\boldsymbol{v}_c = 0$ is satisfied and so also

$$\dot{\boldsymbol{\xi}} = R\,[\boldsymbol{\Omega} \times \boldsymbol{n}]. \tag{5}$$

We make the change of variables

$$\boldsymbol{\xi} = R\,\boldsymbol{\xi}', \qquad \omega = \frac{rJ}{jRT}\,\omega', \qquad \boldsymbol{\Omega} = \frac{1}{T}\,\boldsymbol{\Omega}', \qquad \boldsymbol{v}_c = \frac{R}{T}\,\boldsymbol{v}_c', \qquad t = Tt',$$

$$\boldsymbol{F}_{\text{fric}} = \frac{\gamma J}{RT^2}\,\boldsymbol{F}_{\text{fric}}', \qquad N = \frac{\gamma J}{RT^2}N', \qquad \gamma = \frac{R^2(M+m)}{J}, \qquad T = \sqrt{\frac{R}{g}},$$

to reduce the system (2) to the following dimensionless form omitting the prime symbol

$$\ddot{\boldsymbol{\xi}} = \boldsymbol{F}_{\text{fric}} + N\boldsymbol{n} - \boldsymbol{k}, \tag{6a}$$

$$\dot{\boldsymbol{\Omega}} = \boldsymbol{Q} - \gamma\,[\boldsymbol{n} \times \boldsymbol{F}_{\text{fric}}], \tag{6b}$$

$$N = \langle \boldsymbol{n}, \boldsymbol{k} \rangle - \langle \dot{\boldsymbol{\xi}}, \dot{\boldsymbol{n}} \rangle \tag{6c}$$

$$\boldsymbol{F}_{\text{fric}} = -\mu N\,\frac{\boldsymbol{v}_c}{|\boldsymbol{v}_c|}, \qquad \boldsymbol{v}_c = \dot{\boldsymbol{\xi}} - [\boldsymbol{\Omega} \times \boldsymbol{n}], \tag{6d}$$

here $\boldsymbol{Q} = -\dot{\boldsymbol{\omega}}$ is a control vector. Note that since the moment of inertia of a homogeneous spherical shell is $J = 2MR^2/3$, then

$$\gamma = \frac{(M+m)R^2}{J} = \frac{3}{2}\frac{(M+m)R^2}{MR^2} \geq \frac{3}{2}.$$

## 3 Motion Control of the Spherobot

We assume now that there are no slipping between the spherobot and the surface $\Sigma$, and no twisting along the surface normal $\boldsymbol{n}$, i.e. $\langle \boldsymbol{\Omega}, \boldsymbol{n} \rangle = 0$. Expressing the friction force $\boldsymbol{F}_{\text{fric}}$ from the Eq. (6a), substituting it into (6b), and taking into account the relation $\dot{\boldsymbol{\xi}} = [\boldsymbol{\Omega} \times \boldsymbol{n}]$, yields

$$(1+\gamma)\,\dot{\boldsymbol{\Omega}} - \gamma\boldsymbol{n}\,\langle \dot{\boldsymbol{\Omega}}, \boldsymbol{n} \rangle = \boldsymbol{Q} - \gamma\,[\boldsymbol{n} \times \boldsymbol{k}], \qquad \dot{\boldsymbol{\xi}} = [\boldsymbol{\Omega} \times \boldsymbol{n}]. \tag{7}$$

Note that the Eq. (7) are nonlinear in variables $\boldsymbol{\Omega}$ and $\boldsymbol{\xi}$ because the surface normal $\boldsymbol{n}$, in general, depends on the contact point between the robot and the surface $\Sigma$. Let us denote the position of the contact point as $\boldsymbol{\xi}_c$. It is convenient to rewrite the system (7) in terms of the variables $\boldsymbol{\Omega}$ and $\boldsymbol{\xi}_c$. Using the relation $\boldsymbol{\xi} = \boldsymbol{\xi}_c + \boldsymbol{n}$, we obtain the following equations:

$$\boldsymbol{\Pi}\,\dot{\boldsymbol{\Omega}} = \boldsymbol{Q} - \gamma\,[\boldsymbol{n} \times \boldsymbol{k}], \qquad \boldsymbol{\Pi} = (1+\gamma)I - \gamma\,\boldsymbol{n}\,\boldsymbol{n}^\top, \tag{8a}$$

$$(I + \boldsymbol{\Gamma})\,\dot{\boldsymbol{\xi}}_c = [\boldsymbol{\Omega} \times \boldsymbol{n}], \qquad \boldsymbol{\Gamma} = \frac{\partial \boldsymbol{n}}{\partial \boldsymbol{\xi}_c}. \tag{8b}$$

A reference path to be followed can be described by a smooth function $\boldsymbol{\zeta} = \boldsymbol{\zeta}(t)$ with respect to the time. Our aim is to find a control input $\boldsymbol{Q}$ such that the robot tracks the reference trajectory as close as possible. In order to do it, we suppose that the input $\boldsymbol{Q}$ can be composed as $\boldsymbol{Q} = \boldsymbol{Q}_0 + \boldsymbol{u}$. The first component $\boldsymbol{Q}_0$ is a nominal control, which can be found from the Eq. (8a) as

$$\boldsymbol{Q}_0 = \boldsymbol{\Pi}_0\,\dot{\boldsymbol{\Omega}}_0 + \gamma\,[\boldsymbol{n}_0 \times \boldsymbol{k}], \qquad \boldsymbol{\Pi}_0 = (1+\gamma)I - \gamma\,\boldsymbol{n}_0\,\boldsymbol{n}_0^\top, \tag{9}$$

where the angular velocity $\boldsymbol{\Omega}_0$ and the normal $\boldsymbol{n}_0$ are calculated along the path $\boldsymbol{\zeta}$. The second component $\boldsymbol{u}$ is a corrective control applied to compensate possible deviations of the spherobot from a reference path caused by uncertainties of initial values and unmodeled external disturbances. We choose a corrective control as a linear state-feedback control law

$$\boldsymbol{u} = \boldsymbol{\Theta}_\Omega \left(\boldsymbol{\Omega} - \boldsymbol{\Omega}_0\right) + \boldsymbol{\Theta}_\zeta \left(\boldsymbol{\xi}_c - \boldsymbol{\zeta}\right), \tag{10}$$

where the feedback matrices $\boldsymbol{\Theta}_\Omega$ and $\boldsymbol{\Theta}_\zeta$ may be nonstationary.

Let us define the position error $\boldsymbol{\varepsilon}_\zeta = \boldsymbol{\xi}_c - \boldsymbol{\zeta}$ and the angular velocity error $\boldsymbol{\varepsilon}_\Omega = \boldsymbol{\Omega} - \boldsymbol{\Omega}_0$, then the tracking error system can be written as

$$\boldsymbol{\Pi}_0 \, \dot{\boldsymbol{\varepsilon}}_\Omega = \gamma \left[ \langle \dot{\boldsymbol{\Omega}}_0 \cdot \boldsymbol{n}_0 \rangle I + \boldsymbol{n}_0 \dot{\boldsymbol{\Omega}}_0^\top - [\boldsymbol{k}]_\times^\top \right] \boldsymbol{\Gamma}_0 \boldsymbol{\varepsilon}_\zeta + \boldsymbol{u} + \boldsymbol{v}_1, \tag{11a}$$

$$\left(I + \boldsymbol{\Gamma}_0\right) \dot{\boldsymbol{\varepsilon}}_\zeta = \left([\boldsymbol{\Omega}_0]_\times \boldsymbol{\Gamma}_0 - \dot{\boldsymbol{\Gamma}}_0\right) \boldsymbol{\varepsilon}_\zeta + [\boldsymbol{n}_0]_\times^\top \boldsymbol{\varepsilon}_\Omega + \boldsymbol{v}_2, \tag{11b}$$

here the matrix $\boldsymbol{\Gamma}_0$ is calculated along the path $\boldsymbol{\zeta}$, the exogenous disturbances $\boldsymbol{v}_1$, $\boldsymbol{v}_2$ stand instead of nonlinear terms in the system.

Note that the values $\boldsymbol{\varepsilon}_\Omega$ and $\boldsymbol{\varepsilon}_\zeta$ are related by an addition relation. Consider the condition of no twisting along the surface normal and linearize it along $\boldsymbol{\zeta}$, then:

$$\langle \boldsymbol{\Omega} \cdot \boldsymbol{n} \rangle = \left(\boldsymbol{\Omega}_0 + \boldsymbol{\varepsilon}_\Omega + \ldots\right)^\top \left(\boldsymbol{n}_0 + \boldsymbol{\Gamma}_0 \boldsymbol{\varepsilon}_\zeta + \ldots\right) = \boldsymbol{\Omega}_0^\top \boldsymbol{n}_0 + \boldsymbol{\Omega}_0^\top \boldsymbol{\Gamma}_0 \boldsymbol{\varepsilon}_\zeta + \boldsymbol{\varepsilon}_\Omega^\top \boldsymbol{n}_0 + \cdots = 0.$$

Taking into account that the first term vanishes and ignoring the high order terms, we obtain the following relation:

$$\boldsymbol{n}_0^\top \boldsymbol{\varepsilon}_\Omega + \boldsymbol{\Omega}_0^\top \boldsymbol{\Gamma}_0 \boldsymbol{\varepsilon}_\zeta = 0. \tag{12}$$

Because the control goal is to minimize the deviation between the actual and reference trajectories in the presence of uncertainties, we choose two performance indices

$$J_1(\boldsymbol{\Theta}) = \sup_{\boldsymbol{\varepsilon}_0, \boldsymbol{v}} \frac{\max \left\{ \sup\limits_{0 \le t \le T} \left| \boldsymbol{\varepsilon}_\Omega \right|_2, \sup\limits_{0 \le t \le T} \left| \boldsymbol{\varepsilon}_\zeta \right|_2 \right\}}{\sqrt{\boldsymbol{\varepsilon}_0^\top \boldsymbol{R} \boldsymbol{\varepsilon}_0 + \|\boldsymbol{v}\|^2}}, \tag{13}$$

$$J_2(\boldsymbol{\Theta}) = \sup_{\boldsymbol{\varepsilon}_0, \boldsymbol{v}} \frac{\sup\limits_{0 \le t \le T} \left| \boldsymbol{u} \right|_2}{\sqrt{\boldsymbol{\varepsilon}_0^\top \boldsymbol{R} \boldsymbol{\varepsilon}_0 + \|\boldsymbol{v}\|^2}}, \tag{14}$$

where $\boldsymbol{\Theta} = \left(\boldsymbol{\Theta}_\Omega, \boldsymbol{\Theta}_\zeta\right)$, $\boldsymbol{\varepsilon} = \text{column}\left(\boldsymbol{\Omega} - \boldsymbol{\Omega}_0, \boldsymbol{\xi}_c - \boldsymbol{\zeta}\right)$, $\boldsymbol{\varepsilon}_0 = \boldsymbol{\varepsilon}(0)$, and $\boldsymbol{R} = \boldsymbol{R}^\top > 0$ is the weighting matrix, $\boldsymbol{v}$ is an unmodeled external disturbances. The first functional characterizes the maximal deviation of the robot with respect to the reference trajectory, while the second one characterizes the maximal corrective control input. These criteria are competing, i.e. the greater control input acting to the robot, the smaller its deviation with respect to the reference path. Thus, to determine the feedback matrices $\boldsymbol{\Theta}_\Omega$ and $\boldsymbol{\Theta}_\zeta$, it is required to minimize in Pareto sense two functionals $J_1$ and $J_2$.

Next, we obtain the conditions under which the system (11) can be resolved with respect to $\dot{\varepsilon}_\Omega$ and $\dot{\varepsilon}_\zeta$. Obviously, we should require that the matrices $\boldsymbol{\Pi}_0$ and $(I+\boldsymbol{\Gamma}_0)$ are nondegenerate. First, we consider the matrix $\boldsymbol{\Pi}_0$ and note that the relation

$$\det\Big((1+\gamma)I - \gamma\, \boldsymbol{n}_0\, \boldsymbol{n}_0^\top\Big) = \gamma^3 \det\left(\frac{1+\gamma}{\gamma} I - \boldsymbol{n}_0\, \boldsymbol{n}_0^\top\right) = 0$$

is valid if and only if the value $(1+\gamma)/\gamma$ is an eigenvalue of the matrix $\boldsymbol{n}_0\, \boldsymbol{n}_0^\top$. However, it is easy to show that $\lambda_{1,2}(\boldsymbol{n}_0\, \boldsymbol{n}_0^\top) = 0$ and $\lambda_3(\boldsymbol{n}_0\, \boldsymbol{n}_0^\top) = 1$, hence the matrix $\boldsymbol{\Pi}_0$ is nondegenerate for any $\gamma$. Thus the Eq. (11a) can always be resolved with respect to $\dot{\varepsilon}_\Omega$.

Now consider the matrix $(I+\boldsymbol{\Gamma}_0)$. It will be degenerate if and only if the value $\lambda = -1$ is an eigenvalue of the matrix $\boldsymbol{\Gamma}_0$. Let us calculate the characteristic polynomial of $\boldsymbol{\Gamma}_0$:

$$\det\Big(\lambda I - \boldsymbol{\Gamma}_0\Big) = \lambda(\lambda^2 + 2H_0\lambda + K_0) = 0,$$

where $H_0$ and $K_0$ are the average and Gaussian curvatures of the surface $\Sigma$, respectively, calculated along the curve $\boldsymbol{\zeta}$. It is easy to check that $\lambda = -1$ is the root of the characteristic equation if the condition $2H_0 = 1 + K_0$ is satisfied. Hence the matrix $(I+\boldsymbol{\Gamma}_0)$ is nonsingular if

$$1 + K_0 \neq 2H_0. \tag{15}$$

Thus, under the condition (15), the Eq. (11b) can be resolved with respect to $\dot{\varepsilon}_\zeta$ and, therefore, the system (11) is regular.

Finally, let us show that if no twisting along the surface normal $\boldsymbol{n}$, then the corrective control (10) can be rewritten as a linear feedback with respect to $\varepsilon_\zeta$ and $\dot{\varepsilon}_\zeta$. To do this, we discard the term $\boldsymbol{v}_2$ in the Eq. (11b) and add the relation (12), after which we get the system:

$$\begin{aligned}
[\boldsymbol{n}_0]_\times^\top \varepsilon_\Omega &= (I+\boldsymbol{\Gamma}_0)\,\dot{\varepsilon}_\zeta - ([\boldsymbol{\Omega}_0]_\times \boldsymbol{\Gamma}_0 - \dot{\boldsymbol{\Gamma}}_0)\varepsilon_\zeta,\\
\boldsymbol{n}_0^\top \varepsilon_\Omega &= -\boldsymbol{\Omega}_0^\top \boldsymbol{\Gamma}_0 \varepsilon_\zeta,
\end{aligned}$$

Since the matrix $([\boldsymbol{n}_0]_\times, \boldsymbol{n}_0)$ has the full column rank, the solution can be written as:

$$\varepsilon_\Omega = [\boldsymbol{n}_0]_\times (I+\boldsymbol{\Gamma}_0)\,\dot{\varepsilon}_\zeta - [\boldsymbol{n}_0]_\times([\boldsymbol{\Omega}_0]_\times \boldsymbol{\Gamma}_0 - \dot{\boldsymbol{\Gamma}}_0)\varepsilon_\zeta - \boldsymbol{n}_0 \boldsymbol{\Omega}_0^\top \boldsymbol{\Gamma}_0 \varepsilon_\zeta,$$

therefore, the relation (10) can be represented as $\boldsymbol{u} = \widetilde{\boldsymbol{\Theta}}_1\, \dot{\varepsilon}_\zeta + \widetilde{\boldsymbol{\Theta}}_2\, \varepsilon_\zeta$, where

$$\widetilde{\boldsymbol{\Theta}}_1 = \boldsymbol{\Theta}_\Omega [\boldsymbol{n}_0]_\times (I+\boldsymbol{\Gamma}_0), \quad \widetilde{\boldsymbol{\Theta}}_2 = \boldsymbol{\Theta}_\zeta - \boldsymbol{\Theta}_\Omega\Big([\boldsymbol{n}_0]_\times([\boldsymbol{\Omega}_0]_\times \boldsymbol{\Gamma}_0 - \dot{\boldsymbol{\Gamma}}_0) - \boldsymbol{n}_0 \boldsymbol{\Omega}_0^\top \boldsymbol{\Gamma}_0\Big).$$

In conclusion, let us say a few words about the control of the spherobot's motion in case with sliding in some parts of the motion trajectory. On these parts the movement of a spherobot is described by the equations (6), so the control

(10), in generally, does not guarantee that the robot will follow to the reference path. Nevertheless, this control law should be recommended to use, since in the case of relatively short sliding parts with a corresponding deviation from the reference path, this control ensures that the robot approaches the reference path upon subsequent transition to "pure" (without slipping) rolling.

## 4 Optimal Control of Maximal Output Deviations over a Finite Horizon

In this section, we present some results which will be applied for solving the problems indicated in the previous section. More details can be found in [11,12].

### 4.1 Maximal Output Deviations

Consider a linear time-variant system on finite horizon described by the equations

$$\begin{aligned} \dot{x} &= A(t)x + B(t)v, \quad x(t_0) = x_0, \\ z &= C(t)x, \end{aligned} \tag{16}$$

where $x \in \mathbb{R}^{n_x}$ is the state, $v \in L_2([t_0, t_1], \mathbb{R}^{n_v})$ is the disturbance, $z \in \mathbb{R}^{n_z}$ is the output. The generalized $H_2$ norm or the maximal output deviation under external and initial disturbances over finite-time horizon $[t_0, t_1]$ is defined as follows

$$J_{0,v} = \sup_{x_0,\, v \in L_2} \frac{\sup\limits_{t \in [t_0, t_1]} |z(t)|}{\sqrt{x_0^\top R x_0 + \|v\|_2^2}}, \tag{17}$$

where the weighting matrix $R = R^\top > 0$ is a measure of the relative importance of the uncertainty in initial conditions versus the uncertainty in the external disturbance. A "smaller" size of $R$ reflects greater uncertainty in the initial condition. In the particular cases, when the initial state of the system (16) is zero or the external disturbances are absent, the maximal output deviation is defined as

$$J_v = \sup_{v \in L_2} \frac{\sup\limits_{t \in [t_0, t_1]} |z(t)|}{\|v\|_2}, \qquad J_0 = \sup_{x_0 \neq 0} \frac{\sup\limits_{t \in [t_0, t_1]} |z(t)|}{\sqrt{x_0^\top R x_0}}.$$

**Theorem 1** ([12]). *The maximal output deviation under external and initial disturbances over the horizon* $[t_0, t_1]$ *for system* (16) *is given by the formula*

$$J_{0,v} = \sup_{t \in [t_0, t_1]} \lambda_{\max}^{1/2} \left[ C(t) Y(t) C^\top(t) \right], \tag{18}$$

*where* $\lambda_{\max}(M)$ *is the maximum eigenvalue of matrix* $M$*, the matrix function* $Y(t)$ *is the solution to the matrix differential equation*

$$\dot{Y} = A(t)Y + YA^\top(t) + B(t)B^\top(t), \qquad Y(t_0) = R^{-1}. \tag{19}$$

*In the particular case, when* $x_0 = 0$, *the maximal output deviation* $J_v$ *under external disturbances is provided by* (18), (19) *for* $Y(t_0) = 0$. *In other particular case, when* $v(t) \equiv 0$, *the maximal output deviation* $J_0$ *under initial disturbances is given by* (18), (19) *for* $Y(t_0) = R^{-1}$ *and* $B(t) \equiv 0$.

### 4.2 Synthesis of the Optimal Control

Now consider a linear time-variant controlled system

$$\begin{aligned} \dot{x} &= A(t)x + B(t)v + B_u(t)u, \quad x(t_0) = x_0 \\ z &= C(t)x + D(t)u, \end{aligned} \tag{20}$$

where $u$ is the control input. Synthesis of the optimal control as linear non-stationary feedback $u(t,x) = \Theta(t)x$, $t \in [t_0, t_1]$ minimizing the maximum output deviation of the closed-loop system is based on the following theorem.

**Theorem 2** ([12]). *The state-feedback matrices* $\Theta(t)$ *of the optimal control are computed as* $\Theta(t) = Z(t)Y^{-1}(t)$, *where* $Z(t)$, $Y(t)$ *are solutions to the following semidefiniteprogramming problem: find* $\gamma^2$ *subject to constraints*

$$\begin{aligned} &\dot{Y}(t) = A(t)Y(t) + Y(t)A^{\top}(t) + B_u(t)Z(t) + Z^{\top}(t)B_u^{\top}(t) + B(t)B^{\top}(t), \\ &\begin{pmatrix} Y(t) & \star \\ C(t)Y(t) + D(t)Z(t) & \gamma^2 I \end{pmatrix} \geq 0, \quad Y(t_0) = R^{-1}. \end{aligned} \tag{21}$$

To compute the optimal non-stationary state-feedback we discretize (21) by using discrete-time grid $t_k = t_{k-1} + h$, $k = 1, \ldots, N$, where $h = (t_1 - t_0)/N$. Then Eq. (21) get the form

$$\begin{aligned} &\min \gamma^2 : \\ &Y_{k+1} - Y_k - h\left(A_k Y_k + Y_k A_k^{\top} + B_{u,k} Z_k + Z_k^{\top} B_{u,k}^{\top} + B_k B_k^{\top}\right) = 0, \\ &\begin{pmatrix} Y_k & \star \\ C_k Y_k + D_k Z_k & \gamma^2 I \end{pmatrix} \geq 0, \quad Y_0 = R^{-1}; \quad k = 0, \ldots, N-1, \end{aligned} \tag{22}$$

## 5 Computer Simulation of Spherobot Dynamics

Before proceeding to computational experiments, we note some features of the mathematical model being studied. First, we note that this model is correct only when the condition of continuous motion of the spherobot is fulfilled

$$N = \langle \boldsymbol{n} \cdot \boldsymbol{k} \rangle - \langle \dot{\boldsymbol{\xi}} \cdot \dot{\boldsymbol{n}} \rangle \geqslant 0. \tag{23}$$

Secondly, we note that on some time intervals the motion occurs with slipping, and the dynamics of the spherobot is described by the system (6), on other time intervals the motion occurs without slipping, and the dynamics of the spherical robot is described by the system (7). The change of driving mode is determined as follows. Let the regime without slip first take place. Then the speed of the point of contact of the spherobot with the surface is $\boldsymbol{v}_c = \dot{\boldsymbol{\xi}} - [\boldsymbol{\Omega} \times \boldsymbol{n}] = 0$ and, moreover, the force of friction $\boldsymbol{F}_{\text{fric}}$ is the force of static friction, i.e. the inequality $|\boldsymbol{F}_{\text{fric}}| \leqslant \kappa N$, where $N$ is defined according to (23). The transition to the sliding motion mode occurs when this condition is violated, in this case the friction force $\boldsymbol{F}_{\text{fric}}$ is determined by the formula (4).

For numerical integration of the motion equations, instead of the (4) formula, it is advisable to use a slightly different formula

$$\boldsymbol{F}_{\text{fric}} = -\mu N \frac{\boldsymbol{v}_c}{|\boldsymbol{v}_c| + \varepsilon}, \tag{24}$$

where $\varepsilon$ is a small positive number (for example, $\varepsilon = 10^{-4}$). The moment of time at which the velocity $\boldsymbol{v}_c$ vanishes should be associated with the condition $|\boldsymbol{v}_c| < \varepsilon$. If the last one is satisfied at some time and, in addition, the inequality

$$|\boldsymbol{F}_{\text{fric}}| \leqslant \mu N, \qquad \boldsymbol{F}_{\text{fric}} = \ddot{\boldsymbol{\xi}} - N\boldsymbol{n} + \boldsymbol{k} \tag{25}$$

is hold on, then the transition to motion without slip occurs, otherwise there is no change in the motion mode.

We now describe the numerical experiments carried out. Let us choose the following value of the parameter $\gamma = 10.5$. First, consider the motion of a spherobot on a plane along a straight line given by the following equations:

$$x_0(t) = \alpha\, t, \qquad y_0(t) = z_0(t) = 0. \tag{26}$$

Thus, the reference path is the $Ox$ axis and spherobot's velocity is a constant $\alpha$. We choose the initial conditions as

$$\boldsymbol{\xi}(0) = \big(0,\, 0.3,\, 1\big), \qquad \dot{\boldsymbol{\xi}}(0) = \big(\alpha + 0.2,\, 0.2,\, 0\big), \qquad \boldsymbol{\Omega}(0) = \big[\boldsymbol{k} \times \dot{\boldsymbol{\xi}}(0)\big]$$

and set $\alpha = 1.0$. The deviation from the reference trajectory will be estimated as $\Delta(t) = \|\zeta(t) - \zeta_0(t)\|$. Figures 2 – 4 show the results of numerical simulation of the spherobot approaching the motion along a straight line (26) for two values of the coefficient of sliding friction: $\kappa = 0.8$ (red line) and $\kappa = 0.3$ (blue line). In Fig. 2 the desired trajectory of movement is shown in black, and the trajectory of movement of the spherobot center at two different coefficients of friction is shown in red and blue colors. Figures 3 and 4 show the path tracking error and the velocity at the contact point. Under the action of the controller, the position error converges rapidly, and the maximum displacement error does not exceed 0.4. From the simulation results, it can be seen that when the friction coefficient is high enough, the spherical robot rolls on the horizontal plane without sliding and the desired trajectory is reached somewhat faster than in the case of a lower friction coefficient.

Now consider the case of plane motion along a circle given by the equations

$$x_0(t) = \rho\, \cos\alpha t, \qquad y_0(t) = \rho\, \sin\alpha t, \qquad z_0(t) = 0 \tag{27}$$

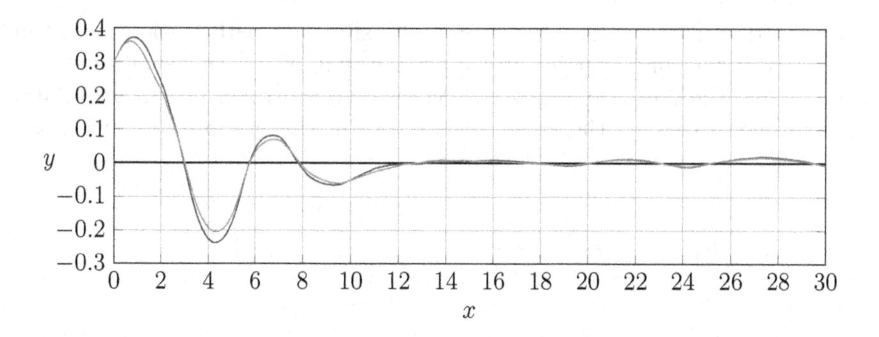


**Fig. 2.** Reference path and real motion trajectory.

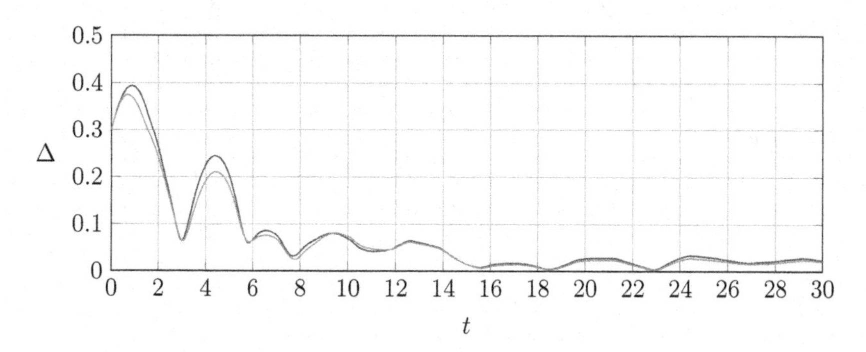


**Fig. 3.** Deviation of the spherobot's center from the reference path.

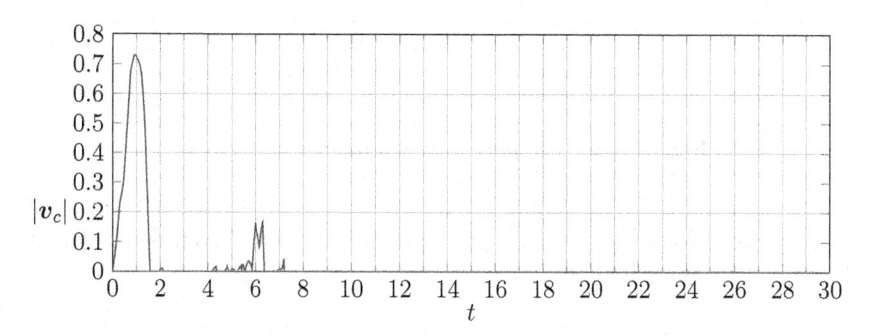


**Fig. 4.** Time histories of the spherobot's velocity at the contact point

with initial value

$$\boldsymbol{\xi}(0) = \big(\rho + 0.5,\ 0,\ 1\big), \qquad \dot{\boldsymbol{\xi}}(0) = \big(0.2,\ \alpha\rho - 0.2,\ 0\big), \qquad \boldsymbol{\Omega}(0) = \big[\boldsymbol{k} \times \dot{\boldsymbol{\xi}}(0)\big].$$

The results of numerical simulation are presented in Figs. 5, 6 and 7. In Fig. 5 the desired trajectory of movement is shown in black, and the trajectories of the spherobot center with friction coefficient $\kappa = 0.8$ and $\kappa = 0.5$, respectively, are shown in red and blue colors. Figures 6 and 7 show the path tracking error and the velocity at the contact point. It can be seen that even with a small coefficient of friction and the presence of slip, the synthesized controller provides a solution to the problem, that is, the robot's trajectory of the motion approaches to a given circle.

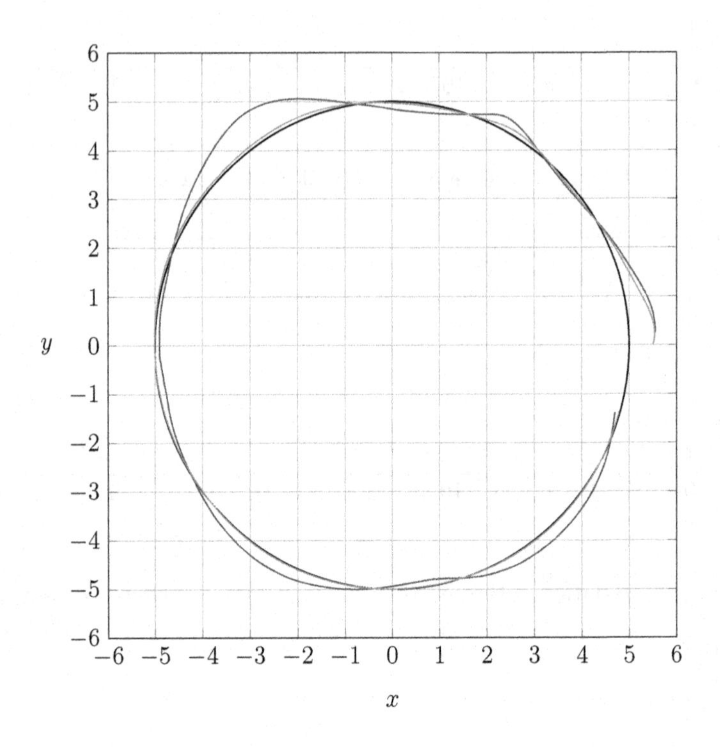


**Fig. 5.** Reference path and real motion trajectory.

Finally, we consider the spherobot's motion on surface given by the equation $2z(1 + x^2) = 1$ along the line

$$x_0(t) = t, \qquad y_0(t) = 0, \qquad z_0(t) = \frac{1}{2}\frac{1}{1+t^2}$$

with initial value

$$\boldsymbol{\xi}(0) = \big(x_0(-5),\, 0,\, z_0(-5)\big), \quad \boldsymbol{\Omega}(0) = \big(0.0,\, 2.2,\, 0.0\big), \quad \dot{\boldsymbol{\xi}}(0) = \big[\boldsymbol{\Omega}(0) \times \boldsymbol{n}_0\big].$$

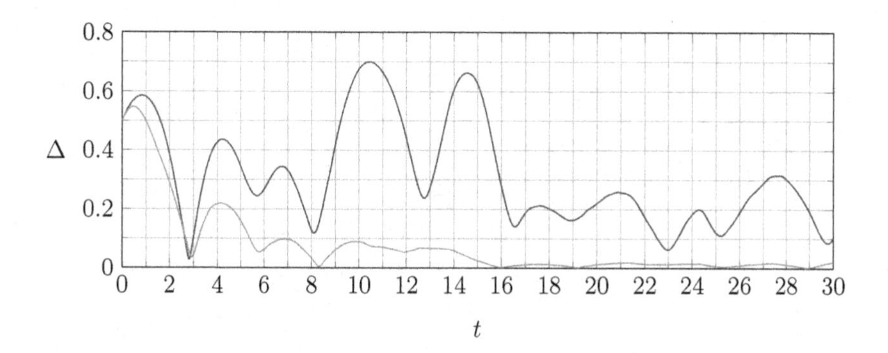


**Fig. 6.** Deviation of the spherobot's center from the reference path.

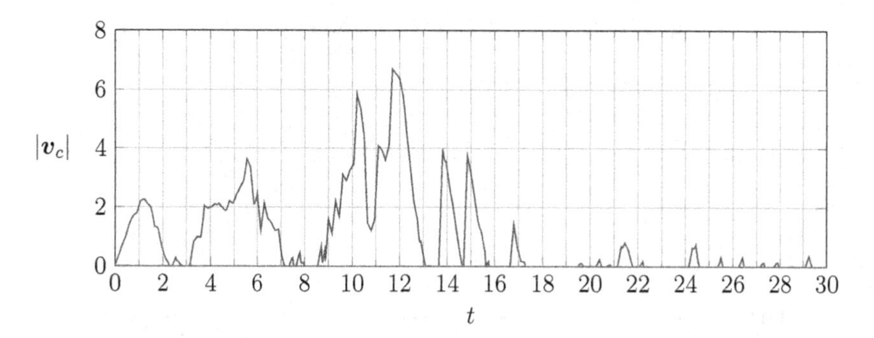


**Fig. 7.** Time histories of the spherobot's velocity at the contact point.

In Fig. 8 the desired trajectory is shown in black, and the real motion trajectory is shown in blue, the friction coefficient is $\kappa = 0.8$. Figures 9 and 10 show the time histories of the velocity at the contact point and the projection of the angular velocity $\boldsymbol{\Omega}$ to the surface normal. From these figures it follows that in the initial section of the trajectory there is a sliding motion. In addition, along the entire trajectory of motion, the spherical robot rolls with spinning around the surface normal. However, despite the presence of slipping and spinning, the synthesized controller can track the given path quickly.

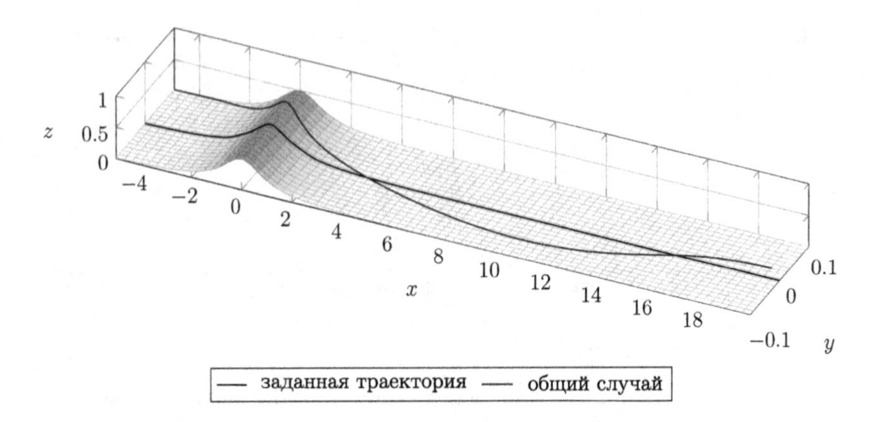


**Fig. 8.** Reference path and real motion trajectory.

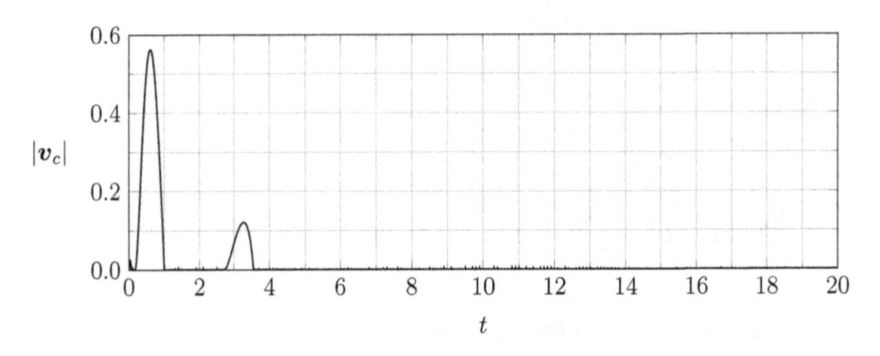


**Fig. 9.** Time histories of the spherobot's velocity at the contact point.

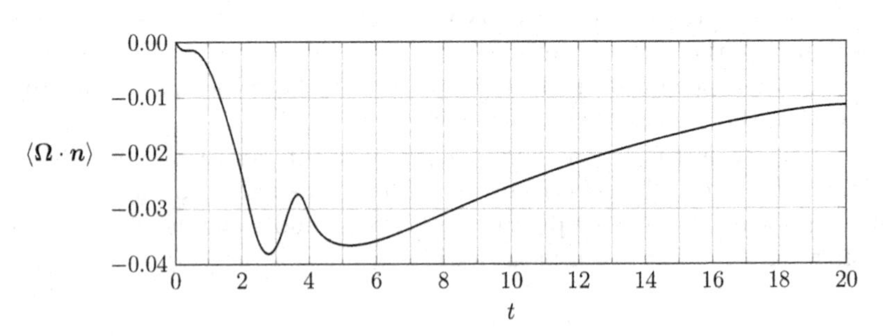


**Fig. 10.** Time histories of the projection of the angular velocity $\boldsymbol{\Omega}$ to the surface normal.

## 6 Conclusion

In this paper, a mathematical model of the movement of the two-thin walled spherical robot along an arbitrary uneven surface is presented. To address the control problem ensuring the movement of the robot in a neighborhood of the desired trajectory of motion, the path-tracking controller is designed based on linearization near a reference trajectory with additional assumptions of no slipping and spinning along normal to a surface. The robot's position error and its angular velocity error are the performance indices. The multi-objective optimal control minimizing the performance indices is synthesized using the LMI approach and generalized $H_2$ norm. The correctness of the operation of the synthesized control laws is demonstrated by numerical experiments.

## References


1. Balandin, D.V., Komarov, M.A., Osipov, G.V.: A motion control for a spherical robot with pendulum drive. J. Comput. Syst. Sci. Int. **52**(4), 650–663 (2013). https://doi.org/10.1134/S1064230713040047
2. Hogan, F.R., Forbes, J.R., Barfoot, T.D.: Rolling stability of a power-generating tumbleweed rover. J. Spacecr. Rocket. **51**(6), 1895–1905 (2014). https://doi.org/10.2514/1.A32883
3. Hogan, F., Forbes, J.: Modeling of spherical robots rolling on generic surfaces. Modeling of spherical robots rolling on generic surfaces. Multibody Syst. Dyn. **35**(1), 91–109 (2015). https://doi.org/10.1007/s11044-014-9438-3
4. Ho, C.-H., Wu, C.-K., Tu, J.-Y., Hsu, S.-K.: Conceptual design of spherical vehicle system for future transportation. J. Autom. Control Eng. **3**(3), 191–195 (2015). https://doi.org/10.12720/joace.3.3.191-195
5. Bai, Y., Svinin, M., Yamamoto, M.: Adaptive trajectory tracking control for the ball-pendulum system with time-varying uncertainties. In: Proceeding of International Conference on Intelligent Robots and Systems (IROS, 2017), Vancouver, BC, Canada, 2017, pp. 2083–2090 (2017). https://doi.org/10.1109/IROS.2017.8206026
6. Ivanova, T.B., Kilin, A.A., Pivovarova, E.N.: Controlled motion of a spherical robot with feedback. J. Dyn. Control Syst. **24**(3), 497–510 (2018). https://doi.org/10.1007/s10883-017-9387-2
7. Chernousko, F.L., Bolotnik, N.N.: Dynamics of mobile systems with controlled configuration. Fizmatlit, Moscow (2022). (in Russian)
8. Kim, I.-K., Kim, J.-S., Goh, C.-S.: Theoretical and practical study of spherical motor without the winding overhang. In: Proceeding of International Electric Machines and Drives Conference (IEMDC 2007), Antalya, Turkey, 2007, pp. 1361–1365 (2007). https://doi.org/10.1109/IEMDC.2007.383627
9. Li, Z., Jun, W.: The principle of work and essential technology of magnetic suspended spherical motor. In: 3d International Conference on Measuring Technology and Mechatronics Automation, Shanghai, China, 2011, pp. 898–901 (2011). https://doi.org/10.1109/ICMTMA.2011.796
10. Lim, C.K., Yan, L., Chen, I.-M., Yang, G., Lin, W.,: A novel approach in generating 3-DOF motions. In: Proceedings of the IEEE International Conference Mechatronics and Automation, 2005, Niagara Falls, ON, Canada, 2005, pp. 1485–1490 (2005). https://doi.org/10.1109/ICMA.2005.1626775
11. Balandin, D.V., Kogan, M.M.: Multi-objective generalized $H_2$ control. Automatica **99**(1), 317–322 (2017)
12. Balandin, D.V., Biryukov, R.S., Kogan, M.M.: Finite-horizon multi-objective generalized $H_2$ control with transients. Automatica **106**(8), 27–34 (2019)

# Mathematical Model of Processing Batches of Raw Materials Taking into Account Ripening Process

Albert Egamov(✉)

Lobachevsky University of Nizhny Novgorod, Nizhny Novgorod, Russia
albert.egamov@itmm.unn.ru

**Abstract.** At optimizing sugar production, as a rule, the question arises about the schedule of processing of various varieties of sugar beet. Due to seasonal temperature conditions, various root crop diseases and other reasons, there is no complete information about the degradation of sugar beet varieties during storage. Therefore, it is not possible to obtain an optimal processing schedule a priori, maximizing the yield of the final product (of sugar). It is necessary to find such strategies (algorithms) for processing batches of raw materials that would be quasi-optimal, if certain pre-known conditions for the parameters of batches. In this article, the mathematical model of processing sugar beetis complicated by taking into account the ripening process. At the same time, new non-standard methods come to the fore as optimal strategies, the study of which allows us to derive qualitatively new quasi-optimal strategies. In addition to theoretical interest, finding a new type of optimal strategies shows the direction for searching for quasi-optimal strategies of practical interest under similar conditions. The specifics of the task allow us to use effectively computer technique and parallel programming methods to find the values of the objective function of each strategy.

**Keywords:** Mathematical modelling · Sugar beet processing · Optimal schedule · Ripening process

## 1 Introduction

Sugar is a strategic product in all countries of the world. Its role in the daily life of a person is quite large, especially in Russia in the summer, when its sufficient

The article was carried out under the contract No SSZ-1771 dated 22.04.2021 on the implementation of R&D on the topic: "Creation of high-tech sugar production on the basis of JSC "Sergach Sugar Plant", within the framework of the Agreement on the provision of subsidies from the federal budget for the development of cooperation between the Russian educational organization of higher education and the organization of the real sector of the economy in order to implement a comprehensive project to create high-tech production No. 075-11-2021-038 of 24.06.2021 (IGC 000000S407521QLA0002).


D. Balandin et al. (Eds.): MMST 2023, CCIS 1914, pp. 190–205, 2024.
https://doi.org/10.1007/978-3-031-52470-7_16

quantity on the shelves of stores is difficult to assess due to high demand, sometimes turning into a rush. In addition, sugar production is very energy-intensive, so it is desirable to optimize sugar beet processing processes at a sugar factory by pre-constructing a mathematical model of production [1–4]. During storage, different varieties (batches) of sugar beet lose their production value (share of sugar) in root crops in different ways, so the order of processing batches clearly affects the input of the sugar, and, therefore, other things being equal, the output of the him. Also during storage, various batches of raw materials over time, generally speaking, reduce their production value in different ways (for example, for sugar beet, production value is the proportion of sugar in sugar beet). Therefore, the optimal schedule for processing batches of raw materials can significantly increase the final yield of the final product [5–8]. The yield of the finished product (sugar) depends on many quantities [9,10], on the percentage of dirt on beets, nitrate content, damage during transportation for processing, processing temperature, and so on. In this paper [11], the mathematical model takes into account a one-day production shutdown. But the main influence on the yield of the final product is the percentage of sucrose in incoming root.

The optimization problem is posed according to the presented mathematical model, which can be reduced to a well-known linear programming problem. It is the assignment problem. The Hungarian algorithm is used as the main method for its solution [12,13]. To solve the problem numerically, you can use the algorithmic language "Python". The subroutine "Hungarian algorithm" is located in the standard libraries of subroutines "scy.py" and "Munkres" [14]. In addition, the specifics of the task allow us to use parallel programming methods to find the values of the objective functions of each of the strategies.

## 2 Mathematical Model and Auxiliary Statements

### 2.1 Mathematical Model

Suppose that for a unit of time (period, stage) the plant processes a fixed quantity (one batch) of raw materials of weight $M$. The value of $M$ is determined by the production power of the plant. There are raw materials (sugar beet of different varieties), and one batch raw materials of one variety are processed at each unit of time. We denote $n$ is the number of beet batches that the plant processes during the "season". We number the batches of raw materials in some way from 1 to $n$. We denote $p_{i0}$ is the sucrose content in the $i$th batch of beets at the beginning of processing, $p_{i1}$ is the sucrose content in the $i$th batch of beets after the first unit of time (the first stage of storage and processing), $p_{ij}$, $j = \overline{1, n-1}$, is the content of the sucrose in the ith batch of beets after the $j$th period (the $j$th stage of storage), etc.

It is convenient to determine the parameters $p_{ij}$ through the initial sugar content $p_{i0}$ and degradation coefficients $b_{ij}$, $j = \overline{1, n-1}$, $b_{ij} = p_{ij}/p_{i\,j-1}$. The corresponding formulas for $p_{ij}$ have the following form

$$p_{ij} = p_{i0} b_{i1} b_{i2} ... b_{ij}, \quad i = \overline{1, n}, \quad j = \overline{1, n-1}. \tag{1}$$

We will formulate the task of finding the optimal sequence of processing of the available batches of raw materials, which ensures the maximum yield of the final product (sugar) for the entire season of processing of raw materials by the sugar plant. The sequence of processing batches of raw materials will be described by permutation $\sigma$ of natural numbers from 1 to $n$. When implementing this sequence, the total output of products after processing all batches within units of time will be proportional to the following value

$$S = \sum_{j=1}^{n} p_{\sigma(j)\, j-1}. \tag{2}$$

The optimization task is to find such a sequence of numbers from 1 to (the sequence of processing of existing batches of raw materials), at which the sugar yield calculated by formula (1) will be maximal. A similar problem is considered in [5,7,15].

The Hungarian algorithm cannot be applied in practice, since its application is based on full knowledge of the state matrix in the assignment problem (in this instance matrix $P$ with elements $p_{ij}$). In other words, for its application already at the beginning of the processing of raw materials, it is necessary to know all the degradation coefficients of all batches for all periods. Therefore, in order to determine suitable strategies, it is necessary to search for other heuristic algorithms.

The strategy implies an unambiguous choice of the batch sent for processing at the beginning of each stage, based on knowledge about the sugar content in unprocessed batches. The name of the strategy, as a rule, coincides with the name of the algorithm that it uses.

It turns out [5–7,15], with some intuitive constraints on parameters, there are simple strategies: batch processing algorithms that will be optimal under these constraints and will turn into quasi-optimal ones for similar parameters. There are two basic strategies, which are greedy and thrifty strategies using algorithms of the same name. The greedy algorithm consists of priority processing the beet batch that can give the greatest output of products at the moment. The thrifty algorithm consists of primary processing the batch with the lowest production value. If there are several of these, the batch with the lowest degradation coefficient for the last period is taken, if the period is the first, then the degradation coefficients corresponding to these batches are found from some other, for example, empirical considerations. For more information about this, see [5].

Below it will be shown which strategies can be made from a combination of greedy and thrifty strategies.

### 2.2 Auxiliary Lemmas

**Lemma 1.** *Consider the function $F(x) = x^{\alpha} - x^{\beta}$, $0 < \alpha < \beta$, $\beta - \alpha \geq 1$. Then, for $x > 0$, there is unique point of the local extremum will be the maximum point $x^*$*

$$x^* = \left(\alpha\beta^{-1}\right)^{\frac{1}{\beta-\alpha}}.$$

*Additionally*

$$x^* \leq \frac{n-2}{n-1}, \quad \text{for} \quad \beta \leq n-1. \tag{3}$$

**Remark.** The inequality (3) remains true if, instead of the inequality $\beta - \alpha \geq 1$, we take the inequality $\beta \leq n-2$. The proof is similar.

The proof of this Lemma 1 is given at the end of the article in the Appendix section. Lemma 2 follows directly from definition of the maximum point.

**Lemma 2.** *Let $x^*$ is maximum point of continuously differentiable function $F(x)$. The next statements are right.*
*If $x_1 < x_2 \leq x^*$, then $F(x_1) < F(x_2)$.*
*If $x^* \leq x_1 < x_2$, then $F(x_1) > F(x_2)$.*

## 3 Basic Results. Optimal Strategies in the Ripening Process

### 3.1 The Necessary Conditions for an Optimal Schedule

In the course of the study, the initial model gradually became more complicated. For example, in the mathematical model of processing, the ripening process began to be taken into account.

"The ripening process is the process of bringing the removed unripe fruits (root crops) in storages, warehouses or specially equipped chambers to the state of consumer ripeness" [16]. Suppose that the ripening process ends for all batches at the same time (at ihe same period) and the withering process begins "immediately" after him. Number $\nu$ is the natural number which stands the stage of the beginning of the process of withering sugar beet, $\nu \in [[n/4]; [[n/2]]$ and $\nu \geq 2$, $\nu$ is known a priori.

Consider the mathematical model in which there is a process of ripening. If we take into account the ripening process $b_{ij} > 1$, at $j \leq \nu - 1$, and $b_{ij} < 1$, at $j \geq \nu$, where $\nu$ is the number of the stage of the beginning of the process of withering sugar beet.

Suppose that at the stages from the 1st stage to the $(\nu-1)$th stage, the process of ripening takes place. After $\nu$th stage the process of withering begins, $\nu \in N$, for all batches of raw materials. Let's introduce the notation $b_i$, $\widehat{b}_i$, $i = \overline{1,n}$. It is assumed that the parameters $b_i$ depend only on the batch number of the raw material and do not depend on number of the stage of processing.

$$b_{ij} = b_i > 1,\, i = \overline{1,n},\, j = \overline{1,\nu-1}; \quad b_{ij} = \widehat{b}_i < 1,\, i = \overline{1,n},\, j = \overline{\nu,n-1}. \tag{4}$$

They can also be called the ripening coefficients and the withering coefficients respectively.

In addition, it is assumed that all varieties of beets have the same initial sugar content, that is

$$p_{i0} = p_0, \quad i = \overline{1,n}. \tag{5}$$

In addition, we will assume that the parties of raw materials can be numbered so that the following chains of inequalities are fulfilled.

$$1 < b_1 < b_2 < ... < b_{n-1} < b_n, \quad (6)$$

$$1 > \widehat{b}_1 > \widehat{b}_2 > ... > \widehat{b}_{n-1} > \widehat{b}_n \geq \frac{n-2}{n-1}, \quad (7)$$

This means that the higher the batch's ripening coefficient, the lower the degradation coefficient subsequently. In the new notation, considering the formulation (1), the formula (2) will be rewritten as

$$S = p_0(1 + \sum_{i=2}^{\nu} b_{\sigma(i)}^{i-1} + \sum_{i=\nu+1}^{n} b_{\sigma(i)}^{\nu-1} \widehat{b}_{\sigma(i)}^{i-\nu}). \quad (8)$$

The task is to finding such a permutation $\sigma$, for which the sugar yield $S^*$ (the value of the function $S$, see (8) will be maximum.

**Theorem 1.** *Let the conditions (4), (5) be fulfilled, the degradation coefficients satisfy the conditions (6), (7). Then the optimal batch schedule for maximizing the objective function (8) before the transition period $\nu$ goes in ascending order of the degradation coefficients, and then in descending order.*

The proof of this Theorem is located in the Appendix. From Theorem 1 follows, that during the the ripening period up to the $\nu$ stage, the degradation coefficients of batches with optimal permutation increase and then decrease. This is a necessary condition for an optimal schedule, taking into account the ripening process and the condition (6), (7). Thus, the following conclusions can be drawn:

1) $\gamma(\nu) = n$,
2) either $\gamma(1) = 1$ or $\gamma(n) = 1$.
3) The number of remaining ways decreased from $n!$ (if the brute force method had been used) up to $C_{n-1}^{\nu-1}$, a polynomial of the order of $\nu - 1$, (selection of $n - 1$ batches for processing during the ripening process).

More precisely, we can talk about the optimal plan only by knowing the dependence of the parameters $\widehat{b}_i$ on $b_i$, for example, by setting the dependence $\widehat{b}_i = G(b_i)$.

### 3.2 Thrifty/Greedy Strategy

If we prove that the inequality

$$b_{\gamma(\nu-1)} < b_{\gamma(n)} \quad (9)$$

is true for the optimal schedule, then the chain of inequalities is valid:

$$b_{\gamma(\nu)} > b_{\gamma(\nu+1)} > ... > b_{\gamma(n)} > b_{\gamma(\nu-1)} > ... > b_{\gamma(1)}.$$

Now these conditions completely define optimal permutation. This is only possible, if $\gamma(k) = k$, $k \leq \nu - 1$; $\gamma(k) = n + \nu - k$, $k \geq \nu$.

Let's find out what additional relation must be fulfilled in order for the inequality (9) to be true. If in the optimal plan swap $\gamma(\nu-1)$-th and $\gamma(n)$-th batch then considering (8) the difference of the objective functions will be written like this

$$0 \leq S^* - S^*_{\nu-1 \leftrightarrow n} = p_0(b^{\nu-2}_{\gamma(\nu-1)} + b^{\nu-1}_{\gamma(n)}\widehat{b}^{n-\nu}_{\gamma(n)}) - p_0(b^{\nu-2}_{\gamma(n)} + b^{\nu-1}_{\gamma(\nu-1)}\widehat{b}^{n-\nu}_{\gamma(\nu-1)}) \Longleftrightarrow$$

$$b^{\nu-2}_{\gamma(\nu-1)} - b^{\nu-1}_{\gamma(\nu-1)}\widehat{b}^{n-\nu}_{\gamma(\nu-1)} \geq b^{\nu-2}_{\gamma(n)} - b^{\nu-1}_{\gamma(n)}\widehat{b}^{n-\nu}_{\gamma(n)}. \quad (10)$$

Let the equality

$$\widehat{b}_i = b_i^{-z}, \quad i = \overline{1, n}, \quad (11)$$

is valid, where $z$ is some fixed positive number. Definition (11) does not contradict the conditions imposed on above on $\widehat{b}_i$. Really, from the simultaneous fulfillment of the conditions (6), (11), the chain of inequalities (7) is right and $\widehat{b}_i < 1$, $i = \overline{1, n}$, because $z$ is a positive number.

From the inequality (10) taking into account (11) we get

$$b^{\nu-2}_{\gamma(\nu-1)} - b^{\nu-1-zn+z\nu}_{\gamma(\nu-1)} \geq b^{\nu-2}_{\gamma(n)} - b^{\nu-1-zn+z\nu}_{\gamma(n)}. \quad (12)$$

Let's find out under what condition on $z$, the inequality of $\nu - 1 - zn + z\nu > \nu - 2$ is valid. Really

$$\nu - 1 - zn + z\nu > \nu - 2 \Longleftrightarrow (\nu - n)z > -1 \Longleftrightarrow z < 1/(n - \nu).$$

Thus, for $z \in (0; 1/(n-\nu))$ for the function $f_4(x) = x^{\nu-2} - x^{\nu-1-zn+z\nu}$ the conditions of Lemma 1 are satisfied. Inequality (12) is rewritten as $f_4(b_{\gamma(\nu-1)}) \geq f_4(b_{\gamma(n)})$ and the inequality (9) is true according Lemma 2.

That is, the next theorem is valid:

**Theorem 2.** *Let the conditions (4) (5) be satisfied, the degradation coefficients satisfy the conditions (6), (11). Then the optimal order of the batches is as follows: the batches corresponding to* $b_1, b_2, \ldots, b_{\nu-1}, b_n, b_{n-1}, \ldots, b_{\nu+1}, b_\nu$.

Figure 1 shows the directed graph, the vertices of which are batches of raw materials with corresponding degradation coefficients. Direction of arrows (edges) from a batch with a large value of $b_i$ to a batch with a smaller one. Projections of the graph vertices on the time axis show the optimal schedule for processing sugar beet batches.

This strategy is called thrifty/greedy strategy, because with first stage to $\nu - 1$ stage the one uses thrifty algorithm, and by $\nu$ stage to $n$ stage greedy algorithm is used. There is strategy, which with the first stage to the $\nu - 1$ stage uses greedy algorithm, and by the $\nu$ stage to the $n$ stage thrifty algorithm is used. The such strategy is called greedy/thrifty one. These strategies were discussed in detail in [7]. There are also conditions when the thrifty/greedy strategy becomes optimal, as well as quasi-optimal.

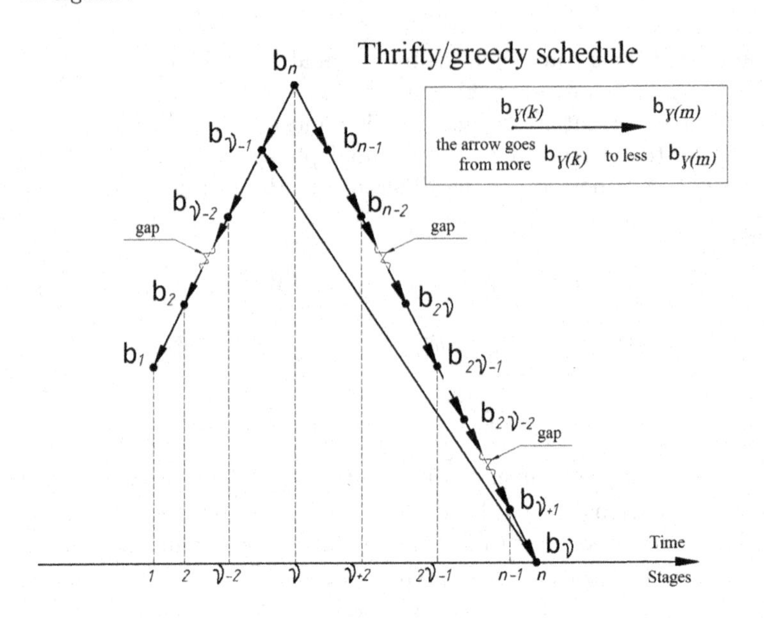


**Fig. 1.** Thrifty/greedy schedule is optimal schedule

Let's prove, that at the conditions of Theorem 2, the resulting optimal strategy is thrifty/greedy.

At the first stage, the sugar content of all batches is the same, so the thrifty algorithm selects the batch with the lowest degradation coefficient $b_1$ for processing. At each stage of ripening before the $j$th stage, $2 \leq j \leq \nu$, the sugar content of the $i$th batch is equal to $p_{ij} = p_0 b_i^{j-1}$. It is clear, than the smaller $b_i$, than the smaller $p_{ij}$ for all $j = \overline{1, n-1}$. The batch with a lower $b_i$ is selected for processing with the first to $\nu - 1$ stages. Therefore sugar content $p_{ij}$ be also the smallest in this moment and it means, that thriftily algoritmi used in this part of processing.

Before each $(\nu + j_0)$th stage (stage of withering), $0 \leq j_0 \leq n - \nu$, the sugar content of the $i$th batch is equal to

$$p_{ij} = p_0 b_i^{\nu-1} \widehat{b}_i^{j_0} = p_0 b_i^{\nu-1-j_0 z}. \tag{13}$$

Since $z < 1(n-\nu)$, $j_0 z < j_0/(n-\nu) \leq 1$ is right. Then $\nu - 1 - j_0 z > \nu - 1 - j_0/(n-\nu) \geq \nu - 2 \geq 0$. Therefore, $\nu - 1 - j_0 z > 0$, that means the power in formula (13) is positive. As a result of this, the more $b_i$, the more $p_{ij}$ for all $j = \overline{\nu, n}$. The batch with a greatest $b_i$ is selected for processing with the $\nu$ to $n$ stages. Therefore sugar content $p_{ij}$ be also the greatest in this moment and it means, that greedy algorithm used in this part of processing.

It is proved, that the optimal strategy will be the "thrifty/greedy" strategy.

### 3.3 CTG – Optimal Schedule

Let's imagine another variant of the dependence of $\widehat{b}_i$ on $b_i$, in which you can also offer a clear optimal plan. Let

$$\widehat{b}_i = \frac{\varphi_i}{b_i}, \quad i = \overline{1, n}, \tag{14}$$

where $\varphi_i$ are positive constants and the chain of inequalities be valid

$$1 > \varphi_1 \geq \varphi_2 \geq ... \geq \varphi_{n-1} \geq \varphi_n. \tag{15}$$

Let's prove the next theorem.

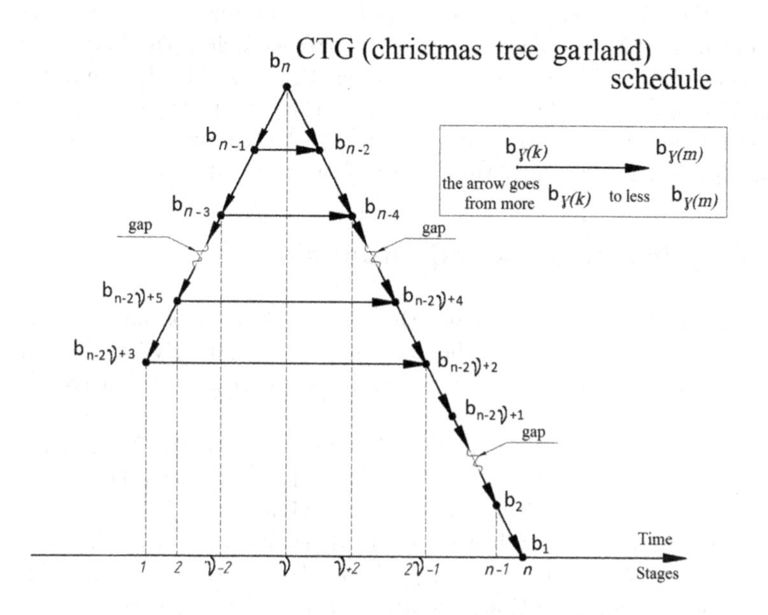


**Fig. 2.** CTG – optimal schedule

**Theorem 3.** *Let the conditions (4), (5) be satisfied, the degradation coefficients satisfy the conditions (6), (14) and (15). Then the optimal order of the batches is as follows: in turn for processing the batches corresponding to the coefficients* $b_{n-2\nu+3}$, $b_{n-2\nu+5}$, ..., $b_{n-1}$, $b_n$, $b_{n-2}$, $b_{n-4}$..., $b_{n-2\nu+4}$, $b_{n-2\nu+2}$, $b_{n-2\nu+1}$,..., $b_2$, $b_1$ *(see Fig. 2).*

The proof of this Theorem is located in the Appendix.

Figure 2 shows an directed graph whose vertices are batches of raw materials with corresponding degradation coefficients. Direction of arrows (edges) from a batch with a large value of $b_i$ to a batch with a smaller one. Projections of the graph vertices on the time axis show the optimal schedule for processing sugar beet batches. This graph illustrates an optimal schedule and is similar to a Christmas tree with a garland, so this algorithm can be called – CTG algorithm (Christmas tree garland).

From Theorem 3 follows description of optimal schedule:
$\gamma(k) = n - 2\nu + 2k + 1$, $k = \overline{1, \nu - 1}$;
$\gamma(k) = n + 2\nu - 2k$, $k = \overline{\nu, 2\nu - 1}$;
$\gamma(k) = n - k + 1$, $k = \overline{2\nu - 1, n}$.

### 3.4 CTG-Strategy

Let's describe the CTG strategy. Before each stage, the batches are "arranged" in ascending order of sugar content at the moment. The CTG algorithm (see Fig. 2) at the first stage sends a batch for processing, which occupies in the arrangement $n - 2\nu + 3$ place, if you order batches before the first stage, at the second stage sends for processing the batch that occupies in the arrangement $n - 2\nu + 4$ place, etc. At the beginning of the $\nu - 1$ stage, a batch is being processed, which is currently in the $n - \nu + 1$ place in order of increasing sugar content. Next, a greedy algorithm is used from the $\nu$ stage.

## 4 Numerical Results and Discussion

As mentioned above, since initially the matrix of states is not fully known, it is not possible to directly use theoretical methods for solving the assignment problem, therefore it is necessary to look for such strategies that could be used in practice.

The following strategies is knew, and namely, greedy; thrifty; greedy/thrifty; thrifty/greedy; strategy of type T(k)G, $k = \overline{1, n - \nu + 1}$; CTG-strategy. The first four strategies have already been discussed above.

Various combinations of thrifty and greedy strategy are possible. For example there is a group of strategies called "T(k)G" (based on T(k)G-algorithm).

Here $k$ is a parameter of each algorithm T(k)G, natural number. The essence of the T(k)G algorithm is as follows. At each of the first $\nu - 1$ stages of the processing process, the sugar content of all remaining batches is calculated. They being put in order and "numbered" in ascending order of sugar content. At this stage, the $k$th batch is selected for processing. At the next stage, this procedure is repeated. Starting with the $\nu$ stage is used greedy algorithm.

The thrifty/greedy strategy is, in fact, a strategy T(1)G. It is clear that the CTG algorithm is not an algorithm of the type T(k)G algorithm for some $k$.

We will solve the optimization problem numerically with the help of computer technology. Let's analyze 4 principal cases.

1. The ripening process is not taken into account, $\nu = 5$.
2. The ripening process is not taken into account, $\nu = 7$.
3. The ripening process is taken into account, $\nu = 5$.
4. The ripening process is taken into account, $\nu = 7$.

For each case, 6 variants of acceptable segments of the initial sugar content and decomposition coefficients are selected. Based on them, the variables $p_{i0}$ and $b_{ij}$, $i = \overline{1,n}$, $j = \overline{1,n-1}$, are generated randomly by means of a uniform distribution. There is no dependence between the coefficients $b_{ij}$.

For each case and variant, 25 experiments are conducted in which the values of the objective function for all the strategies described above are found. After that, their averaged values are calculated (the arithmetic mean of twenty-five values of the objective function for each strategy). It is this part of the program that can be "parallelized". Then they are compared.

We will assume that the first place in all experiments is occupied by the maximal strategy using, algorithm that finds the maximum value in this experiment. Here it is used for comparison. In practice, its application is impossible.

For the next best three strategies, the value of the averaged objective function and the place are listed in the corresponding table. In the tables among T(k)G-strategies take into account only the best strategy of them.

In the first and second cases, the ripening process is not taken into account. The obtained data are reflected in the first and second tables. Two segments are specified in the first column: the permissible segment of the initial sugar content and the permissible segment of degradation coefficients.

In the third and fourth tables, the ripening process to be considered, three segments are specified in the first column: the permissible segments of initial sugar content, the coefficients of ripening and the coefficients of wilting (Table 1).

**Table 1.** Comparison of objective functions for different strategies $\nu = 5$. The ripening process is not taken into account.

| Acceptable segments | Maximum value $\langle S^* \rangle$ | Place of greedy strategy | Value $\langle S \rangle$ | Place of Best $TkG$-strategy | Value $\langle S \rangle$ | Place of $CTG$-strategy | Value $\langle S \rangle$ |
|---|---|---|---|---|---|---|---|
| [0.14; 0.16] [0.97; 0.99] | 1.985 | 2 | 1.967 | 3 | 1.966 | 4 | 1.965 |
| [0.14; 0.16] [0.94; 0.96] | 1.632 | 2 | 1.622 | 3 | 1.620 | 4 | 1.619 |
| [0.14; 0.16] [0.92; 0.96] | 1.553 | 2 | 1.531 | 3 | 1.529 | 4 | 1.527 |
| [0.16; 0.16] [0.97; 0.99] | 2.119 | 4 | 2.096 | 2 | 2.097 | 3 | 2.097 |
| [0.16; 0.16] [0.94; 0.96] | 1.738 | 3 | 1.723 | 4 | 1.723 | 2 | 1.723 |
| [0.16; 0.16] [0.85; 0.96] | 1.378 | 2 | 1.327 | 4 | 1.324 | 3 | 1.326 |

Numerical experiments have shown that when the following conditions are met simultaneously (Table 2):

**Table 2.** Comparison of objective functions for different strategies $\nu = 7$. The ripening process is not taken into account.

| Acceptable segments | Maximum value $\langle S^* \rangle$ | Place of greedy strategy | Value $\langle S \rangle$ | Place of Best $TkG$-strategy | Value $\langle S \rangle$ | Place of $CTG$-strategy | Value $\langle S \rangle$ |
|---|---|---|---|---|---|---|---|
| [0.14; 0.16] [0.97; 0.99] | 1.986 | 2 | 1.968 | 3 | 1.967 | 4 | 1.964 |
| [0.14; 0.16] [0.94; 0.96] | 1.640 | 2 | 1.628 | 3 | 1.625 | 4 | 1.620 |
| [0.14; 0.16] [0.92; 0.96] | 1.551 | 2 | 1.531 | 3 | 1.527 | 4 | 1.518 |
| [0.16; 0.16] [0.97; 0.99] | 2.119 | 3 | 2.095 | 2 | 2.098 | 4 | 2.094 |
| [0.16; 0.16] [0.94; 0.96] | 1.739 | 2 | 1.725 | 3 | 1.724 | 4 | 1.723 |
| [0.16; 0.16] [0.85; 0.96] | 1.388 | 2 | 1.335 | 3 | 1.326 | 4 | 1.320 |

**Table 3.** Comparison of objective functions for different strategies $\nu = 5$. Accounting for the ripening process.

| Acceptable segments | Maximum value $\langle S^* \rangle$ | Place of greedy strategy | Value $\langle S \rangle$ | Place of Best $TkG$-strategy | Value $\langle S \rangle$ | Place of $CTG$-strategy | Value $\langle S \rangle$ |
|---|---|---|---|---|---|---|---|
| [0.14; 0.16] [1.01; 1.03] [0.97; 0.99] | 2.261 | 4 | 2.237 | 2 | 2.238 | 3 | 2.237 |
| [0.14; 0.16] [1.04; 1.06] [0.94; 0.96] | 2.254 | 4 | 2.234 | 3 | 2.235 | 2 | 2.236 |
| [0.14; 0.16] [1.04; 1.08] [0.92; 0.96] | 2.270 | 4 | 2.232 | 2 | 2.234 | 3 | 2.233 |
| [0.16; 0.16] [1.01; 1.03] [0.97; 0.99] | 2.415 | 4 | 2.388 | 2 | 2.389 | 3 | 2.388 |
| [0.16; 0.16] [1.04; 1.06] [0.94; 0.96] | 2.394 | 4 | 2.370 | 3 | 2.370 | 2 | 2.371 |
| [0.16; 0.16] [1.04; 1.18] [0.85; 0.96] | 2.612 | 4 | 2.487 | 2 | 2.497 | 3 | 2.488 |

**Table 4.** Comparison of objective functions for different strategies $\nu = 7$. Accounting for the ripening process.

| Acceptable segments | Maximum value $\langle S^* \rangle$ | Place of greedy strategy | Value $\langle S \rangle$ | Place of Best *TkG*-strategy | Value $\langle S \rangle$ | Place of *CTG*-strategy | Value $\langle S \rangle$ |
|---|---|---|---|---|---|---|---|
| [0.14; 0.16] [1.01; 1.03] [0.97; 0.99] | 2.372 | 4 | 2.344 | 3 | 2.348 | 2 | 2.348 |
| [0.14; 0.16] [1.04; 1.06] [0.94; 0.96] | 2.550 | 4 | 2.521 | 3 | 2.527 | 2 | 2.527 |
| [0.14; 0.16] [1.04; 1.08] [0.92; 0.96] | 2.620 | 4 | 2.566 | 3 | 2.569 | 2 | 2.571 |
| [0.16; 0.16] [1.01; 1.03] [0.97; 0.99] | 2.540 | 4 | 2.513 | 2 | 2.514 | 3 | 2.513 |
| [0.16; 0.16] [1.04; 1.06] [0.94; 0.96] | 2.704 | 3 | 2.677 | 4 | 2.676 | 2 | 2.677 |
| [0.16; 0.16] [1.04; 1.18] [0.85; 0.96] | 3.359 | 4 | 3.166 | 2 | 3.183 | 3 | 3.179 |

1) close (equal) values of the initial sugar content,
2) taking into account the ripening process,
3) a small variance of degradation coefficients,

The CTG-strategy gives either the best value of the objective function among heuristic strategies, or the value of the objective function for CTG-strategy is inferior to the second of the best values, but by no more than 0.001 (see Table 3 and Table 4).

In Sect. 3.2, the conditions under which a thrifty/greedy strategy is optimal were established. Under these conditions, $p_0 = 0.16$, $z = 1/15$ and $z = 1/31$ (see (11)) were tested similar to the computer calculations described above. Even then the deviation of the objective function value of the CTG-strategy from the objective function value of the optimal, thrifty/greedy strategy does not exceed 0.05. In half of the studied options, the CTG-strategy ranks second among the well-known strategies, losing also to the maximal strategy.

Thus, according to numerical results, at the presented conditions, the CTG-strategy is preferable from a practical point of view than any of the strategies when taking into account the ripening process.

## 5 Conclusion

This article presents the necessary conditions for the optimality of the schedule for processing batches of raw materials, taking into account the ripening process at matematical model. It is shown that the optimality of the schedule largely depends on the relationship between the coefficients $b_i$ and $\widehat{b}_i$. Theoretical studies give a idea of building a new quasi-optimal schedule and, accordingly, a new quasi-optimal strategy. A nontrivial heuristic algorithm was found, which was not previously considered in articles.

# A Appendix

**Proof of Lemma** 1. Derivative $F'(x) = \alpha x^{\alpha-1} - \beta x^{\beta-1} = x^{\alpha-1}(\alpha - \beta x^{(\beta-\alpha)})$, therefore, for $x > 0$, this derivative is zero only when $x = x^*$. It is not difficult to see that $x^* < 1$. For $0 < x < x^*$, inequality $F'(x) > 0$ is true, for $x > x^*$, inequality $F'(x) < 0$ is true, that is, $x^*$ is the maximum point. By Bernoulli's inequality [17]

$$\left(\frac{n-2}{n-1}\right)^{\beta-\alpha} = \left(1 - \frac{1}{n-1}\right)^{\beta-\alpha} \geq 1 - \frac{\beta-\alpha}{n-1} \geq \frac{\alpha}{\beta} = (x^*)^{\beta-\alpha}.$$

Another inequality in the line above is true, since

$$1 - \frac{\beta-\alpha}{n-1} \geq \frac{\alpha}{\beta} \iff (n-\beta-1)(\beta-\alpha) \geq 0.$$

Therefore the condition (3) is true and the Lemma 1 is right. The proof of the lemma is completed.

**Proof of Theorem** 1. Suppose, under these conditions, $\gamma$ is the optimal schedule is a permutation of natural numbers from 1 to $n$. Let $S^*_{k\leftrightarrow m}$ denotes the value of the function $S$, see (8), if in the optimal schedule, swap $k$ and $m$ parties in places. It is clear, that $S^* - S^*_{k\leftrightarrow m} \geq 0$.

Suppose that the parties with numbers $k$ and $k+1$ are swapped in the optimal schedule $\gamma$.

Case 1: $k = 1$, then given (8) the difference of the objective functions is written as

$$0 \leq S^* - S^*_{1\leftrightarrow 2} = p_0(1 + b_{\gamma(2)}) - p_0(1 + b_{\gamma(1)}) \iff b_{\gamma(1)} \leq b_{\gamma(2)}.$$

From assumption (6) follows $b_{\gamma(1)} < b_{\gamma(2)}$.

Case 2: $k = \overline{2, \nu-1}$, then according to (8) the difference of the objective functions is written as

$$0 \leq S^* - S^*_{k\leftrightarrow k+1} = p_0(b^{k-1}_{\gamma(k)} + b^k_{\gamma(k+1)}) - p_0(b^{k-1}_{\gamma(k+1)} + b^k_{\gamma(k)})$$

$$\iff b^{k-1}_{\gamma(k)} - b^k_{\gamma(k)} \geq b^{k-1}_{\gamma(k+1)} - b^k_{\gamma(k+1)} \iff f_1(b_{\gamma(k)}) \geq f_1(b_{\gamma(k+1)}).$$

The function $f_1(x) \equiv x^{k-1} - x^k$ fits the condition of Lemma 1. Hence according Lemma 2, the inequality $b_{\gamma(k)} \leq b_{\gamma(k+1)}$ is valid. From assumption (6) follows $b_{\gamma(k)} < b_{\gamma(k+1)}$.

Case 3: $k = \nu$, the transition from ripening to withering, the difference of objective functions will be written as

$$0 \le S^* - S^*_{\nu\leftrightarrow\nu+1} = p_0(b^{\nu-1}_{\gamma(\nu)} + b^{\nu-1}_{\gamma(\nu+1)}\widehat{b}_\nu) - p_0(b^{\nu-1}_{\gamma(\nu+1)} + b^{\nu-1}_{\gamma(\nu)}\widehat{b}_{\gamma(\nu)})$$

$$\iff b^{\nu-1}_{\gamma(\nu)}(1 - \widehat{b}_{\gamma(\nu)}) \ge b^{\nu-1}_{\gamma(\nu+1)}(1 - \widehat{b}_{\gamma(\nu+1)}).$$

Suppose that $b_{\gamma(\nu)} \le b_{\gamma(\nu+1)}$, then

$$1 \ge \left(\frac{b_{\gamma(\nu)}}{b_{\gamma(\nu+1)}}\right)^{\nu-1} \ge \frac{1-\widehat{b}_{\gamma(\nu+1)}}{1-\widehat{b}_{\gamma(\nu)}} \Rightarrow 1-\widehat{b}_{\gamma(\nu)} \ge 1-\widehat{b}_{\gamma(\nu+1)} \iff \widehat{b}_{\gamma(\nu+1)} \ge \widehat{b}_{\gamma(\nu)}.$$

It follows from the inequalities (6), (7), that $b_{\gamma(\nu)} > b_{\gamma(\nu+1)}$. But there is the assumption $b_{\gamma(\nu)} \le b_{\gamma(\nu+1)}$. We got a contradiction. Therefore the inequality $b_{\gamma(\nu)} > b_{\gamma(\nu+1)}$ is valid.

Case 4: $k = \overline{\nu+1, n-1}$, then given (8) the difference of the objective functions is written as

$$0 \le S^* - S^*_{k\leftrightarrow k+1} = p_0(b^{\nu-1}_{\gamma(k)}\widehat{b}^{k-\nu}_{\gamma(k)} + b^{\nu-1}_{\gamma(k+1)}\widehat{b}^{k-\nu+1}_{\gamma(k+1)} - b^{\nu-1}_{\gamma(k+1)}\widehat{b}^{k-\nu}_{\gamma(k+1)} - b^{\nu-1}_{\gamma(k)}\widehat{b}^{k-\nu+1}_{\gamma(k)})$$

$$\iff b^{\nu-1}_{\gamma(k)}(\widehat{b}^{k-\nu}_{\gamma(k)} - \widehat{b}^{k-\nu+1}_{\gamma(k)}) \ge b^{\nu-1}_{\gamma(k+1)}(\widehat{b}^{k-\nu}_{\gamma(k+1)} - \widehat{b}^{k-\nu+1}_{\gamma(k+1)}). \tag{16}$$

Denote $f_2(x) = x^{k-\nu} - x^{k-\nu+1}$. The estimate $k - \nu + 1 \le (n-1) - 2 + 1 < n - 1$ is correct. It is clear that the function $f_2(x)$ is suitable for the conditions of Lemma 1. We suppose that $b_{\gamma(k)} < b_{\gamma(k+1)}$, then considering (6), (7), we have $\widehat{b}_{\gamma(k)} > \widehat{b}_{\gamma(k+1)} \ge \frac{n-2}{n-1}$ and according to Lemma 2 the inequality $f_2(\widehat{b}_{\gamma(k)}) < f_2(\widehat{b}_{\gamma(k+1)})$ is right. Then the inequality (16) is not true. We got a contradiction. Therefore $b_{\gamma(k)} > b_{\gamma(k+1)}$. That was what needed to be proved.

After considering these cases, it becomes clear that during the ripening period up to the $\nu$ stage, the degradation coefficients of batches with optimal permutation increase and then decrease. This is a necessary condition for an optimal schedule, taking into account the ripening process and the conditions (6), (7).

The proof of the theorem is completed.

**Proof of Theorem 3.** From (6) and (15) follows the validity of inequalities in the chain (7). The conditions of withering $\widehat{b}_i < 1$, $i = \overline{1, n}$, is also fulfilled. The correctness of the definition (14) is justified.

In the optimal schedule, we swap places $\gamma(\nu - j)$-th and $\gamma(\nu + j)$-th batches, $j = \overline{1, \nu - 1}$. The difference of the objective functions (see 8) will be written like this

$$0 \le S^* - S^*_{(\nu-j)\leftrightarrow(\nu+j)} = p_0(b^{\nu-j-1}_{\gamma(\nu-j)} + b^{\nu-1}_{\gamma(\nu+j)}\widehat{b}^j_{\gamma(\nu+j)}) - p_0(b^{\nu-j-1}_{\gamma(\nu+j)} + b^{\nu-1}_{\gamma(\nu-j)}\widehat{b}^j_{\gamma(\nu-j)})$$

$$\iff b^{\nu-j-1}_{\gamma(\nu-j)}(1 - \varphi^j_{\nu-j}) \ge b^{\nu-j-1}_{\gamma(\nu+j)}(1 - \varphi^j_{\nu+j}) \iff$$

$$\left(\frac{b_{\gamma(\nu-j)}}{b_{\gamma(\nu+j)}}\right)^{\nu-j-1} \ge \frac{1-\varphi^j_{\nu+j}}{1-\varphi^j_{\nu-j}}. \tag{17}$$

Suppose that $b_{\gamma(\nu-j)} < b_{\gamma(\nu+j)}$, then from (17) it follows that

$$\frac{1-\varphi_{\nu+j}^{j}}{1-\varphi_{\nu-j}^{j}} < 1 \iff \varphi_{\nu-j}^{j} < \varphi_{\nu+j}^{j} \iff \varphi_{\nu-j} < \varphi_{\nu+j}.$$

Contradiction with the chain (15). Hence,

$$b_{\gamma(\nu-j)} > b_{\gamma(\nu+j)}, \quad j = \overline{1, \nu-1}. \tag{18}$$

It follows, in particular, that $\gamma(\nu-j) \neq 1$, $j = \overline{1, \nu-1}$, which means $\gamma(n) = 1$. The optimal schedule is shown on Fig. 2.

We will prove that the optimal schedule is completely determined by the additional conditions (18). It is not difficult to see that from the chain $\gamma(n) < \gamma(n-1) < ... < \gamma(2\nu-1)$ it follows that $\gamma(2\nu-1) \leq n-2\nu+2$. Equality is achieved when $\gamma(k) = n-k+1$, $k = \overline{2\nu-1, n}$. On the other hand there are two sequences whose members are different: $n = \gamma(\nu) > \gamma(\nu+1) > ... > \gamma(2\nu-1)$, and $n = \gamma(\nu) > \gamma(\nu-1) > ... > \gamma(1) > \gamma(2\nu-1)$, so between the number $\gamma(\nu)$ and the number $\gamma(2\nu-1)$ are at least $(\nu-2)+(\nu-1) = 2\nu-3$ distinct natural numbers, and their $\gamma(\nu) - \gamma(2\nu-1) \leq n-(n-2\nu+2)-1 = 2\nu-3$.

The proof of the theorem is completed.

## References

1. Pirogova, O., Nuzhdin, R., Pivovar, B.: Business analysis of factors and conditions for the development of sugar production organizations. In: E3S Web of Conferences, vol. 244, p. 10056 (2021)
2. Rudoy, D., et al.: Mathematical modeling in the agro-industrial complex: basic problems and models construction. In: E3S Web of Conferences, vol. 381, p. 01082 (2023)
3. Jiao, Z., Higgins, A.J., Prestwidge, D.B.: An integrated statistical and optimisation approach to increasing sugar production within a mill region. Comput. Electron. Agric. **48**, 170–181 (2005)
4. Junqueira, R., Morabito, R.: Modeling and solving a sugarcane harvest front scheduling problem. Int. J. Prod. Econ. **231**(1), 150–160 (2019)
5. Balandin, D., Egamov, A., Kuzenkov, O., Pristavchenko, O., Vildanov, V.: Mathematical modelling and optimization of scheduling for processing beet in sugar production. In: Balandin, D., Barkalov, K., Meyerov, I. (eds.) MMST 2022. CCIS, vol. 1750, pp. 227–238. Springer, Cham (2022). https://doi.org/10.1007/978-3-031-24145-1_19
6. Balandin, D.V., Kuzenkov, O.A., Egamov, A.I.: Evaluation of the efficiency of sugar beet processing strategies based on the data of the Sergach sugar factory. In: IOP Conference Series: Earth and Environmental Science, vol. 1206, p. 012046 (2023)
7. Balandin, D.V., et al.: Mathematical model and combined quasi-optimal algorithm of sugar beet processing. Bull. Voronezh State Univ. Ser.: Syst. Anal. Inf. Technol. **2**, 62–76 (2023). [in Russian]
8. Balandin, D.V., Kuzenkov, O.A., Vildanov, V.K.: A software module for constructing an optimal schedule for processing raw materials. Mod. Inf. Technol. IT-Educ. **17**(2), 442-452 (2021). [in Russian]. https://doi.org/10.25559/SITITO.17.202102

9. Kukhar, V.N., et al.: Methods for assessing the technological properties of sugar beet using indicators of the content of potassium, sodium and $\alpha$-amine nitrogen determined in beetroot and its processing products. Sugar **1**, 18–36 (2019). [in Russian]
10. Zelepukin, Yu.I., Zelepukin, S.Yu.: Assessment of sugar beet quality. Sugar **11**, 31–35 (2021). [in Russian]. https://doi.org/10.24412/2413-5518-2021-11-31-35
11. Egamov, A.I.: Stability of the optimal schedule for perishable product processing. In: E3S Web of Conferences, vol. 395, p. 03007 (2023). https://doi.org/10.1051/e3sconf/202339503007
12. Rainer, B., Mauro, D., Silvano, M.: Assignment Problems, 382 p. Society for Industrial and Applied Mathematics, Philadelphia (2009)
13. Bunday, B.: Basic Linear Programming, 163 p. London (1984)
14. Johansson, R.: Numerical Python: Scientific Computing and Data Science Applications with Numpy, SciPy and Matplotlib, 2nd edn., 723 p. Apress (2018)
15. Egamov, A.I.: Estimation of average losses of quasi-optimal strategies without taking into account the ripening process. In: XXII International Scientific and Practical Conference "Actual problems of science and education in the conditions of modern challenges", 11 July 2023, pp. 40–47. Pressroom, Publishing House, Moscow (2023). [in Russian]
16. Agricultural dictionary-reference. Editor-in-chief: A.I. Gaister. Moscow Leningrad: State Publishing House of collective farm and state farm literature "Selkhozgiz", 1280 p. (1934). [in Russian]
17. Bronstein, I.N., Semendyaev, K.A.: Handbook of Mathematics for Engineers and Students of Higher Education Institutions, 13th edn., 544 p. Nauka, Moscow (1985). [in Russian]

# Artificial Intelligence and Supercomputer Simulation

# Isolation of ECG Sections Associated with Signs of Cardiovascular Diseases Using the Transformer Architecture

Alexey Petukhov, Denis Rodionov, Denis Karchkov(✉), Viktor Moskalenko, Alexander Nikolskiy, and Nikolai Zolotykh

Lobachevsky State University of Nizhny Novgorod, 603022 Nizhny Novgorod, Russia
karchkov.denis@mail.ru
http://www.unn.ru

**Abstract.** The problem of explainability of artificial intelligence is extremely relevant at the present time. The most acute need for this tool is felt in the areas of practical application of intelligent systems. One of these areas is medicine. Extracting knowledge and understanding the reason for the model's prediction will allow us to provide in a human-readable form the main factors used by the model to predict. The article will consider a method for highlighting areas of the electrocardiogram associated with signs of cardiovascular diseases using the transformer machine learning architecture. Visualization of the attention of the transformer to the ECG signal allows you to highlight the segments of the electrocardiogram, which may show signs of various cardiovascular diseases. For the diagnosis of inferior myocardial infarction, the algorithm for labeling the data extracted from the trained model identified Q and R waves, for anterior septal myocardial infarction, Q and T waves, as well as the ST segment, and for left ventricular hypertrophy, the ST segment. Areas of attraction of "attention" of the neural network are shown graphically and displayed in the article.

**Keywords:** analysis of the electrocardiogram signal · artificial intelligence in medicine · explainable artificial intelligence · transformer · transformer attention mechanism

## 1 Introduction

We live in an amazing era where computerization, machine learning and artificial intelligence are rapidly advancing. At the heart of machine learning is the ability to train a computer to recognize patterns or dependencies based on examples, rather than setting rigid programming rules [1,2].

An electrocardiogram (ECG) is a commonly used method for measuring the physiological activity of the heart and diagnosing various cardiovascular diseases, including arrhythmia, myocardial infarction, and others [3].

Supported by Lobachevsky State University of Nizhny Novgorod.


D. Balandin et al. (Eds.): MMST 2023, CCIS 1914, pp. 209–222, 2024.
https://doi.org/10.1007/978-3-031-52470-7_17

It is worth noting that in recent years, deep learning has achieved results in many areas, in particular in computer vision. Unlike classical methods, in deep learning there is no need for explicit feature extraction by the researcher, which is an advantage. Features are extracted automatically and implicitly using data mining [4,10,11].

In the process of optimizing the parameters of a neural network, artificial intelligence optimizes a mathematically defined objective function that determines the quality of solutions. However, artificial intelligence often derives general patterns and rules, but their internal representation may not be clear to humans. Therefore, there is a need for interpretability, that is, the representation of the neural network in a simpler and more understandable form, for example, dimensionality reduction. The goal of "explainable artificial intelligence" is to extract knowledge from a neural network so that people can use this knowledge to understand important patterns derived by artificial intelligence [5,6].

## 2 Solutions Overview

The growth of digital ECGs is significant, and there are promising studies that use traditional machine learning methods to detect cardiac abnormalities [12, 13]. However, deep neural networks (DNNs) remain the preferred methods for ECG analysis. To detect abnormalities, researchers use raw or preprocessed ECG signals [14,15] or convert them into images using wavelet or Fourier transform [16,17]. Several reviews cover deep learning approaches for ECG classification [18,19].

Most studies use partial and end-to-end deep learning approaches with convolutional layers and residual connections. Some of these methods have demonstrated cardiologist-level accuracy for certain cardiac disorders. One of the key challenges in medical industry research is the limited amount of data available, often due to privacy and security concerns. As a result, only a few works with large datasets are used for ECG analysis [20,21]. Additionally, extracting labels from text-based diagnostic reports may introduce additional errors [22,23]. Typically, studies analyzing 12-lead ECGs use open datasets such as PTB-XL [24]. It consists of thousands of fixed-parameter ECGs, preprocessed and annotated by a small group of cardiologists who may make similar annotation errors. The small number of samples in public datasets limits the number of violations that neural networks can effectively classify. Thus, deep neural networks trained on these datasets for common disorders may not perform as well on other datasets.

Various studies show that the same deep neural networks trained on different datasets produce significantly different results on their respective test sets, even for the most common heart diseases such as atrial fibrillation [25–27]. This difference in predictive performance across different ECG datasets reduces the robustness of the networks and hinders their widespread clinical application.

Detection of myocardial infarction in the PTB-XL database is a problem that has been studied in detail before. Researcher Rajendra Acharya [28] was one of the first to develop a convolutional neural network for automatic detection of myocardial infarction. This was done on noisy and denoised electrocardiograms, without manual extraction of characteristic information. The achieved accuracy rate is 93.53%.

In the work "A New Pattern Recognition Method for Detection and Localization of Myocardial Infarction Using T-Wave Integral and Total Integral as Extracted Features from One Cycle of ECG Signal" by researchers Naser Safdarian, N.J. Dabanloo, and Gholamreza Attarodi [29] used two characteristics: T-wave integral and general integral to localize and detect myocardial infarction. Using these two characteristics and various statistical learning methods such as artificial neural networks, probabilistic neural networks, nearest neighbors, multilayer perceptrons and naive Bayes classifiers, they were able to achieve an accuracy of 94%.

Javad Kojuri in his work "Prediction of acute myocardial infarction with artificial neural networks in patients with non-diagnostic electrocardiogram" [30] developed two different types of artificial neural networks: with radial basis functions (RBF) with a classification accuracy of 83% and a multilayer perceptron (MLP) to detect myocardial infarction with 96% accuracy.

Li Sun in the study "ECG Analysis Using Multiple Instance Learning for Myocardial Infarction Detection" [31] states that supervised learning methods have had limited success and applies multiple instance learning (MIL) for automatic classification of electrocardiograms. They demonstrate that their proposed LTMIL algorithm outperforms previous supervised learning approaches in terms of classification performance. As a result of the study, an accuracy of 90% was achieved.

The work "A novel electrocardiogram parameterization algorithm and its application in myocardial infarction detection" [32] proposes a new characteristic of the electrocardiogram, which is obtained using the best 20-order polynomial approximation of a given ECG signal. Researcher Bin Liu, who developed a classification model using this feature, achieved an accuracy of 94.4% in detecting myocardial infarction.

A team of researchers led by Mohammad Kachuee in "ECG Heartbeat Classification: A Deep Transferable Representation" [33] used transfer learning from a model trained for arrhythmia classification to create a model capable of detecting myocardial infarction. For this purpose, a deep convolutional neural network model is used, which has shown a classification accuracy of 95.9% for detecting myocardial infarction.

## 3 Problem Statement

Consider the problem of detecting and predicting various cardiovascular diseases based on ECG. The input data is the PTB-XL database [8], consisting of 21,799 ECGs in 12 leads, lasting 10 s with a sampling frequency of 500 Hz. Thus, each data example is a two-dimensional array of $12 \times 5000$ numbers - ECG signal intensities for some lead at some point in time. Table 1 presents the cardiovascular diseases we are considering.

**Table 1.** Description of input data.

| Disease code | Type of disease | Number of examples |
|---|---|---|
| IMI | Inferior myocardial infarction | 2676 |
| ASMI | Anterior septal myocardial infarction | 2357 |
| LVH | Left ventricular hypertrophy | 2132 |
| PVC | Premature contraction of the ventricles | 1143 |

For each disease, it is necessary to set the task of binary classification. Figure 1 shows an example of a 12-lead electrocardiogram with a duration of 10 s.

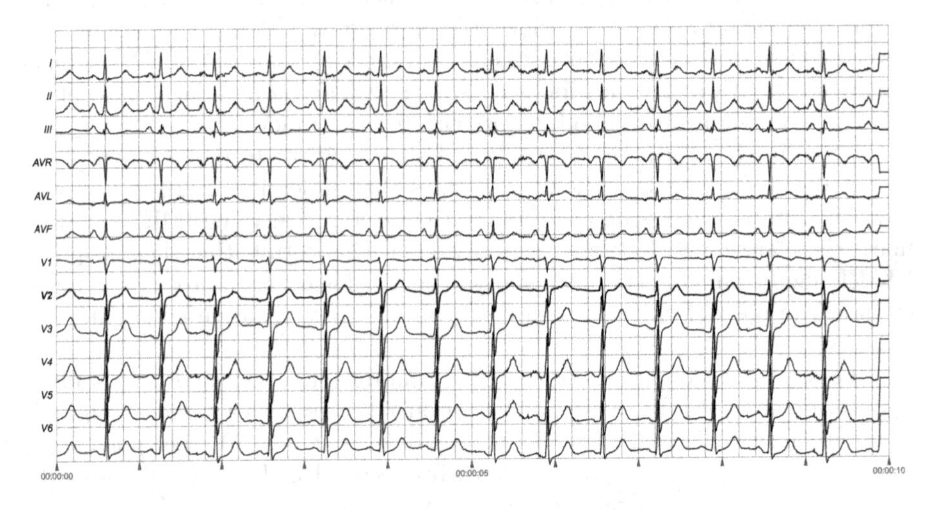


**Fig. 1.** 12 lead ECG example.

Commonly important ECG regions associated with signs of cardiovascular disease include the QRS, ST, and T complexes, as well as other characteristics that may indicate abnormalities. To isolate these areas, various signal analysis methods, including machine learning methods, can be used. Further, using the trained machine learning model, we will highlight the parts of the original ECG signal that were the most important for the model for diagnosing each of the above cardiovascular diseases. Thus, we will extract knowledge from the neural network and provide it in a human-readable format. Such markup of the ECG signal will help people understand in which particular areas of the electrocardiogram signs of each of the selected diagnoses are observed.

### 3.1 Method for Solving the Problem

To solve the problem of binary classification of each disease, we will use such a machine learning model architecture as a transformer. The architecture was first introduced in the work of researcher Vaswani Ashish "Attention is all you need" [9] and has become widely applicable in various machine learning tasks.

The purpose of knowledge extraction, as described in [7], is to provide in a human-readable form the reasons for the model's prediction. For the diagnosis of cardiovascular diseases, such reasons are the areas of the original ECG signal, on which a sign of the diagnosis is observed. To extract knowledge from the trained model and provide it in a human-readable form, a method will be used that uses the attention of the transformer to the ECG signal.

The internal attention mechanism used by the transformer allows this model to weight different parts of the original ECG signal. Removing the attention values of an additional artificial so-called token responsible for classification to non-overlapping sections of the original ECG signal allows obtaining weights indicating the importance of these sections for the disease under study.

Each $12 \times 5000$ input data sample is divided into non-overlapping segments of 10 points in length along the time coordinate. Such segments from all 12 leads are collected in the so-called token of 120 numbers. Thus, the original ECG signal is converted into 500 tokens, each of which has a size of 120 numbers. In addition, an artificial classification token is added. After 3 layers of the encoder, only the classification token is passed to the fully connected layer responsible for the final prediction of the model. In the encoder layers, the number of heads equal to the number of leads is used in order to visualize the attention of the transformer to each lead separately in the knowledge extraction method. Figure 2 shows the processing of the initial data and the architecture of the model used.

### 3.2 Model Quality Metrics

The confusion matrix for binary classification consists of four elements: true positive, false negative, false positive and true negative. These elements represent the number of correctly and incorrectly classified examples of each class.

- True Positive (TP): Number of examples correctly classified as positive;
- False Negative (FN): Number of examples incorrectly classified as negative;
- False Positive (FP): Number of examples incorrectly classified as positive;
- True Negative (TN): Number of examples correctly classified as negative.

The confusion matrix allows you to evaluate the performance of the classifier and calculate various metrics, such as precision, recall.

$$precision = \frac{TP}{TP + FP} \tag{1}$$

$$recall = \frac{TP}{TP + FN} \tag{2}$$

The quality of the binary classification will be assessed using the F1 score, which is a measure that combines precision and recall of the model [34]. F1-measure is calculated using the following formula:

$$F1 = 2 * \frac{precision * recall}{precision + recall} \quad (3)$$

The F1-measure is a good measure when both precision and recall are important. A high F1-score indicates a good balance between precision and recall of the classification model. The Accuracy metric for binary classification is calculated using the following metrics:

$$Accuracy = 2 * \frac{TP + TN}{TP + TN + FP + FN} \quad (4)$$

Accuracy measures what proportion of all model predictions are correct and can be interpreted as the overall performance of the classifier.

# 4 Experimental Results

The essence of the practical experiments is to test the capabilities of the "Transformer" architecture to determine the following diagnoses: premature ventricular contraction (PVC), inferior myocardial infarction (IMI), anteroseptal myocardial infarction (ASMI) and left ventricular hypertrophy (LVH). At the same time, the use of a specialized token defined by the solution architecture will allow us to obtain information about the signal sections that attracted the attention of the model.

## 4.1 Premature Contraction of the Ventricles

Premature ventricular contraction (PVC), also known as ventricular extrasystoles, is a type of arrhythmia in which additional electrical impulses occur in the heart's ventricles, causing additional ventricular contractions. This behavior of the heart muscle can be easily tracked on an electrocardiogram without special knowledge.

To assess the quality of the trained model on the test sample, we use the confusion matrix (Table 2) and the F1-score.

**Table 2.** Confusion matrix for localization of premature ventricular contractionl.

| | | Predicted class | |
|---|---|---|---|
| | | Negative | Positive |
| True class | Negative | 4080 | 52 |
| | Positive | 53 | 176 |

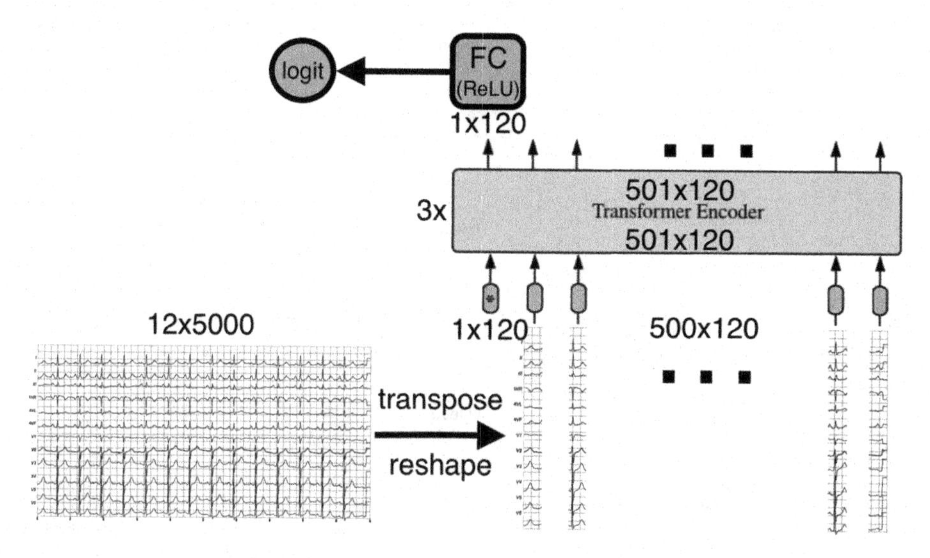


**Fig. 2.** Modified architecture Transformer to control the attention of the model.

When choosing the final class as the boundary for the model prediction values, such a boundary was chosen that the accuracy would be equal to the recall. Thus, the quality of the trained model corresponds to the value of 0.77 F1-Score. An accuracy of 98% was achieved.

Let us consider the results of applying the knowledge extraction method in signal processing with a diagnosis of premature ventricular contraction (Fig. 3). It can be seen that the red dots, indicating the attention of the model, are concentrated in areas of abnormal contraction following the completed cardiocycle. In particular, the areas after the peak of ventricular contraction (R wave) and the beginning of myocardial recovery (beginning of T wave) attracted the attention of the network.

## 4.2 Inferior Myocardial Infarction

Inferior myocardial infarction is a condition in which the lower part of the heart muscle, called the myocardium, suffers damage or death due to a lack of blood supply. It is caused by a disease of the coronary arteries, which supply oxygen-rich blood to the heart. This disease is life-threatening and requires immediate medical attention.

In this case, the F1-Score was equal to 0.65. An accuracy of 91% was obtained for this diagnosis. The confusion matrix is presented in Table 3.

Let us present the results of applying the knowledge extraction method that uses the attention of the transformer to the ECG signal in Fig. 4. The initial ECG signal in the selected leads is shown in blue, the attention of the model is shown in red, the bright red dots highlight the ECG areas associated with the signs of a given cardiovascular disease for a given ECG signal.

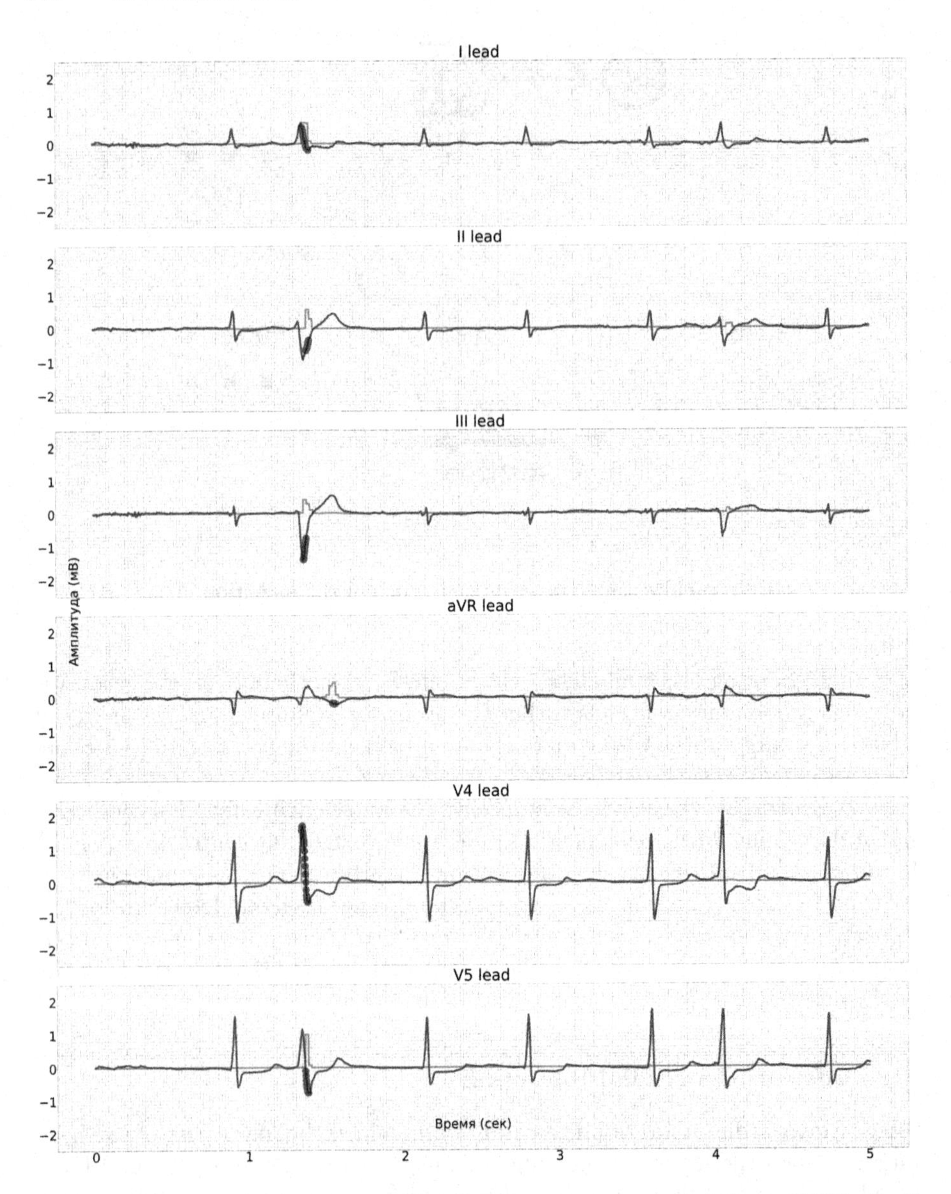


**Fig. 3.** The result of applying the knowledge extraction method for an ECG signal with a diagnosis of ventricular extrasystole.

**Table 3.** Confusion matrix for the problem of localization of inferior myocardial infarction on the ECG signal.

| | | Predicted class | |
|---|---|---|---|
| | | Negative | Positive |
| True class | Negative | 3639 | 186 |
| | Positive | 187 | 349 |

The results of applying the method of extracting knowledge from the trained model allow us to conclude that for lower myocardial infarction, Q and R waves are important parts of the ECG signal, where a symptom of the diagnosis is observed.

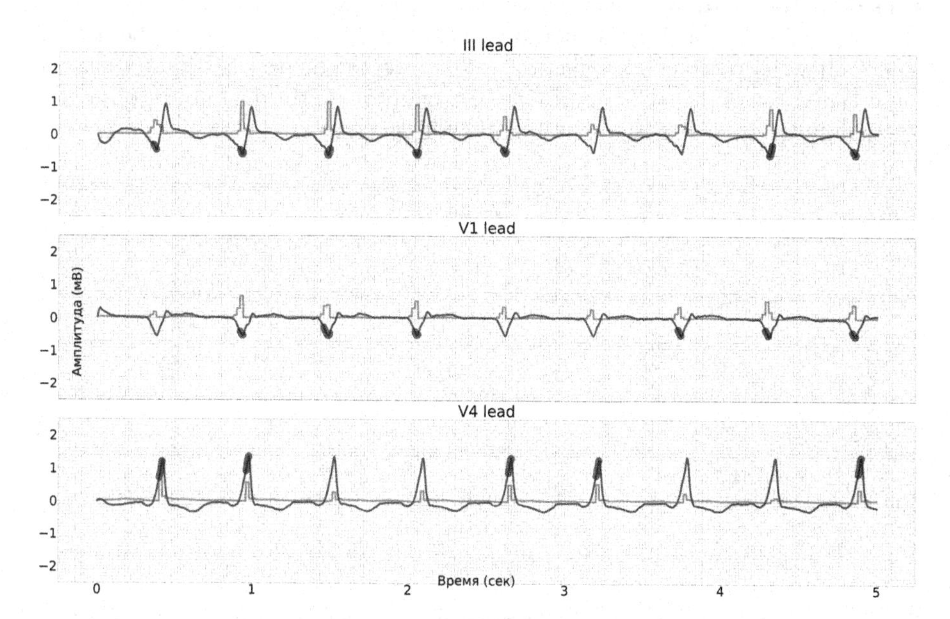


**Fig. 4.** The result of applying the knowledge extraction method for an ECG signal with a diagnosis of inferior myocardial infarction.

### 4.3 Anterior Septal Myocardial Infarction

Anterior septal myocardial infarction (ASMI) is a form of myocardial infarction characterized by damage to the anterior segment of the heart. In ASMI, obstruction in the heart's blood supply occurs in the anterior interventricular space (between the right and left ventricles of the heart). With a complete blockage of the blood supply to this part of the heart, necrosis of the cells of the heart muscle occurs, including due to a lack of oxygen.

Thus, the quality of the trained model corresponds to the value of 0.76 F1 measure, accuracy 95%. The confusion matrix is presented in Table 4.

**Table 4.** Confusion matrix of prediction model of anterior septal myocardial infarction.

| | | Predicted class | |
|---|---|---|---|
| | | Negative | Positive |
| True class | Negative | 3777 | 112 |
| | Positive | 113 | 359 |

The results of applying the knowledge extraction method using the attention of the transformer to the ECG signal are shown in Fig. 5. The blue color shows the original ECG signal in the selected leads, the red color shows the attention of the model, the bright red dots highlight the ECG areas associated with the signs of a given cardiovascular disease for a given ECG signal.

The results of applying the methods of extracting knowledge from the trained model allow us to conclude that for anterior septal myocardial infarction, the important areas of the ECG signal, where a symptom of the diagnosis is observed, are mainly Q and T waves, as well as the ST segment.

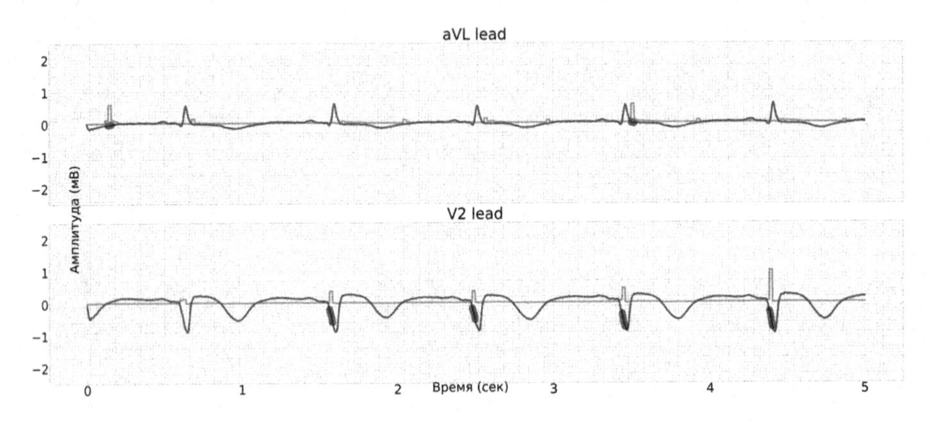


**Fig. 5.** The result of applying the knowledge extraction method on records with a diagnosis of anterior septal myocardial infarction.

## 4.4 Left Ventricular Hypertrophy

Left ventricular hypertrophy (LVH) is a condition in which the musculature of the left ventricle of the heart becomes thicker and stronger than normal. This is often the body's response to increased stress on the heart, such as high blood pressure or heart failure.

The main causes of LVH are hypertension (arterial hypertension), heart valve disease, hypertrophic cardiomyopathy, and other heart conditions. These conditions cause an increase in the load on the heart, which leads to thickening of the walls of the left ventricle.

For the task of predicting left ventricular hypertrophy, the value of the F1-Score was 0.68, accuracy 94%. The confusion matrix is in Table 5.

The results of applying the knowledge extraction method using the attention of the transformer to the ECG signal are shown in Fig. 6. The initial ECG signal in the selected leads is shown in blue, the attention of the model is shown in red, the bright red dots highlight the ECG areas associated with the signs of a given cardiovascular disease for a given ECG signal.

**Table 5.** Confusion matrix of prediction model of left ventricular hypertrophy.

| | | Predicted class | |
|---|---|---|---|
| | | Negative | Positive |
| True class | Negative | 3799 | 135 |
| | Positive | 136 | 291 |

The results of applying the method of extracting knowledge from the trained model allow us to conclude that for left ventricular hypertrophy, R waves are important parts of the ECG signal, where a symptom of the diagnosis is observed.

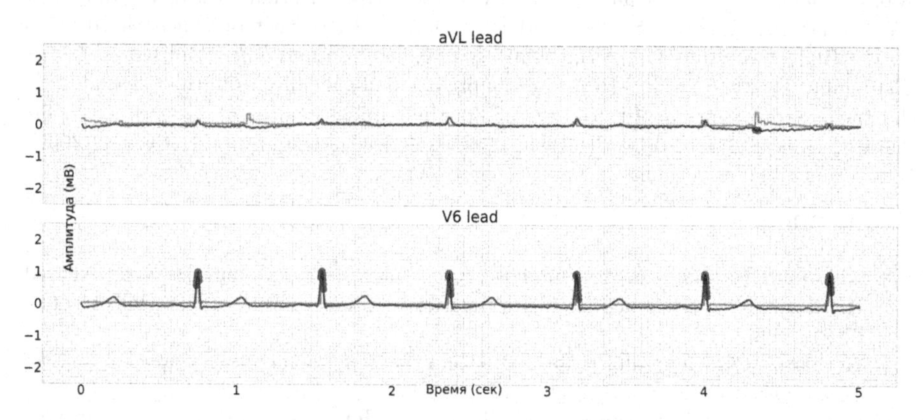


**Fig. 6.** The result of applying the knowledge extraction method for an ECG signal with a diagnosis of left ventricular hypertrophy.

## 5 Conclusion

In the course of the study, such an architecture as a transformer was chosen as the architecture of the machine learning model that solves the problem of diagnosing selected cardiovascular diseases in the form of a binary classification problem for each diagnosis independently. The number of trainable model parameters is 377881. For the studied diagnoses: premature ventricular contraction (PVC), inferior myocardial infarction (IMI), anterior septal myocardial infarction (ASMI), left ventricular hypertrophy (LVH) - F1-score was 0.77, 0.65, 0.76 and 0.68, respectively. Moreover, the accuracy of determining these diagnoses is 91%, 95%, 94% and 98%, respectively. The article presents the architecture of the model used, as well as error matrices, learning curves in the form of figures for each disease under study.

A method for extracting knowledge from a trained machine learning model is implemented, using the attention of the transformer to the ECG signal. The results of applying this method for each of the studied diagnoses were presented.

The obtained results are consistent with the segments of the cardiogram accepted by specialists, which may show signs of the studied cardiovascular diseases. Thus, the implemented methods of knowledge extraction will allow us to recognize important areas of the ECG for other, less studied cardiovascular diseases.

The results of applying knowledge extraction methods from the trained model can be used to study such cardiovascular diseases. Further work will be devoted to the study of methods not only for the selection of areas, but also for extracting from the model an explanation of specific problems in these selected areas of the ECG.

**Acknowledgements.** The work was supported by project 0729-2021-013, which is carried out within the framework of the State task for the performance of research work by laboratories that have passed the competitive selection within the framework of the national project "Science and Universities", in respect of which the decision of the Budget Commission of the Ministry of Education and Science of Russia (dated September 14, 2021 No. BC-P/23) on the provision of subsidies from the federal budget for the financial support of the state task for the implementation of research work. The work was also supported by the federal academic leadership program "Priority 2030".

## References

1. Abiodun, O.I., Jantan, A., Omolara, A.E., Dada, K.V., Mohamed, N.A., Arshad, H.: State-of-the-art in artificial neural network applications: a survey. Heliyon **4**(11) (2018)
2. Gorban', A.N.: Generalized approximation theorem and computational capabilities of neural networks. Sibirskii zhurnal vychislitel'noi matematiki **1**(1), 11–24 (1998)
3. Center for Interventional Arrhythmology Homepage. http://www.aritm.ru/diagnoz/ekg.htm. Accessed 21 Aug 2023
4. Ronneberger, O., Fischer, P., Brox, T.: U-net: convolutional networks for biomedical image segmentation. In: Navab, N., Hornegger, J., Wells, W.M., Frangi, A.F. (eds.) MICCAI 2015, Part III. LNCS, vol. 9351, pp. 234–241. Springer, Cham (2015). https://doi.org/10.1007/978-3-319-24574-4_28
5. Bhatt, U., Weller, A., Moura, J. M.: Evaluating and aggregating feature-based model explanations (2020). https://doi.org/10.48550/arXiv.2005.00631
6. Zhang, Y., Tiňo, P., Leonardis, A., Tang, K.: A survey on neural network interpretability. IEEE Trans. Emerg. Top. Comput. Intell. **5**(5), 726–742 (2021)
7. Yang, C., Liu, J., Shi, C.: Extract the knowledge of graph neural networks and go beyond it: an effective knowledge distillation framework. In: Proceedings of the Web Conference, pp. 1227–1237 (2021)
8. Wagner, P., et al.: PTB-XL, a large publicly available electrocardiography dataset. Sci. Data **7**(1), 154–155 (2020)
9. Vaswani, A., et al.: Attention is all you need. In: Advances in Neural Information Processing Systems, vol. 30 (2017)
10. Rodionov, D., Karchkov, D., Moskalenko, V., Nikolsky, A., Osipov, G., Zolotykh, N.: Possibility of using various architectures of convolutional neural networks in the problem of determining the type of rhythm. In: Kryzhanovsky, B., Dunin-Barkowski, W., Redko, V., Tiumentsev, Y. (eds.) NEUROINFORMATICS 2022. SCI, vol. 1064, pp. 362–370. Springer, Cham (2022). https://doi.org/10.1007/978-3-031-19032-2_38

11. Rodionov, D.M., Karchkov, D.A., Moskalenko, V.A., Nikolsky, A.V., Osipov, G.V., Zolotykh, N.Y.: Diagnosis of sinus rhythm and atrial fibrillation using artificial intelligence. Probl. Inform. **1** (2022). https://doi.org/10.24412/2073-0667-2022-1-77-88
12. Alickovic, E., Subasi, A.: Medical decision support system for diagnosis of heart arrhythmia using DWT and random forests classifier. J. Med. Syst. **40**(4), 108–119 (2016)
13. Dohare, A.K., Kumar, V., Kumar, R.: Detection of myocardial infarction in 12 lead ECG using support vector machine. Appl. Soft Comput. **64**, 138–147 (2018)
14. Śmigiel, S., Pałczyński, K., Ledziński, D.: ECG signal classification using deep learning techniques based on the PTB-XL dataset. Entropy **23**(9), 1121–1122 (2021)
15. Baek, Y.S., Lee, S.C., Choi, W., Kim, D.H.: A new deep learning algorithm of 12-lead electrocardiogram for identifying atrial fibrillation during sinus rhythm. Sci. Rep. **11**(1), 12818–12829 (2021)
16. Huang, J., Chen, B., Yao, B., He, W.: ECG arrhythmia classification using STFT-based spectrogram and convolutional neural network. IEEE Access **7**, 92871–92880 (2019)
17. Wang, T., Lu, C., Sun, Y., Yang, M., Liu, C., Ou, C.: Automatic ECG classification using continuous wavelet transform and convolutional neural network. Entropy **23**(1), 119–129 (2021)
18. Hong, S., Zhou, Y., Shang, J., Xiao, C., Sun, J.: Opportunities and challenges of deep learning methods for electrocardiogram data: a systematic review. Comput. Biol. Med. **122**, 103801–103812 (2020)
19. Murat, F., Yildirim, O., Talo, M., Baloglu, U.B., Demir, Y., Acharya, U.R.: Application of deep learning techniques for heartbeats detection using ECG signals-analysis and review. Comput. Biol. Med. **120**, 103726–103737 (2020)
20. Hannun, A.Y., et al.: Cardiologist-level arrhythmia detection and classification in ambulatory electrocardiograms using a deep neural network. Nat. Med. **25**(1), 65–69 (2019)
21. Ribeiro, A.H., et al.: Automatic diagnosis of the 12-lead ECG using a deep neural network. Nat. Commun. **11**(1), 1760–1770 (2020)
22. Weimann, K., Conrad, T.O.: Transfer learning for ECG classification. Sci. Rep. **11**(1), 5251–5261 (2021)
23. Wang, Z., Shah, A.D., Tate, A.R., Denaxas, S., Shawe-Taylor, J., Hemingway, H.: Extracting diagnoses and investigation results from unstructured text in electronic health records by semi-supervised machine learning. PLoS ONE **7**(1), 30412–30422 (2012)
24. Wagner, P., Strodthoff, N., Bousseljot, R., Samek, W., Schaeffter, T.: PTB-XL, a large publicly available electrocardiography dataset (version 1.0.3). PhysioNet (2022). https://doi.org/10.13026/kfzx-aw45
25. Jo, Y.Y., et al.: Explainable artificial intelligence to detect atrial fibrillation using electrocardiogram. Int. J. Cardiol. **328**, 104–110 (2021)
26. Smisek, R., Nemcova, A., Marsanova, L., Smital, L., Vitek, M., Kozumplik, J.: Cardiac Pathologies detection and classification in 12-lead ECG. In: Computing in Cardiology, pp. 1–4. IEEE (2020)
27. Nonaka, N., Seita, J.: In-depth benchmarking of deep neural network architectures for ECG diagnosis. In: Machine Learning for Healthcare Conference, pp. 414–439 (2021)

28. Acharya, U.R., Fujita, H., Oh, S.L., Hagiwara, Y., Tan, J.H., Adam, M.: Application of deep convolutional neural network for automated detection of myocardial infarction using ECG signals. Inf. Sci. **415**, 190–198 (2017)
29. Safdarian, N., Dabanloo, N.J., Attarodi, G.: A new pattern recognition method for detection and localization of myocardial infarction using T-wave integral and total integral as extracted features from one cycle of ECG signal. J. Biomed. Sci. Eng. **7**(10), 818–852 (2014)
30. Kojuri, J., Boostani, R., Dehghani, P., Nowroozipour, F., Saki, N.: Prediction of acute myocardial infarction with artificial neural networks in patients with nondiagnostic electrocardiogram. J. Cardiovasc. Dis. Res. **6**(2), 51–69 (2015)
31. Sun, L., Lu, Y., Yang, K., Li, S.: ECG analysis using multiple instance learning for myocardial infarction detection. IEEE Trans. Biomed. Eng. **59**(12), 3348–3356 (2012)
32. Liu, B., et al.: A novel electrocardiogram parameterization algorithm and its application in myocardial infarction detection. Comput. Biol. Med. **61**, 178–184 (2015)
33. Kachuee, M., Fazeli, S., Sarrafzadeh, M.: ECG heartbeat classification: a deep transferable representation. In: International Conference on Healthcare Informatics, pp. 443–444. IEEE, ICHI (2018). https://doi.org/10.1109/ICHI.2018.00092
34. Chicco, D., Jurman, G.: The advantages of the Matthews correlation coefficient (MCC) over F1 score and accuracy in binary classification evaluation. BMC Genom. **21**(1), 1–13 (2020)

# Synchronous Activity in Small Ensembles of Inhibitory Coupled Phi-Neurons

Alexander Korotkov, Artyom Emelin, Tatiana Levanova(✉), and Grigory Osipov

Lobachevsky University, Nizhny Novgorod 603950, Russia
tatiana.levanova@itmm.unn.ru

**Abstract.** We study the impact of non-identity of elements on patterns of neuron-like activity in a recently proposed simple phenomenological model of the half-center oscillator (HCO). The model consists of two non-identical oscillatory phi-neurons, which are phase equations, interacting via the inhibitory coupling. We investigate different types of synchronized dynamics, such as oscillation death, 0:1 and m:n synchronization in the context of central pattern generators modeling. In our study we focus on bifurcation transitions between these temporal patterns. The obtained results were also generalized to the case of three elements in the ensemble.

**Keywords:** Adler equation · chemical synaptic coupling · half-center oscillator · synchronized activity · bifurcations

## 1 Introduction

One of the important problems in the studies of motor control in neuroscience is the origin of coordinated rhythmic activity in organisms: chewing, respiration, walking, swimming and so on. These types of dynamics can be produced by so called central pattern generators (CPGs), which are groups of neurons that can produce rhythmic patterns even in absence of motor and sensory feedback [1–5].

CPGs' architecture is based on small building blocks called half-center oscillators (HCO) [6–8], which include two units coupled via reciprocal synaptic inhibition. Here each unit may be an individual neuron, as well as a population of neurons in more complex case. During the oscillation, each component switches back and forth between sustained low activity states and high activity states [9]. To the present moment, a great amount of experimental data was obtained that support the half-center hypothesis [10–14].

Despite the fact that in many cases a real circuitry involved in a particular CPG is far from being known, studies that involve mathematical models are essential for obtaining meaningful hypotheses related to some operational principles of biological CPGs [15–18]. A deep study of simplified models arises as a natural first step for a wide range of valuable applications, from medicine to robotics [19–21].


D. Balandin et al. (Eds.): MMST 2023, CCIS 1914, pp. 223–239, 2024.
https://doi.org/10.1007/978-3-031-52470-7_18

Recently in [22] a new simple phenomenological model of the half-center oscillator was proposed. It consists of two identical phi-neurons coupled by mutual inhibitory synapses. Each phi-neuron is a phase oscillators, which in absence of couplings demonstrates regular oscillatory dynamics. This minimal model allows one to reproduce temporal patterns typical for HCO in the simplest manner as it was in early modeling of animal locomotor CPG [23,24].

In this work we go beyond above mentioned studies by investigating patterns of synchronized activity that emerge in the proposed model if elements are not identical. This approach allows us to make a proposed toy model a bit more biologically relevant. Thus, our aim is to study complex impact of inhibitory couplings and non-identity of elements on temporal patterns typical for CPG.

The paper is organized as follows. Section 2 is devoted to HCO which consists of two non-identical phase oscillators with inhibitory couplings. In Subsect. 2.1 we introduce the mathematical model, temporal patterns of synchronized activity depending on different values of control parameters in which are studied in Subsect. 2.2. Bifurcation transitions between these patters are studied in Subsect. 2.3. In Sect. 3 we generalize our results to the case of three coupled elements. Here Subsect. 3.1 is devoted to the model description, while in Subsect. 3.2 and Subsect. 3.3 the cases of symmetric and asymmetric couplings in the motif of three elements are studied in detail. In Sect. 4 we discuss our findings, before we draw our conclusions.

## 2 Two Coupled Non-identical Elements

### 2.1 The Model

Let us consider the model of an HCO consists of two non-identical phi-neurons coupled by mutual inhibitory synaptic couplings. The topology of HCO is presented in Fig. 1.

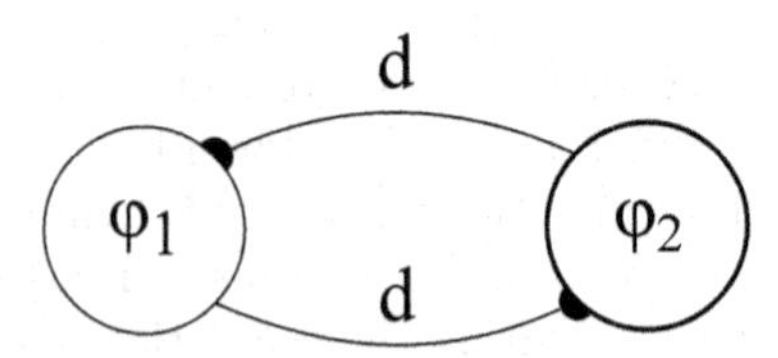


**Fig. 1.** Model of HCO consists of two oscillatory neurons with synaptic inhibitory couplings.

Mathematically this ensemble can be modelled as follows. A single phi-neuron can be described using the Adler equation [25]:

$$\dot{\phi} = \gamma - \sin \phi, \tag{1}$$

where the variable $\phi$ corresponds to the phase of an individual element. Parameter $\gamma$ determines the type of neuron behavior in the following way: $\gamma < 1$ corresponds to unexcited state of the neuron (in this case a stable and unstable equilibria on a unit circle exist), and $\gamma > 1$ – to spiking behavior, when the phase point begins to move counterclockwise along this circle (no equilibria case). The value $\gamma = 1$ is a bifurcation value, at which a saddle-node bifurcation occurs: stable and unstable equilibria merge into one. In this case a neuron still remains unexcited, but now it can generate a single response on the external stimulus.

The coupling between the elements is described by term $I(\phi)$:

$$I(\phi) = \frac{1}{1 + e^{k(\cos(\sigma) - \sin(\phi))}}. \tag{2}$$

This coupling function was first introduced in [26] and then tested in a number of subsequent theoretical studies [18,27,28]. In our study we use its inhibitory version as in [22]. Parameter $k$ is responsible for the switching speed in the coupling function (2). In the present study we set $k = -500$.

Thus, the HCO consist of two phi-neurons with mutual synaptic inhibitory couplings is described by the following system of ODEs:

$$\begin{cases} \dot{\phi}_1 = \gamma_1 - \sin\phi_1 - d \cdot I(\phi_2) \\ \dot{\phi}_2 = \gamma_2 - \sin\phi_2 - d \cdot I(\phi_1) \end{cases} \tag{3}$$

where parameter $d$ corresponds to the strength of inhibitory coupling $I(\phi)$. The phase space of (3) is torus $(\phi_1, \phi_2)$.

We introduce the difference between the elements in the ensemble using parameter $\Delta$. We set both elements generate spiking temporal pattern with different frequencies in the following way: $\gamma_1 = \gamma$ and $\gamma_2 = \gamma + \Delta$, where $\gamma = 1.01$. We choose parameters $d$ and $\sigma$ as control parameters in our study.

In this study we investigate how the non-identity of element affects the main patterns of neuron-like activity of HCO, obtained earlier in [22]: in-phase activity, anti-phase activity, and silent state. In order to do this, we conduct a two-parameter bifurcation analysis on the plane $P = (\sigma, d)$ for various fixed values of parameter $\Delta$. Recall that the parameter $\sigma$ is responsible for the duration of the inhibitory effect, and the parameter $d$ – for the strength of the influence of elements on each other.

It follows from the method of choosing the coupling function (2) that the parameter $\sigma$ can take values from 0 to $\frac{\pi}{2}$. Note that the values of the parameter of coupling strength $d$ for biological reasons should not be chosen too large, since in this case the simulation will not be biologically relevant. Also, the parameter $d$ obviously cannot take negative values. For the convenience of modeling, we assume that $d$ can vary in the range from 0 to 0.5.

### 2.2 Maps of Temporal Patterns

In order to study and classify temporal patterns of synchronous activity in the system under study, we built two-parameter maps of neuron-like activity on the

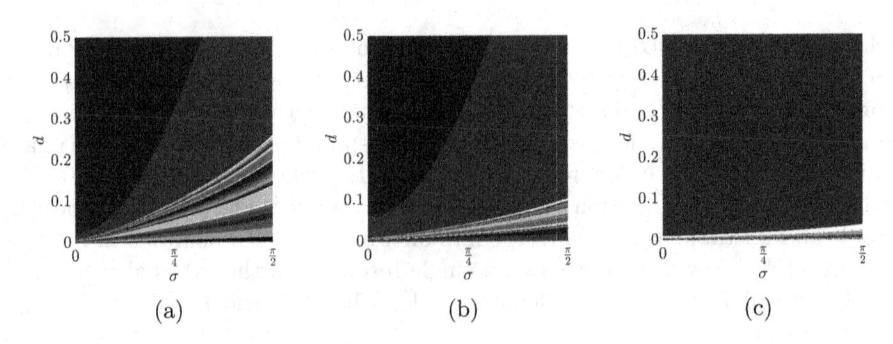


**Fig. 2.** Maps of temporal patterns in the system (3). Regions of different colors correspond to different temporal patterns. Here dark red area corresponds to the activity pattern "both elements are at silent state" (silent state, oscillation death), the red area corresponds to the activity pattern "the second element suppresses activity in the first element" (0:1 synchronization). All other colors correspond to different synchronization patterns. Colors are set randomly. (a) $\Delta = 0.01$, (b) $\Delta = 0.05$, (c) $\Delta = 0.5$. (Color figure online)

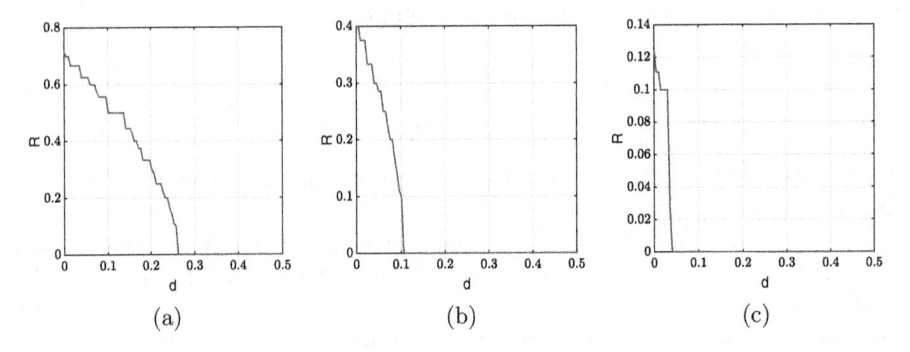


**Fig. 3.** The dependence of the rotation number $R$ on the coupling strength $d$ for a fixed value of $\sigma = \frac{\pi}{2}$. (a) $\Delta = 0.01$, (b) $\Delta = 0.05$, (c) $\Delta = 0.5$.

parameter plane $P = (\sigma; d)$, where $\sigma \in [0;\ \frac{\pi}{2}]$, $d \in [0;\ 0.5]$, see Fig. 2. The parameter plane $P$ was covered with a grid consisting of $500 \times 500$ nodes. Let us briefly describe the algorithm. At each node of $P$, the simulation of the trajectory of the system (3) was started. In the case of non-identical elements the system (3) does not possess multistability, so there is no need to use any inheritance scheme here. Initial conditions at all points of the grid were chosen to be the same. To eliminate the transient process, the system (3) was preliminarily integrated over the time interval from 0 to 500.

As one can see in Fig. 2, in all cases most of the map is covered with regions of different synchronizations. For small values of parameter $\Delta$ the most of the map is divided between the silent state (dark red color) and the 0:1 synchronization state (red color) (see Fig. 2(a)). With increase in the values of non-identity parameter $\Delta$ (Fig. 2(b)) the 0:1 synchronization region begins to expand. Finally,

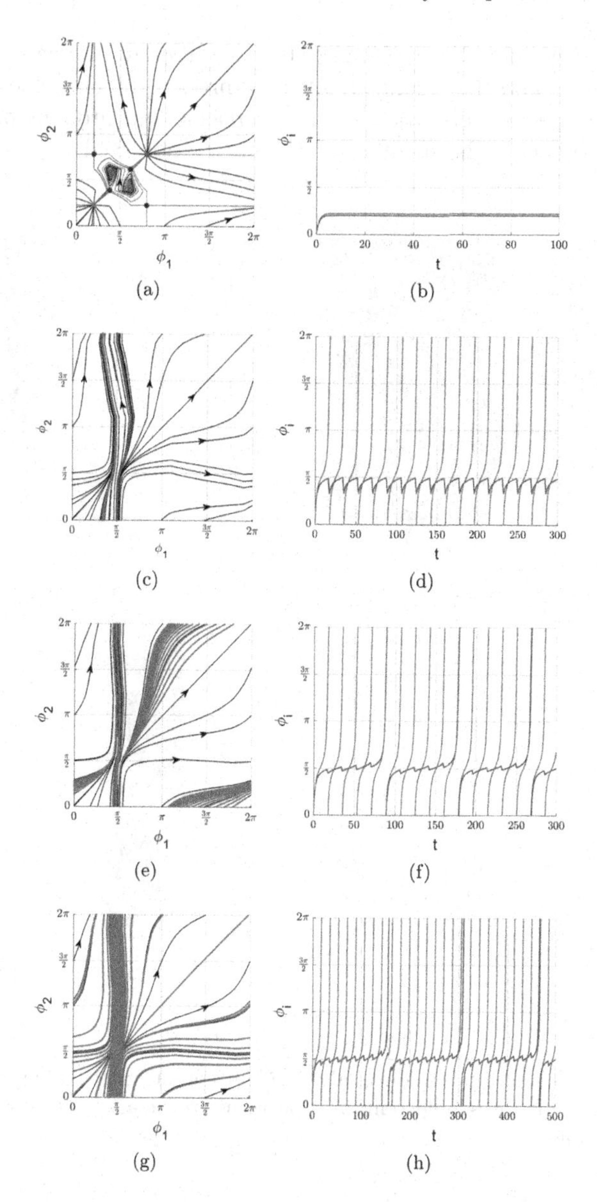

**Fig. 4.** Phase portraits (left panel) and time series (right panel) for different types of synchronization. (a), (b) Silent state (oscillation death), $\sigma = 0.3683$, $d = 0.4199$. (c), (d) The second element suppresses activity in the first element (0:1 synchronization), $\sigma = 1.885$, $d = 0.4199$. (e), (f) 5:1 synchronization, $\sigma = 2.2284$, $d = 0.1566$. (g), (h) Higher-order synchronization, $\sigma = 2.0865$, $d = 0.1638$. The red lines on time series mark the phase of the first element, the green lines – the phase of the second element. The green bold lines in subfigures (c), (e), (g) correspond to the stable limit cycle, the red bold line in subfigure (c) – to the unstable limit cycle. The black lines correspond to regular trajectories of the system. Parameter of non-identity: $\Delta = 0.05$. (Color figure online)

for quite sufficient values of $\Delta$, e.g. $\Delta$=0.05, the difference between the elements becomes too large and the second element suppresses the first element in a wide range of control parameters, with a narrow areas of synchronous patterns $m:n$ (Fig. 2(c)). In this case the silent state can not be observed.

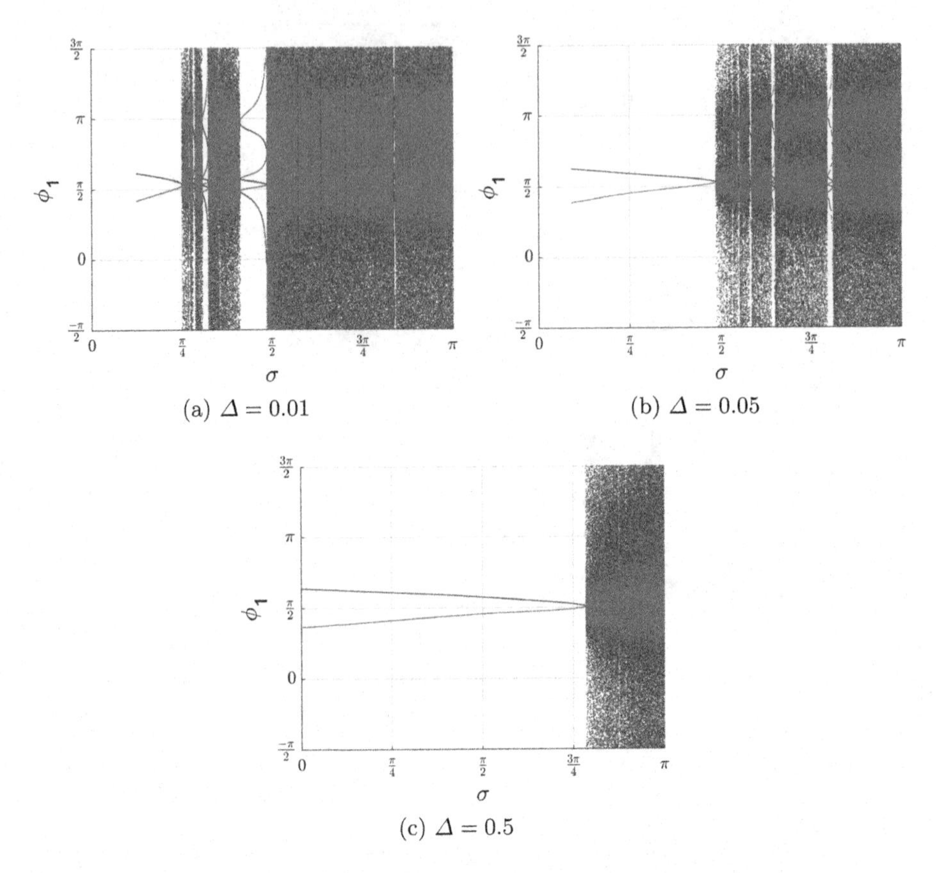


(a) $\Delta = 0.01$ (b) $\Delta = 0.05$ (c) $\Delta = 0.5$

**Fig. 5.** One-parameter bifurcation analysis. Coupling strength value $d = 0.1$. Green color corresponds to the stable temporal patterns, red color – to unstable patterns. (Color figure online)

To demonstrate the transitions between different patterns of synchronous neuron-like activity in more detail, for a fixed parameter $\sigma = \frac{\pi}{2}$ the dependence of the rotation number $R$ on the coupling strength $d$ was plotted (see Fig. 3). The rotation number was determined by the formula $R = \frac{\phi_1(T)-\phi_1(0)}{\phi_2(T)-\phi_2(0)}$, where $T = 10^5$.

Examples of patterns of neuron-like activity and their mathematical images in the phase space of the system are shown in Fig. 4. The presence of $m : n$ synchronization allow one to obtain various complex patterns of activity that are valuable for HCO modelling.

### 2.3 Bifurcation Transitions

Let us have a closer look at each of described regions in Fig. 2 and study the bifurcation transitions between observed types of temporal activity patterns. For this purpose we conduct one-parameter bifurcation analysis for fixed value of coupling strength $d = 0.1$ and three different cases for non-identity parameter $\Delta$ studied in the previous subsection: $\Delta = 0.01$, $\Delta = 0.05$, $\Delta = 0.5$.

Figure 5 presents the Poincare section with the plane $\phi_1 + \phi_2 = 3 * \pi/2$ and shows how attracting sets and temporal patterns corresponding to them change with the increase in the value of $\sigma$. As one can see, quasi-periodic regime in all cases is born as a result of saddle-node bifurcations involving stable and unstable limit cycles. On the left border in Fig. 5(a) and Fig. 5(b) the limit cycle is born as a result of a saddle-node bifurcation on an invariant curve.

Detailed illustration of the bifurcation transitions that occur with the increase in the value of parameter $\sigma$ for fixed $\Delta = 0.01$ one can find in Fig. 6, 7 and 8. For other fixed values of non-identity parameter $\Delta$ ($\Delta = 0.05$ and $\Delta = 0.5$) the bifurcation scenarios are qualitatively similar.

Note that bifurcation transitions include the appearance of one or several heteroclinic trajectories between saddle equilibria. For instance, in Fig. 6c a heteroclinic trajectory between saddle equilibria $S_1$ and $S_2$ exist; in Fig. 6e two heteroclinic trajectories exist: heteroclinic trajectory between saddle equilibria $S_1$ and $S_4$ and heteroclinic trajectory between $S_2$ and $S_3$.

Although the system under study consists of asymmetric non-identical elements, it possesses symmetry, due to which four equilibria are born at once as a result of the saddle-node bifurcation (see, e.g., Fig. 6b). Some of heteroclinic trajectories also appear pairwise (see, e.g., Fig. 6e, Fig. 7a, Fig. 7e). Heteroclinic trajectories, which do not have a symmetric trajectory, are symmetrical to themselves (see, e.g., Fig. 6c, Fig. 7c). The system (3) is invariant under the following replacement: $\phi_i \longrightarrow \pi - \phi_i$, $t \longrightarrow -t$. Thus, the phase portrait will have symmetry with time reversal about the point $(\frac{\pi}{2}, \frac{\pi}{2})$.

In Fig. 8(a) for $\sigma = 0.25$ in the phase space of the system under study two stable equilibria, two unstable equilibria, four saddles, as well as two invariant curves exist. The first invariant curve connects saddle equilibrium $S_1$ and stable node $N_1$. The second invariant curve connects saddle equilibrium $S_2$ and unstable node $N_2$. Later for $\sigma = 0.37$ these eight equilibria merge pairwise, as a result two limit cycles are born: stable (green line) and unstable (red line) ones (see Fig. 8(c)).

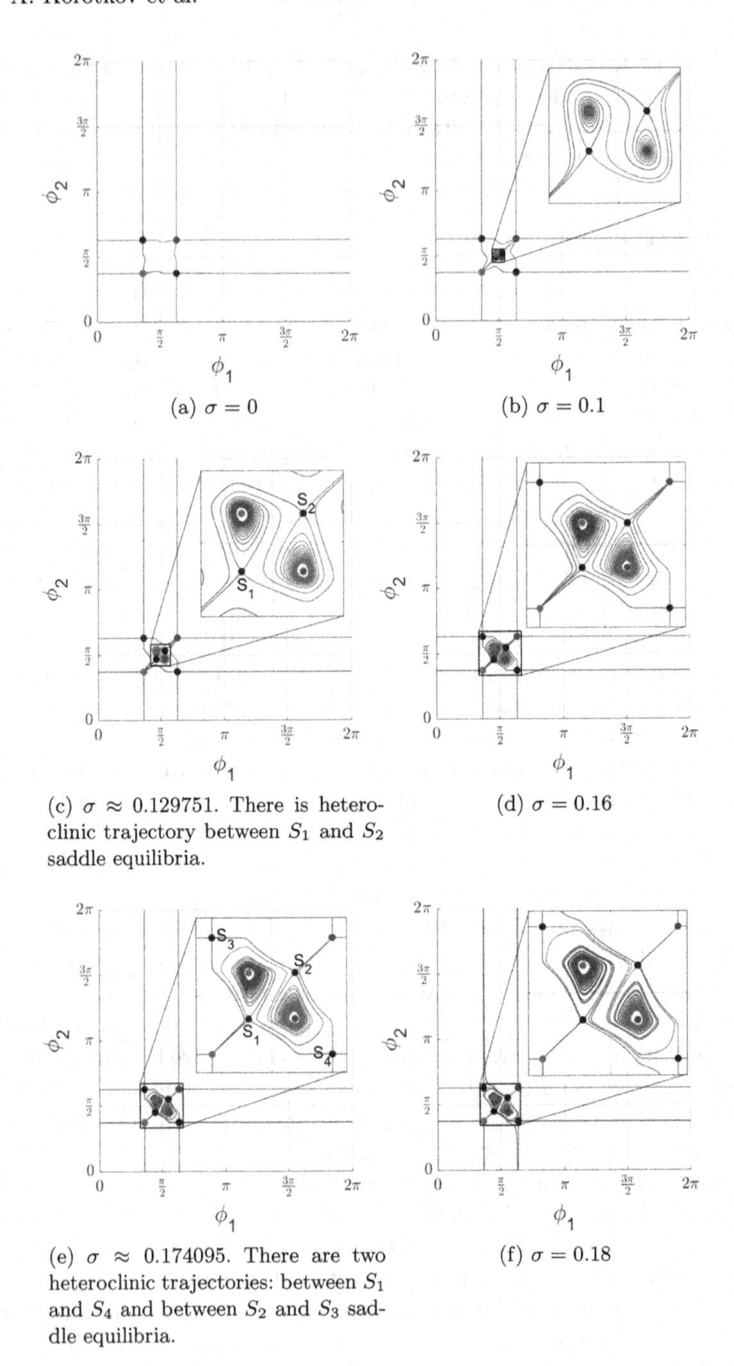


(a) $\sigma = 0$

(b) $\sigma = 0.1$

(c) $\sigma \approx 0.129751$. There is heteroclinic trajectory between $S_1$ and $S_2$ saddle equilibria.

(d) $\sigma = 0.16$

(e) $\sigma \approx 0.174095$. There are two heteroclinic trajectories: between $S_1$ and $S_4$ and between $S_2$ and $S_3$ saddle equilibria.

(f) $\sigma = 0.18$

**Fig. 6.** Phase portraits of system (3). Red dots correspond to unstable equilibria, green dots – to stable equilibria, blue dots – to saddle equilibria. Red lines correspond to unstable separatrices, green lines – to stable separatrices. Parameter values: $d = 0.1$, $\Delta = 0.01$. (Color figure online)

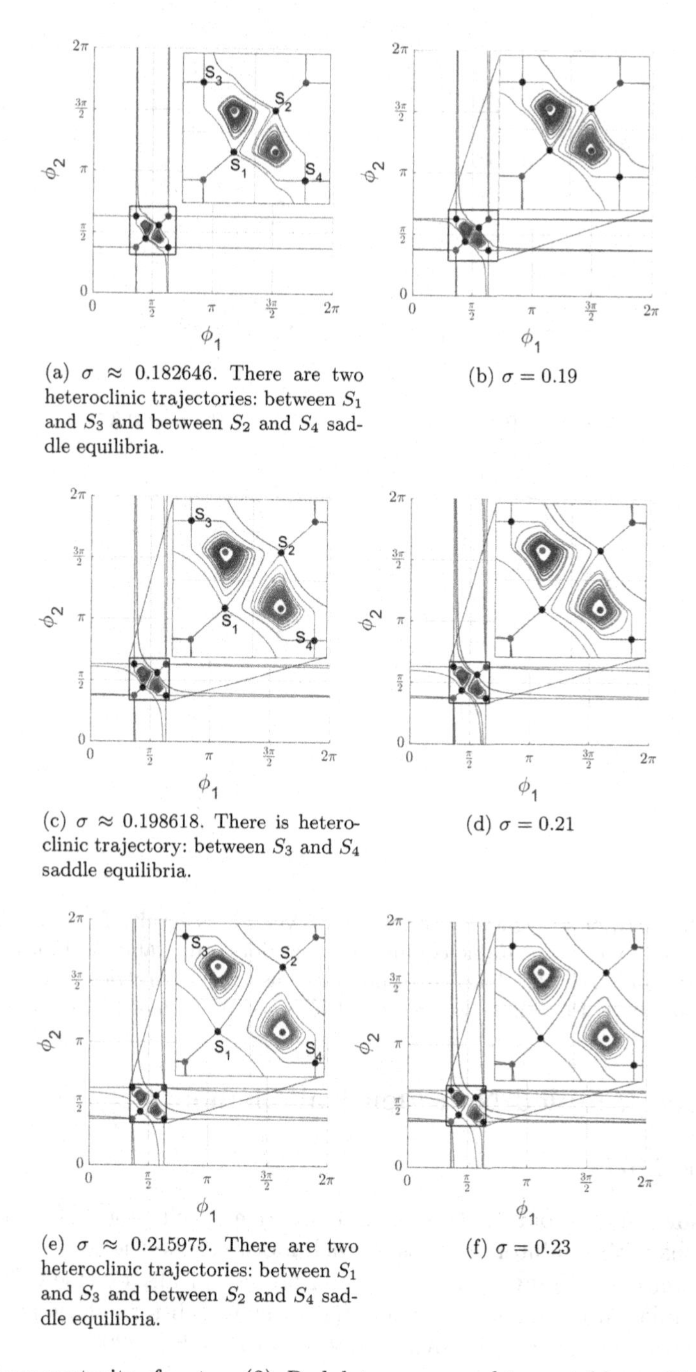


(a) $\sigma \approx 0.182646$. There are two heteroclinic trajectories: between $S_1$ and $S_3$ and between $S_2$ and $S_4$ saddle equilibria.

(b) $\sigma = 0.19$

(c) $\sigma \approx 0.198618$. There is heteroclinic trajectory: between $S_3$ and $S_4$ saddle equilibria.

(d) $\sigma = 0.21$

(e) $\sigma \approx 0.215975$. There are two heteroclinic trajectories: between $S_1$ and $S_3$ and between $S_2$ and $S_4$ saddle equilibria.

(f) $\sigma = 0.23$

**Fig. 7.** Phase portraits of system (3). Red dots correspond to unstable equilibria, green dots – to stable equilibria, blue dots – to saddle equilibria. Red lines correspond to unstable separatrices, green lines – to stable separatrices. Parameter values: $d = 0.1$, $\Delta = 0.01$. (Color figure online)

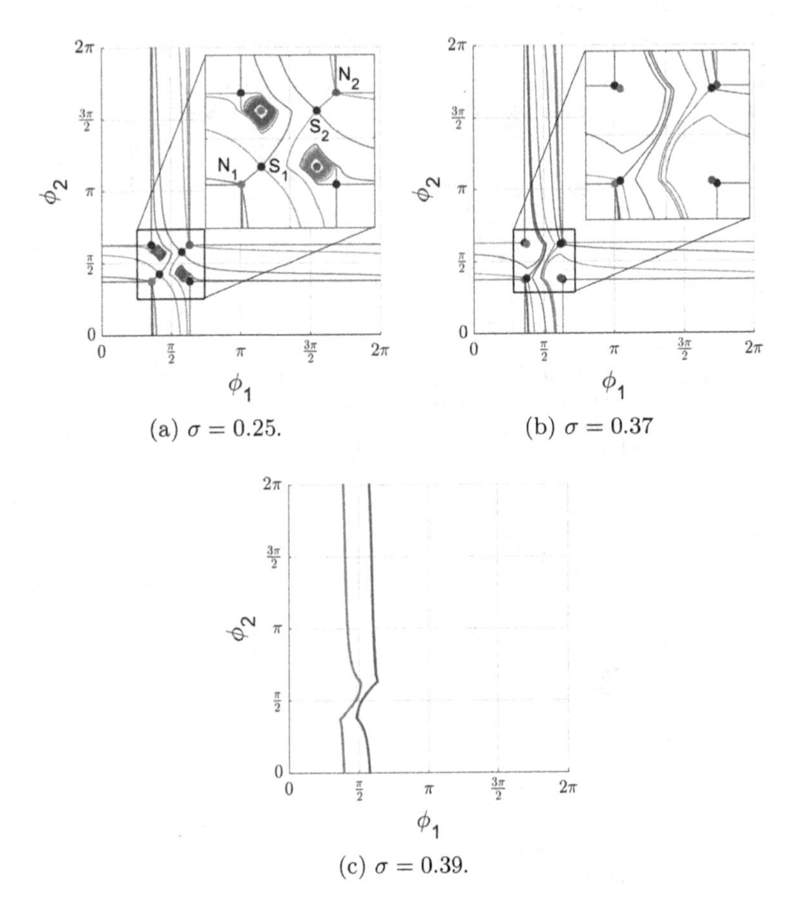


**Fig. 8.** The birth of stable and unstable limit cycles as a result of SNIC bifurcations. Red dots correspond to unstable equilibria, green dots – to stable equilibria, blue dots – to saddle equilibria. Red lines correspond to unstable separatrices, green lines – to stable separatrices. Parameter values: $d = 0.1$, $\Delta = 0.01$. (Color figure online)

## 3 Three Coupled Non-identical Elements

### 3.1 The Model

Let us now employ our HCO model to build a 3-cell motif of non-identical phi-neurons with inhibitory synapses. Such motif can be the part of more complex CPG circuits. Many anatomically and physiologically diverse CPG circuits involve similar three-cell motifs, including the spiny lobster pyloric network, the Tritonia swim circuit and the Lymnaea respiratory CPGs [29].

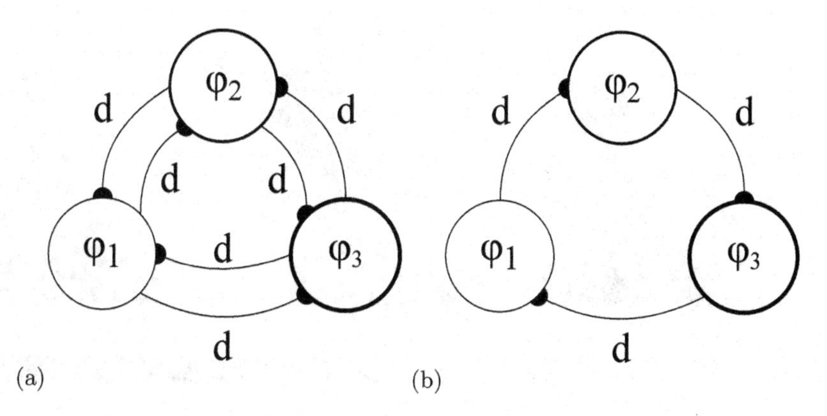


**Fig. 9.** Model of three oscillatory neurons with (a) mutual symmetric inhibitory couplings (b) asymmetric inhibitory couplings.

Here two main topologies of couplings are possible: symmetric and asymmetric (see Fig. 9). The the following subsections we study temporal patterns related to HCO for both described topologies.

### 3.2 Symmetrical Couplings

The governing equations for the motif of three phi-neurons coupled with symmetrical inhibitory couplings (see Fig. 9(a)) can be written as follows:

$$\begin{cases} \dot{\phi}_1 = \gamma_1 - \sin\phi_1 - d \cdot I(\phi_2) - d \cdot I(\phi_3) \\ \dot{\phi}_2 = \gamma_2 - \sin\phi_2 - d \cdot I(\phi_1) - d \cdot I(\phi_3) \\ \dot{\phi}_3 = \gamma_3 - \sin\phi_3 - d \cdot I(\phi_1) - d \cdot I(\phi_2) \end{cases} \tag{4}$$

The couplings between the elements are described by term $I(\phi)$ as in (2). The difference between the elements in the ensemble is introduced using parameter $\Delta$ in following way: $\gamma_1 = \gamma$, $\gamma_2 = \gamma + \Delta$, $\gamma_3 = \gamma + 2\Delta$, where $\gamma = 1.01$. We choose parameters $d$ and $\sigma$ as control parameters in our study. All other parameters of the system (4) are fixed as in the previous Sect. 2.

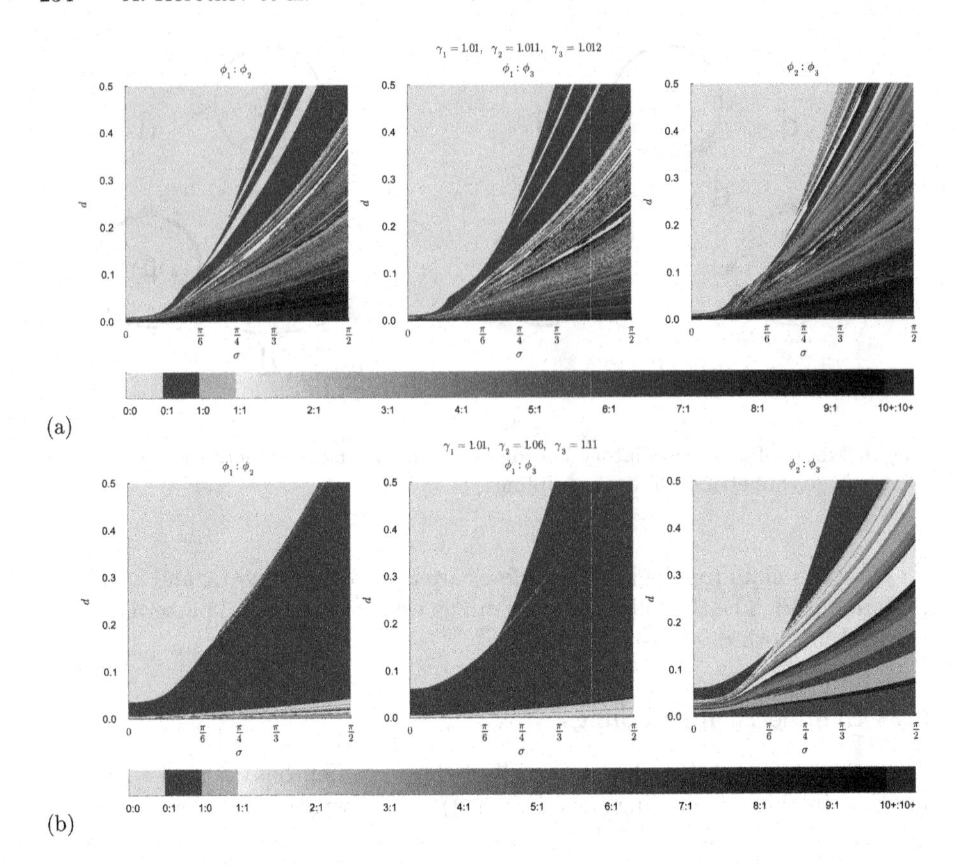


**Fig. 10.** Maps of temporal patterns in the system with symmetric couplings (4). Regions of different colors correspond to different temporal patterns (see colored legend). (a) $\Delta = 0.001$, (b) $\Delta = 0.05$. (Color figure online)

Similar to the case of 2 elements, we built two-parameter maps of neuron-like synchronized activity on the parameter plane $P = (\sigma; d)$, where $\sigma \in [0;\ \frac{\pi}{2}]$, $d \in [0;\ 0.5]$, see Fig. 10. The point of a certain color in the left map corresponds to the phase ratio $\phi_1 : \phi_2$, in the middle map – to the phase ratio $\phi_1 : \phi_3$, in the right map – to the phase ratio $\phi_2 : \phi_3$. All-together these maps should be interpreted as follows. One can choose a certain point at the parameter plane, e.g. point $p_1 = (\pi/4, 0.2)$, which corresponds to the following values of governing parameters: $\sigma = \pi/4$ and $d = 0.2$. For point $p_1$ the ratio between phases of three elements are the following: $\phi_1 : \phi_2 = 0 : 0$, $\phi_1 : \phi_3 = 0 : 0$, $\phi_2 : \phi_3 = 0 : 0$, marked in grey color in all subfigures. This fact means that all elements are suppressed and generate no activity (oscillation death).

Examples of different temporal patterns observed in the system (4) can be found in Fig. 11. These patterns include types of synchronous activity valuable for CPG modelling, such as 'winner-takes-all' pattern (Fig. 11(a)), pattern for

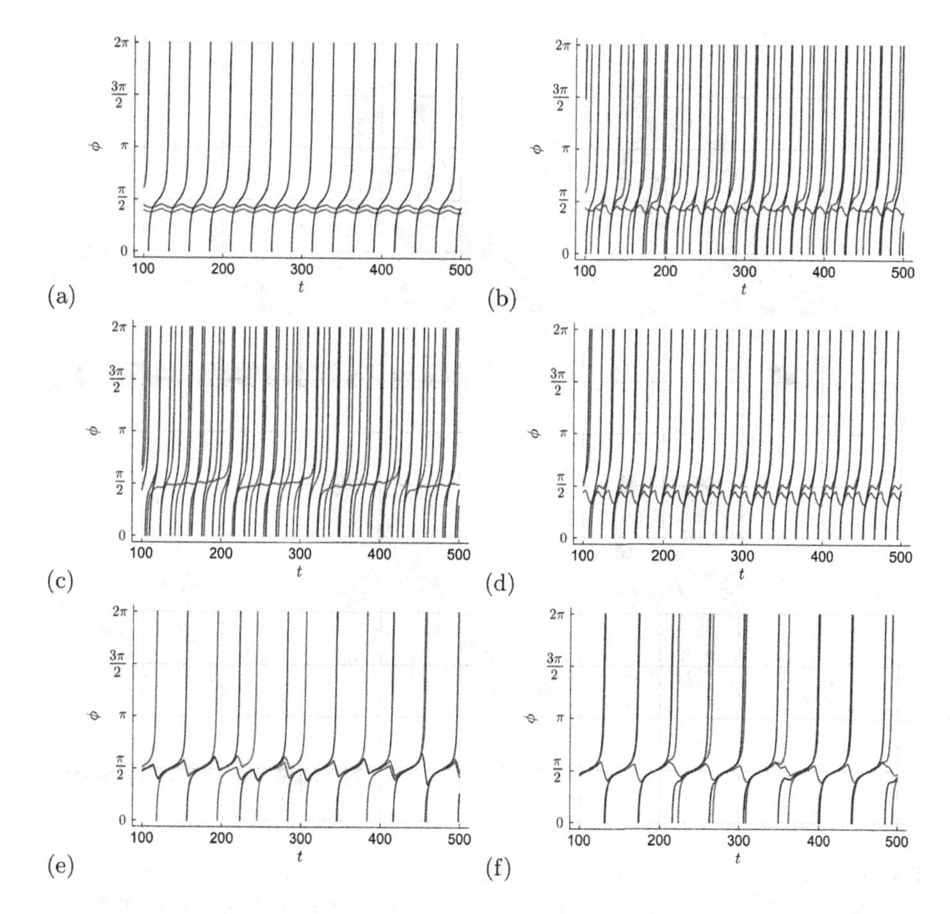


**Fig. 11.** Time series for different types of synchronization in the 3-cell motif with symmetrical inhibitory couplings.

which one element is suppressed and two other active and synchronized in-phase (Fig. 11(d)) or anti-phase (Fig. 11(b)), as well as other more sophisticated types of synchronous rhythms (Fig. 11(c), Fig. 11(e), Fig. 11(f)).

### 3.3 Asymmetrical Couplings

The governing equations for the motif of three phi-neurons coupled with asymmetrical inhibitory couplings (see Fig. 9(b)) can be written as follows:

$$\begin{cases} \dot{\phi}_1 = \gamma_1 - \sin\phi_1 - d \cdot I(\phi_3) \\ \dot{\phi}_2 = \gamma_2 - \sin\phi_2 - d \cdot I(\phi_1) \\ \dot{\phi}_3 = \gamma_3 - \sin\phi_3 - d \cdot I(\phi_2) \end{cases} \tag{5}$$

All parameters of the system (5) are fixed as it was done for the system (4).

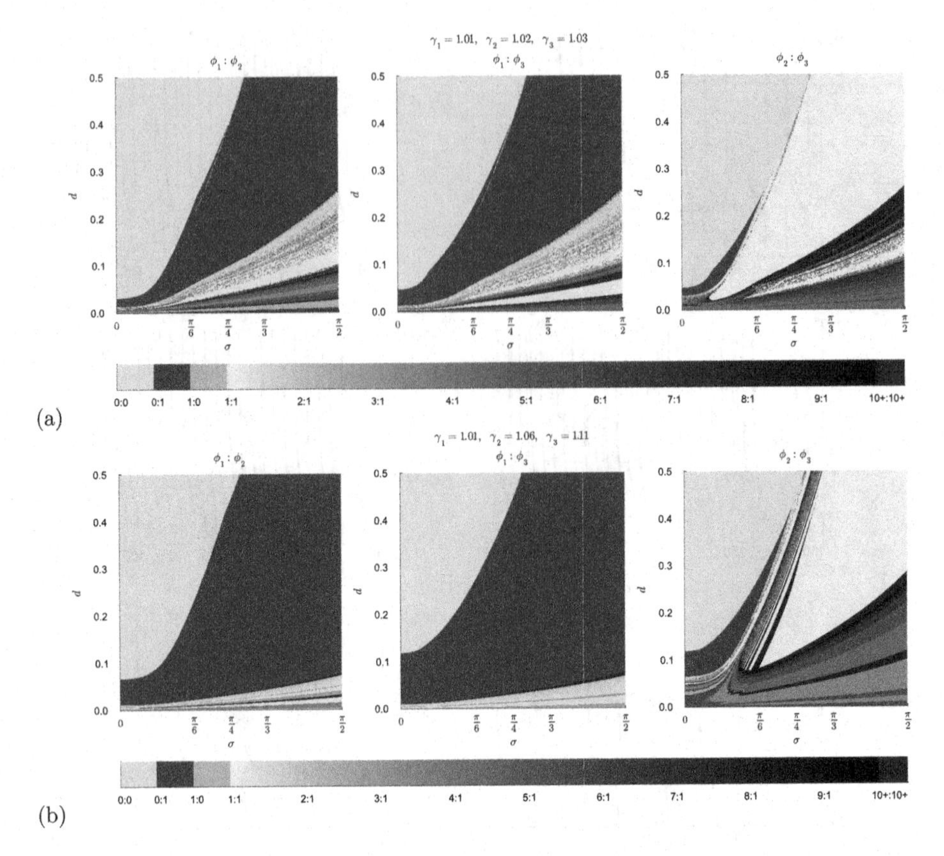


**Fig. 12.** Maps of temporal patterns in the system with asymmetrical couplings (5). Regions of different colors correspond to different temporal patterns (see colored legend). (a) $\Delta = 0.01$, (b) $\Delta = 0.05$. (Color figure online)

Similar to the previous case, we built two-parameter maps of neuron-like synchronized activity on the parameter plane $P = (\sigma; d)$, where $\sigma \in [0;\ \frac{\pi}{2}]$, $d \in [0;\ 0.5]$, see Fig. 12. The point of certain color in the left map corresponds to the phase ratio $\phi_1 : \phi_2$, in the middle map – to the phase ratio $\phi_1 : \phi_3$, in the right map – to the phase ratio $\phi_2 : \phi_3$. Together, the obtained maps should be interpreted as in the previous subsection. For instance, one choose point $p_1 = (\pi/4, 0.2)$, which corresponds to the pair of governing parameter values $\sigma = \pi/4$ and $d = 0.2$. For point $p_1$ phase relations are the following: $\phi_1 : \phi_2 = 0 : 1$ (red colored region), $\phi_1 : \phi_3 = 0 : 1$ (red colored region) and $\phi_2 : \phi_3 = 1 : 1$ (yellow colored region). This fact leads us to the conclusion that for chosen governing parameter values the second and the third elements suppress activity in the first element.

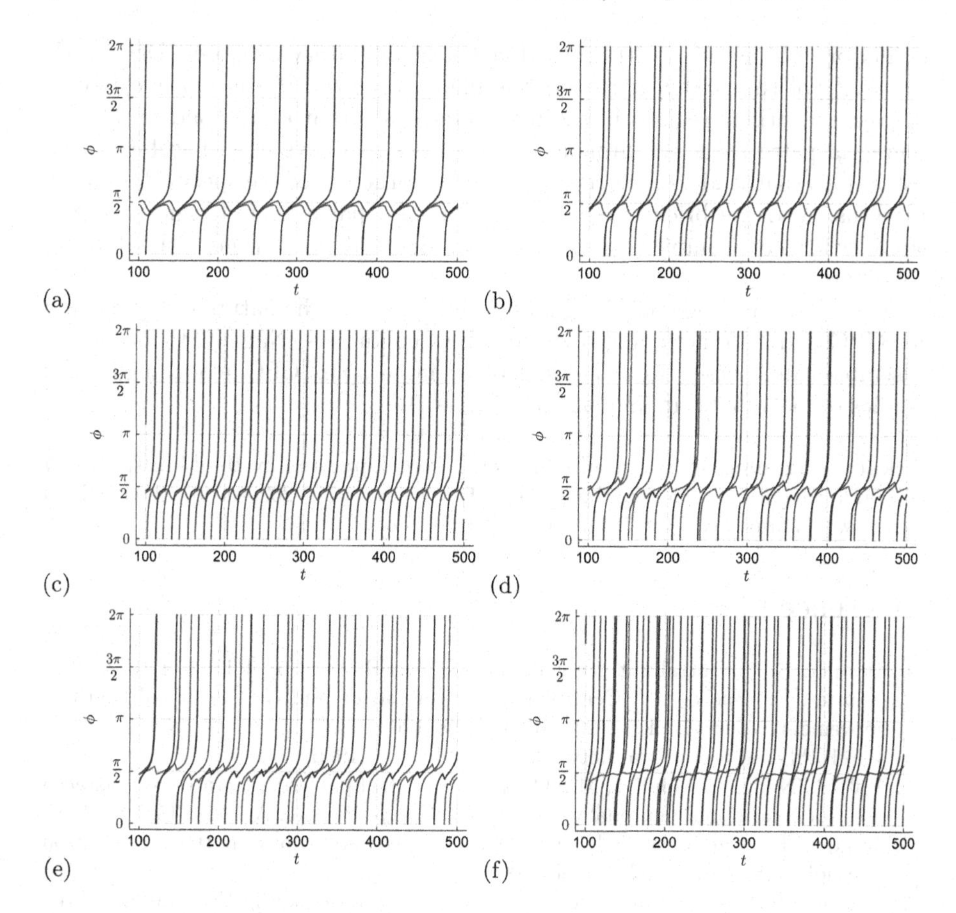


**Fig. 13.** Time series for different types of synchronization in the 3-cell motif with asymmetrical couplings.

Examples of different temporal patterns observed in the system (5) can be found in Fig. 13, including 'winner-takes-all' pattern and various types of complex synchronous dynamics. As in case of symmetrical couplings, here on the parameter plane $P = (\sigma; d)$ a wide region exist where the ensemble exhibit no oscillatory activity. Also, a temporal pattern for which the second and the third elements suppress activity in the first element is stable in a sufficiently wide range of governing parameter values. In case of symmetric couplings such pattern is also presented, but its stability regions on the parameter plane $P$ are noticeably smaller.

## 4 Conclusion

In this study we have investigated in details the case of non-identical elements for earlier proposed phenomenological model of the HCO, which consists of two oscillatory phi-neurons with mutual chemical inhibitory synapses. Despite the simplicity of the model, it reflects the main properties of the biological HCO. We have identified regions in the parameter space that correspond to different temporal patterns of synchronous activity. We also have described bifurcation transitions that lead to the occurrence and destruction of the specified temporal patterns. Also, we briefly describe the temporal patterns and regions of their stability for more complex case of three non-identical phi-neurons with inhibitory couplings.

Our study may help one to gain new insights into the nature of the locomotor CPG and its functioning under different conditions. Namely, the suggested ensemble can be used as a building block in complex CPG modelling, e.g. in studies of locomotion in animals and robots [19].

**Acknowledgements.** The results in Sect. 2 were supported by the Ministry of Science and Education of Russian Federation (Project No. 0729-2020-0036), the results in Sect. 3 were supported by RSF grant # 23-12-00180.

## References

1. Hooper, S.L.: Central pattern generators. Curr. Biol. **10**(5), R176–R179 (2000)
2. Marder, E., Bucher, D.: Central pattern generators and the control of rhythmic movements. Curr. Biol. **11**(23), R986–R996 (2001)
3. Farel, P.B.: Model neural networks and behavior (1986)
4. Bal, T., Nagy, F., Moulins, M.: The pyloric central pattern generator in crustacea: a set of conditional neuronal oscillators. J. Comp. Physiol. A **163**, 715–727 (1988)
5. Marder, E., Calabrese, R.L.: Principles of rhythmic motor pattern generation. Physiol. Rev. **76**(3), 687–717 (1996)
6. Brown, T.G.: The intrinsic factors in the act of progression in the mammal. Proc. R. Soc. London Ser. B Containing Pap. Biol. Character **84**(572), 308–319 (1911)
7. Jankowska, E., Jukes, M., Lund, S., Lundberg, A.: The effect of DOPA on the spinal cord 5. Reciprocal organization of pathways transmitting excitatory action to alpha motoneurones of flexors and extensors. Acta Physiol. Scand. **70**(3–4), 369–388 (1967)
8. Lundberg, A.: Half-centres revisited. In: Regulatory Functions of the CNS Principles of Motion and Organization, pp. 155–167. Elsevier (1981)
9. Rubin, J.: Comparative analysis of half-center central pattern generators (CPGS) (2014)
10. Burke, R., Degtyarenko, A., Simon, E.: Patterns of locomotor drive to motoneurons and last-order interneurons: clues to the structure of the CPG. J. Neurophysiol. **86**(1), 447–462 (2001)
11. Kriellaars, D., Brownstone, R., Noga, B., Jordan, L.: Mechanical entrainment of fictive locomotion in the decerebrate cat. J. Neurophysiol. **71**(6), 2074–2086 (1994)
12. Lafreniere-Roula, M., McCrea, D.A.: Deletions of rhythmic motoneuron activity during fictive locomotion and scratch provide clues to the organization of the mammalian central pattern generator. J. Neurophysiol. **94**(2), 1120–1132 (2005)
13. Rybak, I.A., Shevtsova, N.A., Lafreniere-Roula, M., McCrea, D.A.: Modelling spinal circuitry involved in locomotor pattern generation: insights from deletions during fictive locomotion. J. Physiol. **577**(2), 617–639 (2006)

14. Yakovenko, S., McCrea, D., Stecina, K., Prochazka, A.: Control of locomotor cycle durations. J. Neurophysiol. **94**(2), 1057–1065 (2005)
15. Pais, D., Caicedo-Nunez, C.H., Leonard, N.E.: Hopf bifurcations and limit cycles in evolutionary network dynamics. SIAM J. Appl. Dyn. Syst. **11**(4), 1754–1784 (2012)
16. Zou, W., Senthilkumar, D., Zhan, M., Kurths, J.: Reviving oscillations in coupled nonlinear oscillators. Phys. Rev. Lett. **111**(1), 014101 (2013)
17. Komarov, M., Pikovsky, A.: Dynamics of multifrequency oscillator communities. Phys. Rev. Lett. **110**(13), 134101 (2013)
18. Korotkov, A.G., Levanova, T.A., Zaks, M.A., Maksimov, A.G., Osipov, G.V.: Dynamics in a phase model of half-center oscillator: two neurons with excitatory coupling. Commun. Nonlinear Sci. Numer. Simul. **104**, 106045 (2022)
19. Tsybina, Y.A., et al.: Toward biomorphic robotics: a review on swimming central pattern generators. Chaos Solitons Fractals **165**, 112864 (2022)
20. Zharinov, A.I., Kurganov, D.P., Potapov, I.A., Khoruzhko, M.A., Kazantsev, V.B., Lobov, S.A.: Self-organizing cpgs in the control loop of a biomorphic fish robot. In: 2022 Fourth International Conference Neurotechnologies and Neurointerfaces (CNN), pp. 219–222. IEEE (2022)
21. Mitin, I., et al.: Modeling biomorphic robotic fish swimming: simulations and experiments. In: Ronzhin, A., Meshcheryakov, R., Xiantong, Z. (eds.) ICR 2022. LNCS, vol. 13719, pp. 189–198. Springer, Cham (2022). https://doi.org/10.1007/978-3-031-23609-9_17
22. Emelin, A., Korotkov, A., Levanova, T., Osipov, G.: Motif of two coupled phase equations with inhibitory couplings as a simple model of the half-center oscillator. In: Balandin, D., Barkalov, K., Meyerov, I. (eds.) MMST 2022. CCIS, vol. 1750, pp. 82–94. Springer, Cham (2022). https://doi.org/10.1007/978-3-031-24145-1_7
23. Cohen, A.H., Holmes, P.J., Rand, R.H.: The nature of the coupling between segmental oscillators of the lamprey spinal generator for locomotion: a mathematical model. J. Math. Biol. **13**(3), 345–369 (1982)
24. Buono, P.L., Golubitsky, M.: Models of central pattern generators for quadruped locomotion I. Primary gaits. J. Math. Biol. **42**(4), 291–326 (2001)
25. Adler, R.: A study of locking phenomena in oscillators. Proc. IEEE **61**(10), 1380–1385 (1973)
26. Korotkov, A.G., Kazakov, A.O., Levanova, T.A., Osipov, G.V.: Chaotic regimes in the ensemble of fitzhhugh-nagumo elements with weak couplings. IFAC-PapersOnLine **51**(33), 241–245 (2018)
27. Korotkov, A.G., Kazakov, A.O., Levanova, T.A., Osipov, G.V.: The dynamics of ensemble of neuron-like elements with excitatory couplings. Commun. Nonlinear Sci. Numer. Simul. **71**, 38–49 (2019)
28. Korotkov, A.G., Kazakov, A.O., Levanova, T.A.: Effects of memristor-based coupling in the ensemble of FitzHugh-Nagumo elements. Eur. Phys. J. Spec. Top. **228**(10), 2325–2337 (2019)
29. Lozano, Á., Rodríguez, M., Barrio, R.: Control strategies of 3-cell central pattern generator via global stimuli. Sci. Rep. **6**(1), 23622 (2016)

# The Concept of Hippocampal Activity Restoration Using Artificial Intelligence Technologies

Anna V. Beltyukova, Vyacheslav V. Razin, Nikolay V. Gromov, Margarita I. Samburova, Mikhail A. Mishchenko, Ivan M. Kipelkin, Anton E. Malkov, Lev A. Smirnov, Tatiana A. Levanova(✉), Svetlana A. Gerasimova, and Albina V. Lebedeva

Lobachevsky University, Nizhny Novgorod 603022, Russia
tatiana.levanova@itmm.unn.ru

**Abstract.** Real-time processing and decoding of neural signals play a crucial role in biohybrid neuroprostheses, as they provide feedback to modulate or replace neural function. However, several technological challenges associated with this process remain unsolved. These challenges include the need for complex computation in real-time, handling large volumes of data from hundreds or thousands of channels, and extracting meaningful features to drive stimulation. To address these challenges, deep neural networks (DNN) integrated with biohybrid systems have emerged as a novel strategy. In this paper we propose an approach based on DNN for prediction of hippocampal signals based on received biological input. Proposed study is a first step in the complex task of the development of a neurohybrid chip, which allows one to restore memory functions in the damaged rodent hippocampus.

**Keywords:** hippocampal signals · biohybrid systems · deep neural networks · neuroprosthetics

## 1 Introduction

Diseases and injuries of the nervous system account for about 6.3% of the Global Burden of Disease [1]. These conditions affect more than one billion people worldwide and are expected to worsen in the coming years due to population aging In the context of increasing life expectancy, dementia and memory loss are becoming increasingly pressing issues for the modern healthcare system. These facts, in turn, have a huge impact on both society and the economy [2,3]. Despite significant scientific and technological advancements in recent years, current first-choice therapies still fall short in effectively controlling symptoms and promoting sustainable recovery. For this reason, the restoration of cognitive and motor functions in patients with disabilities has become a global priority in healthcare and scientific research [2–4]. The functional connection between living neurons


D. Balandin et al. (Eds.): MMST 2023, CCIS 1914, pp. 240–252, 2024.
https://doi.org/10.1007/978-3-031-52470-7_19

and artificial synapses and neurons is a key requirement for the development of neuroprostheses, brain-computer interfaces, and the creation of hybrid networks. Significant progress has been made in the field of bioelectronics and neural engineering, enabling the development of devices based on electroceutics, such as neural interfaces and neuroprostheses. These devices have the capability to restore the functionality of nervous tissue [5–8].

The direct interaction of artificial circuits with large neural networks has also stimulated the development of new "neurobiohybrid" (or simply "biohybrid" or "hybrid") systems [9]. These systems hold promise as intriguing clinical solutions for the treatment of brain lesions [10]. One area of particular interest in healthcare is neuromorphic computing systems. These systems are viewed as energy-efficient and capable of real-time processing, making them essential components for next-generation hybrid neurotechnology aimed at brain recovery. In this context, establishing a real-time interface between a computing device and a biological system is crucial [11,12]. A neuromorphic computing roadmap has recently been described demonstrating that current and future neuromorphic systems will be able to process and dynamically learn low power signals [13,14]. It naturally follows from this that complex processing of biological signals can then be built into new types of neuroprostheses.

The main driver of the growth and development of the brain-machine interface in recent years has been the use of increasingly sophisticated machine learning methods to decode neural activity and control various types of prosthetic devices such as prosthetic hands [15–17], cursors [18–23], spellers [24,25], and robots [26–29]. Real-time processing and decoding of neural signals play a crucial role in biohybrid neuroprostheses, as they provide feedback to modulate or replace neural function. However, there are several technological challenges associated with this process. These challenges include the need for complex computation in real-time, handling large volumes of data from hundreds or thousands of channels, and extracting meaningful features to drive stimulation. To address these challenges, integrating Artificial Neural Networks (ANNs) with biohybrid systems has emerged as a novel strategy. ANNs offer adaptive and natural interactions between biological and artificial components. They can be employed for tasks like signal detection, classification, and simulating a biological neural network for replacement experiments.

Our aim is to propose an approach based on DNN for prediction of hippocampal signals based on received biological input. The hippocampus plays a key role in memory formation and memory impairment in neurodegenerative diseases [30]. This study is a first step in the complex task of the development of a neurohybrid chip, which allows one to restore memory functions in the damaged rodent hippocampus.

# 2 The Concept of Restoration the Hippocampal Functions

## 2.1 The Structure and Functions of Hippocampus

The hippocampus is a structure of the limbic system of the brain (olfactory brain), which located deep in the cerebral hemispheres. Namely, it is a part of the medial temporal regions of the hemispheres – one of several interconnected brain regions that make up the hippocampal formation.

Hippocampus has a special neuroanatomy: a combination of a relatively simple organization of the main cellular layers with a highly organized laminar distribution of many input pathways (perforant path). This structure is schematically presented in Fig. 1. Entorhinal cortex provide the major input to the dentate gyrus and a prominent input to the hippocampus and the subiculum. It also contains the main connectional route from the entorhinal cortex to all fields of the hippocampal formation known as perforant pathway. It was shown earlier in biological experiments [31] that dynamics of the intrinsic pathway in the hippocampus (dentate granule cells → CA3 pyramidal cells → CA1 pyramidal cells) could be studied using random impulse train stimulation of excitatory perforant path input. In experiments *in vitro*, i.e. on hippocampal slices, one can obtain sequential excitation of dentate gyrus (DG) → CA3 → CA1 using electrical stimulation.

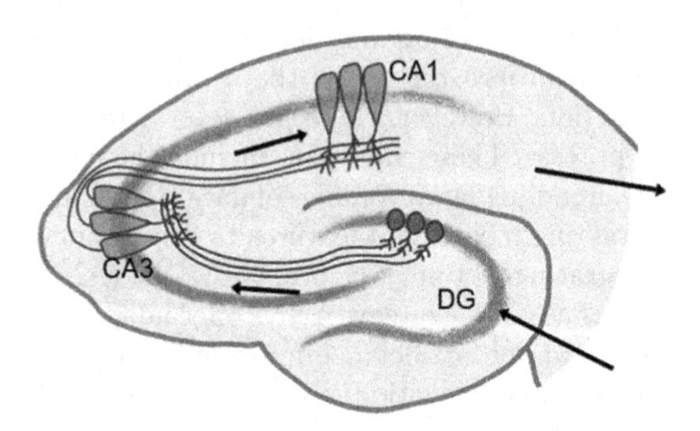


**Fig. 1.** Scheme of laminar structure and perforant pathway in the hippocampus. Arrows mark the direction of signal propagation. Dentate granule cells are pictured in red. Pyramidal cells in CA1 and CA3 are pictured in blue. (Color figure online)

The hippocampus is involved in converting short-term memory into long-term memory, and together with the anterior cingulate cortex, some part of the hippocampus is involved in some form of general attentional control. The hippocampus in animals and humans is responsible for spatiotemporal context-dependent memory [32].

The hippocampus can be effectively used as a neurobiological model system [33] because of its well-studied structure and functions. Many neurodegenerative diseases and injuries lead to impaired hippocampal functionality and decreased cognitive functions [34], which make the hippocampus a valuable object of study in the investigations related to neuroprosthetics.

### 2.2 Advanced Hippocampal Prostheses: A Quick Review

The well-known group of scientists Dr. R.E. Hampson, Dr. S.A. Deadwyler, Dr. T.W. Berger and colleagues in this field have carried out impressive work on the development of a hippocampal neuroprosthesis. The first work on a neural prosthesis was done by this research group in 2011–2012 on rats, although preliminary work was started as early as the 1990s [15–19]. Using a mathematical model with multiple inputs and outputs of MIMO (Multiple Inputs - Multiple Outputs) and training this system with data on electrophysiological activity from the CA3 region of the hippocampus, neuronal activity was predicted at the output in the CA1 region of the hippocampus, which resulted in an increase in working memory in rats in the task of delayed pattern matching (DNMS) [20,21]. So, in 2018, they successfully demonstrated the implementation of this system on people operated on after epilepsy for the first time. They called this system Proof-of-Concept to restore and improve memory function by successfully facilitating memory encoding using the patient's own spatiotemporal neural codes in the patient's hippocampus. The results demonstrated the simplification of memory coding, which is an important feature for the creation of an implantable neural prosthesis necessary to improve human memory [22,23].

### 2.3 Proposed Approach to Hippocampal Signals Restoring

The impressive work on neuroprosthetics of the hippocampus, mentioned above, contributed to the development of this area, including the use of artificial intelligence technology. In this study, we propose Proof-of-Concept for prosthetic system that is able to restore hippocampal functions using artificial intelligence technology. Namely, we propose a deep neural network (DNN) that is able to predict local field potentials in the rat hippocampus slices of the rodent hippocampus. This DNN can be implemented on chip, e.g. memristive chip [35], which can be connected with living slices in order to read and apply stimuli to it. The schematic representation of the restoring system is presented in Fig. 2.

In our work, we have developed an intelligent approach to restoring electrophysiological activity in a slice of the mouse hippocampus instead of the damaged area. First, we carried out a biological experiments series to record field excitatory postsynaptic potentials in hippocampal slices in two hippocampal areas (CA1 and CA3) under normal conditions (the "healthy" slice) during electrical stimulation of the dentate gyrus (DG). The dentate gyrus is the input from the entorhinal cortex in the perforant pathway. The output pathway in the hippocampus passes through the CA1 area and subiculum, returning to the entorhinal cortex. Then we mechanically damaged one of the central hippocampus area (CA3 area) in the slice. This disrupted synaptic transmission in the perforant pathway (the "damaged" slice). After that we continued recording field

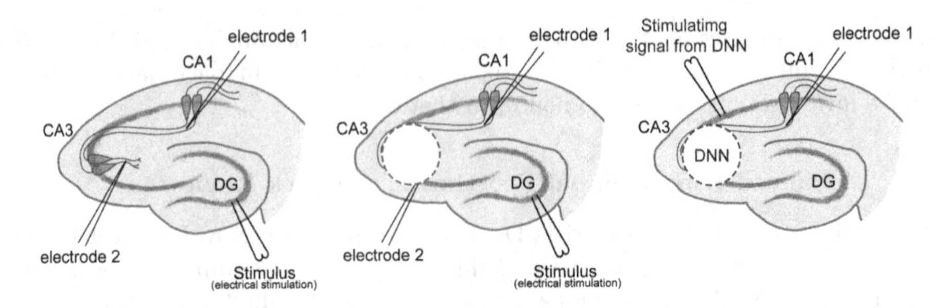


**Fig. 2.** Scheme of experimental protocols for hippocampal functions restoring. Pyramidal cells in CA1 and CA3 are pictured in blue. (Color figure online)

excitatory postsynaptic potentials in each slice from the experimental animals, following the described protocols. Thus, we recorded pairs of fEPSP signals from the CA1 and CA3 regions of the hippocampus (Fig. 3). As a result, we obtained a set of paired fEPSPs at different amplitudes of stimulus. The detailed fEPSP registration protocol is described in Sect. 3.1. Subsequent procedures with the resulting set of signals involve preprocessing, filtering, and their presentation to an artificial neural network for further training and prediction of the output signal in the hippocampus. As a result, the signals predicted by the neural network will be presented as a neuronal stimulus, which replaces the activity in the damaged CA3 hippocampal area in a real-time biological experiment. These works will be carried out in the near future and will serve as experimental proof of the concept developed in this project.

# 3 Hippocampal Signals Data

## 3.1 Data Collection

All experiments were carried out on the $C57BL/6$ mouse line of 2–3-months of age. Field excitatory postsynaptic potential (fEPSP) were recorded on surviving slices of the hippocampus of 400-μm thickness, prepared using a vibratome (MicromHM 650 V; Thermo Fischer Scientific). Slices were permanently placed in artificial cerebrospinal fluid (ACSF) with the composition (mM): $NaCl$ 127, $KCl$ 1, $KH_2PO_4$ 1.2, $NaHCO_3$ 26, D-Glucose 10, $CaCl_2$ 2.4, $MgCl_2$ 1.3 (pH 7.4–7.6; Osmolarity 295 ± 5 mOsm). The solution was constantly saturated with gas mixture containing 95% $O_2$ and 5% $CO_2$. For morphological analysis, a Scientifica SliceScope (Scientifica Ltd) microscope with aerial 4X lens was used. The fEPSP were recorded using glass microelectrodes (Sutter Instruments Model P-97) from the Harvard Apparatus (Capillares 30-0057 GC150F-10, $1.5OD * 0.86IK * 100$ Lmm, Quantity: 225). The position of registration and stimulation electrodes was as follows. Stimulation electrode was positioned in the DG region. Registration electrodes were positioned in CA3 and CA1 regions. The amplitudes for stimulation and than registration of fEPSP were as follows: 100; 200; 300; 400; 500; 1000 μA and duration was 50 μs.

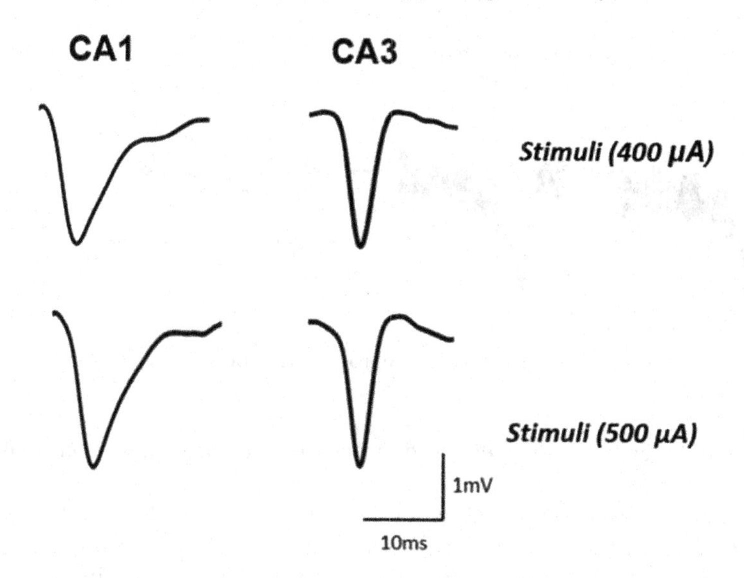


**Fig. 3.** Pairs of signals in hippocampal areas – CA1 and CA3 – for different stimulation amplitudes 400 μA (upper panel) and 500 μA (lower panel).

The electrodes were filled with the ACSF solution. The pipette tip resistance was 4–8 M$\Omega$. The electrodes were installed in the holder of a HEKA EPC 10 USB PROBE 1 preamplifier head connected to the HEKA EPC 10 USB biosignal amplifier. A stainless-steel bipolar electrode was used for electrical stimulation of the dentate gyrus area in the hippocampus. A more detailed description of such protocols is presented in our previous works [36–40]. Data visualization and recording were performed using Patchmacter software.

### 3.2 Data Preprocessing

Obtained data as all biological signals should be preprocessed first [41]. Basic preprocessing pipeline is presented in Fig. 4.

Obtained data contain artifacts from electrical stimuli that should be removed. As the stimulus in all experiments has the same duration and is applied at the same moment, we removed it from our data by excluding 20 time points corresponding to it.

Then the noise was filtered with a Gaussian filter with $\sigma = 35$ and with $radius = 4\sigma$. After that, we normalized the data by converting them to zero mean and unit variance. The data was then separated into inputs and outputs for the models.

In order to obtain train signals we separated all signals into 6 groups depending on the amplitude of external stimulus. All signals inside each group were then averaged except one, which was used as test signal. it allows one to remove the dependence of the signal on minor experimental conditions, such as the incli-

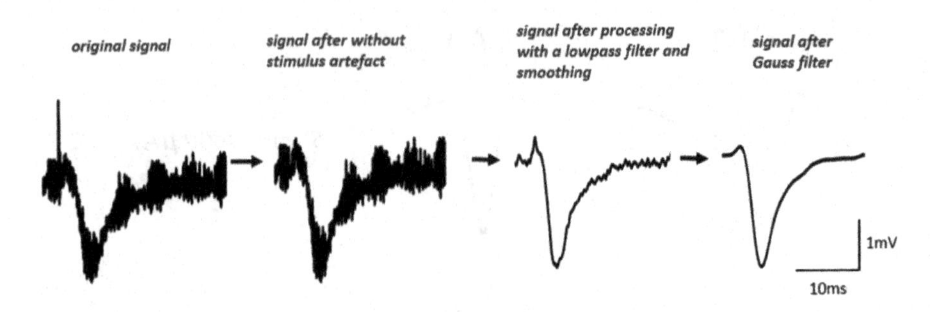


**Fig. 4.** Signal preprocessing pipeline.

nation and exact location of electrode insertion, the exact insertion depth, and others.

The data we used as an input for the model consist of 1000 time samples recorded in CA3 region. We used as an output 1000 time samples recorded in CA1 region. The optimal number of time samples in the input data was determined experimentally during the optimization process.

## 4 DNN Architecture

We proposed a deep learning model to predict of fEPSP signals in CA1 area of hippocampus (CA1 signal) based on fEPSP signals in CA3 area of hippocampus (CA3 signal) as an input.

We used 2-layered LSTM architecture previously tested for time series prediction, including extreme events prediction [42] and epilepsy prediction from EEG data [43]. The model includes a linear input layer of size $1 \times 250$ with the Relu activation function. There are also 2 LSTM cells, each with 250 neurons in hidden state, and the output from these cells are 1000 hidden states from both cells. And the last linear output layer with a weight matrix of size $500 \times 1$. This final output allows us to obtain the predicted values of CA1.

We optimising MSE (Mean Squared Error) loss function, which is:

$$MSE = \frac{1}{n}\sum_{i=1}^{n}(y_i - \hat{y_i})^2, \tag{1}$$

where n is number of training sequences, $y_i$ is correct sample and $\hat{y_i}$ is predicted sample

We define the evaluation metric as combination of squares of slope, half-response length, time rise and time decay with corresponding weights 0.1, 0.5, 0.1, 0.3. Also we normalized values of each part of metric to be in range from 0 to 1. The choice of the evaluation metric was specified by biological background of the problem.

The model was trained on 9 preprocessed records. We used cross validation approach with the following partition scheme: in each iteration 8 records was used for the training set, and 1 for the test set.

## 5 Results

To evaluate the model we took into account the following features of the signal: the beginning and the end of the response and half-response. Taking into account several features of a signal at once during training allows us to improve the quality of the model. The values of obtained evaluation metrics for predicted responses to different stimuli amplitudes are presented in Fig. 5.

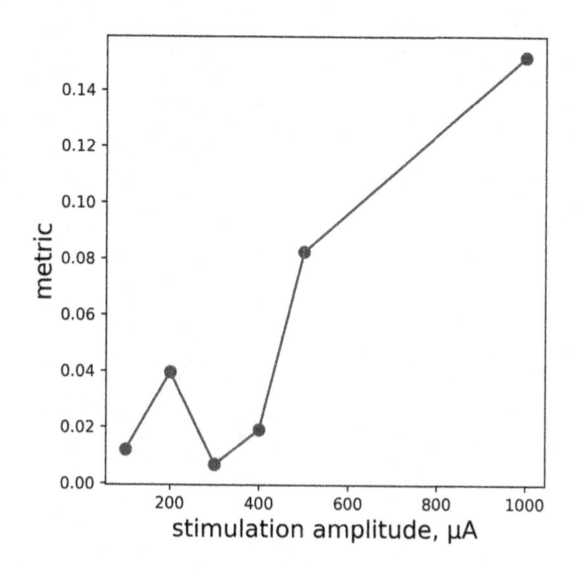


**Fig. 5.** Evaluation metrics for predicted responses to different stimuli amplitudes.

True and predicted value for preprocessed fEPSP signals are shown in Fig. 6. Despite the fact that preprocessed signal still contain noise, it can be clearly seen that LSTM network allows us to obtain a good match of true and predicted signals in terms of response and half-response lengths.

Despite the relatively small amount of data available for training, overfitting is not observed here, since the responses to described stimuli belong to the same distribution.

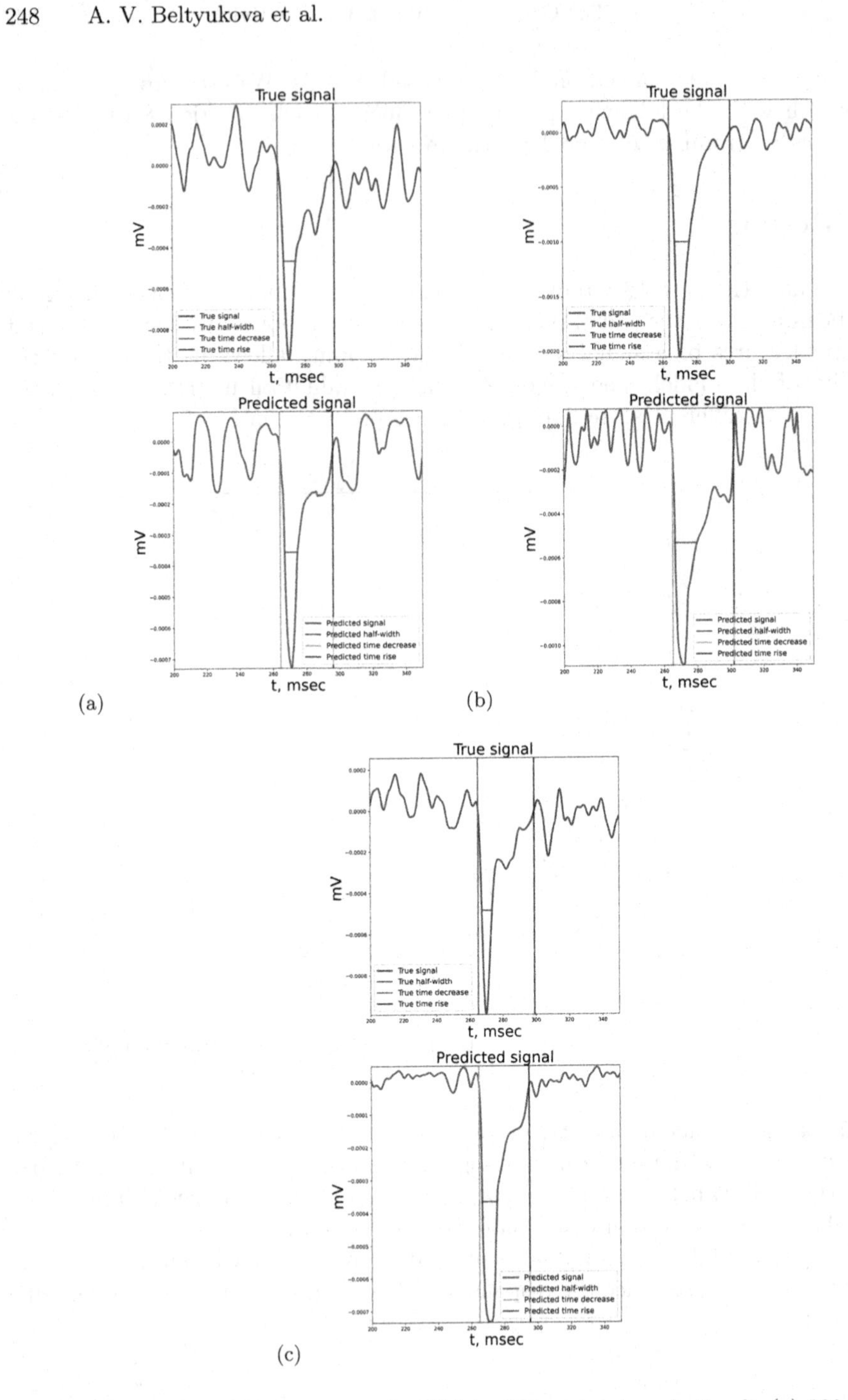


**Fig. 6.** True and predicted response in CA1 to different external stimuli: (a) 200 μA, (b) 300 μA, (c) 500 μA. Borderlines of the response and half-response are marked with vertical lines.

## 6 Conclusion

In the presented study we have proposed the concept of a neurohybrid system that can replace or restore the activity in hippocampal damaged area of the experimental animal brain for subsequent memory functions restoration. The first part of the study - a proof-of-concept - have been performed on hippocampal brain slices (in vitro experiments). Namely, we have proposed an approach based on LSTM deep neural network for prediction of hippocampal signals. In the framework of the proposed approach we obtain averaged fEPSP signals in CA1 and CA3 areas of hippocampus and predict the response in CA1 area based on CA3 input. We have described in detail methods of data collection and preprocessing.

To assess the performance of proposed deep learning model, we have introduced complex evaluation metrics, which allows us to take into account valuable features of the signals. We have calculated training and prediction values for this evaluation metrics for different amplitudes of external stimuli that initiate the sequential excitation of regions in perforant pathway of hippocampus. Training accuracy measures how well the deep learning model fits the features of the response (such as slope, half-width of the response, time rise and time decay) for a given stimulus amplitude. Prediction accuracy measures how well the deep learning model predicts the synaptic parameters for the data that were excluded from the training set. We have compare results of prediction for averaged and non-averaged signals. Averaged signals allow us to deduce a universal form of response for each stimulus. Using non-averaged signals imply that similar job will be done by our deep neural model itself.

At the second stage of the study, we will use transfer learning approach to generalize these results on signals from hippocampus in vivo. Being implemented on a neurohybrid system, such model can provide the restoration of memory functions in the damaged rodent hippocampus. Note that different promising architectures, such as memristor-based chips, can be used as a hardware platform in this task [44].

Although proposed investigation is limited to rodent studies, it is highly likely that our approach will be applicable to other types of acquired brain injury. Such studies are extremely important for the treatment of neurodegenerative diseases associated with memory impairment and have high prospects for future use in practical medicine.

**Acknowledgements.** Data collection and preprocessing was supported by Russian Science Foundation (Project No. 23-75-10099), numerical results (model training and testing) was supported by Ministry of Science and Education of Russian Federation (Contract FSWR-2021-0013).

## References


1. Chin, J.H., Vora, N.: The global burden of neurologic diseases. Neurology **83**(4), 349–351 (2014)
2. WHO: The top 10 causes of death. World Health Organization (2020)
3. Langa, K.M.: Cognitive aging, dementia, and the future of an aging population. In: Future Directions for the Demography of Aging: Proceedings of a Workshop, pp. 249–268. National Academies Press Washington, DC (2018)
4. French, B., et al.: Repetitive task training for improving functional ability after stroke. Cochrane Database Syst. Rev. (11) (2016)
5. Panuccio, G., Semprini, M., Natale, L., Buccelli, S., Colombi, I., Chiappalone, M.: Progress in neuroengineering for brain repair: new challenges and open issues. Brain Neurosci. Adv. **2**, 2398212818776475 (2018)
6. Famm, K.: Drug discovery: a jump-start for electroceuticals (vol 496, pg 159, 2013). Nature **496**(7445), 300 (2013)
7. Berger, T.W., et al.: Restoring lost cognitive function. IEEE Eng. Med. Biol. Mag. **24**(5), 30–44 (2005)
8. Hampson, R., Simeral, J., Deadwyler, S.A.: Cognitive processes in replacement brain parts: a code for all reasons. Toward Replacement Parts, p. 111 (2005)
9. Vassanelli, S., Mahmud, M.: Trends and challenges in neuroengineering: toward "intelligent" neuroprostheses through brain-"brain inspired systems" communication. Front. Neurosci. **10**, 438 (2016)
10. George, R., et al.: Plasticity and adaptation in neuromorphic biohybrid systems. Iscience **23**(10) (2020)
11. Broccard, F.D., Joshi, S., Wang, J., Cauwenberghs, G.: Neuromorphic neural interfaces: from neurophysiological inspiration to biohybrid coupling with nervous systems. J. Neural Eng. **14**(4), 041002 (2017)
12. Sharifshazileh, M., Burelo, K., Sarnthein, J., Indiveri, G.: An electronic neuromorphic system for real-time detection of high frequency oscillations (HFO) in intracranial EEG. Nat. Commun. **12**(1), 3095 (2021)
13. Corradi, F., Indiveri, G.: A neuromorphic event-based neural recording system for smart brain-machine-interfaces. IEEE Trans. Biomed. Circuits Syst. **9**(5), 699–709 (2015)
14. Christensen, D.V., et al.: 2022 roadmap on neuromorphic computing and engineering. Neuromorphic Comput. Eng. **2**(2), 022501 (2022)
15. Chapin, J.K., Moxon, K.A., Markowitz, R.S., Nicolelis, M.A.: Real-time control of a robot arm using simultaneously recorded neurons in the motor cortex. Nat. Neurosci. **2**(7), 664–670 (1999)
16. Velliste, M., Perel, S., Spalding, M.C., Whitford, A.S., Schwartz, A.B.: Cortical control of a prosthetic arm for self-feeding. Nature **453**(7198), 1098–1101 (2008)
17. Hochberg, L.R., et al.: Reach and grasp by people with tetraplegia using a neurally controlled robotic arm. Nature **485**(7398), 372–375 (2012)
18. Wolpaw, J.R., McFarland, D.J., Neat, G.W., Forneris, C.A.: An EEG-based brain-computer interface for cursor control. Electroencephalogr. Clin. Neurophysiol. **78**(3), 252–259 (1991)
19. Serruya, M.D., Hatsopoulos, N.G., Paninski, L., Fellows, M.R., Donoghue, J.P.: Instant neural control of a movement signal. Nature **416**(6877), 141–142 (2002)
20. Wolpaw, J.R., McFarland, D.J.: Control of a two-dimensional movement signal by a noninvasive brain-computer interface in humans. Proc. Natl. Acad. Sci. **101**(51), 17849–17854 (2004)

21. Li, Z., O'Doherty, J.E., Hanson, T.L., Lebedev, M.A., Henriquez, C.S., Nicolelis, M.A.: Unscented Kalman filter for brain-machine interfaces. PLoS ONE **4**(7), e6243 (2009)
22. Gilja, V., et al.: Clinical translation of a high-performance neural prosthesis. Nat. Med. **21**(10), 1142–1145 (2015)
23. Pandarinath, C., et al.: High performance communication by people with paralysis using an intracortical brain-computer interface. Elife **6**, e18554 (2017)
24. Farwell, L.A., Donchin, E.: Talking off the top of your head: toward a mental prosthesis utilizing event-related brain potentials. Electroencephalogr. Clin. Neurophysiol. **70**(6), 510–523 (1988)
25. Sellers, E.W., Kubler, A., Donchin, E.: Brain-computer interface research at the university of South Florida cognitive psychophysiology laboratory: the P300 speller. IEEE Trans. Neural Syst. Rehabil. Eng. **14**(2), 221–224 (2006)
26. Bell, C.J., Shenoy, P., Chalodhorn, R., Rao, R.P.: Control of a humanoid robot by a noninvasive brain-computer interface in humans. J. Neural Eng. **5**(2), 214 (2008)
27. Galán, F., et al.: A brain-actuated wheelchair: asynchronous and non-invasive brain-computer interfaces for continuous control of robots. Clin. Neurophysiol. **119**(9), 2159–2169 (2008)
28. Millán, J.D.R., Galán, F., Vanhooydonck, D., Lew, E., Philips, J., Nuttin, M.: Asynchronous non-invasive brain-actuated control of an intelligent wheelchair. In: 2009 Annual International Conference of the IEEE Engineering in Medicine and Biology Society, pp. 3361–3364. IEEE (2009)
29. Bryan, M., Nicoll, G., Thomas, V., Chung, M., Smith, J.R., Rao, R.P.: Automatic extraction of command hierarchies for adaptive brain-robot interfacing. In: 2012 IEEE International Conference on Robotics and Automation, pp. 3691–3697. IEEE (2012)
30. Plata, A., et al.: Astrocytic atrophy following status epilepticus parallels reduced Ca2+ activity and impaired synaptic plasticity in the rat hippocampus. Front. Mol. Neurosci. **11**, 215 (2018)
31. Berger, T.W., et al.: A hippocampal cognitive prosthesis: multi-input, multi-output nonlinear modeling and VLSI implementation. IEEE Trans. Neural Syst. Rehabil. Eng. **20**(2), 198–211 (2012)
32. Andersen, P.: The Hippocampus Book. Oxford University Press, Oxford (2007)
33. Gergues, M.M., et al.: Circuit and molecular architecture of a ventral hippocampal network. Nat. Neurosci. **23**(11), 1444–1452 (2020)
34. Li, A., Li, F., Elahifasaee, F., Liu, M., Zhang, L., Initiative, A.D.N.: Hippocampal shape and asymmetry analysis by cascaded convolutional neural networks for Alzheimer's disease diagnosis. Brain Imaging Behav. 1–10 (2021)
35. Mikhaylov, A., et al.: Neurohybrid memristive CMOS-integrated systems for biosensors and neuroprosthetics. Front. Neurosci. **14**, 358 (2020)
36. Mishchenko, M.A., Gerasimova, S.A., Lebedeva, A.V., Lepekhina, L.S., Pisarchik, A.N., Kazantsev, V.B.: Optoelectronic system for brain neuronal network stimulation. PLoS One **13**(6), e0198396 (2018)
37. Lebedeva, A., et al.: Integration technology for replacing damaged brain areas with artificial neuronal networks. In: 2020 4th Scientific School on Dynamics of Complex Networks and their Application in Intellectual Robotics (DCNAIR), pp. 158–161. IEEE (2020)
38. Lebedeva, A., et al.: Neuromorphic system development based on adaptive neuronal network to modulate synaptic transmission in hippocampus. In: 2021 Third International Conference Neurotechnologies and Neurointerfaces (CNN), pp. 57–60. IEEE (2021)

39. Gerasimova, S., et al.: A neurohybrid memristive system for adaptive stimulation of hippocampus. Chaos, Solitons Fractals **146**, 110804 (2021)
40. Lebedeva, A., et al.: Development a cross-loop during adaptive stimulation of hippocampal neural networks by an artificial neural network. In: 2022 Fourth International Conference Neurotechnologies and Neurointerfaces (CNN), pp. 82–85. IEEE (2022)
41. Unakafova, V.A., Gail, A.: Comparing open-source toolboxes for processing and analysis of spike and local field potentials data. Front. Neuroinform. **13**, 57 (2019)
42. Gromov, N., Gubina, E., Levanova, T.: Loss functions in the prediction of extreme events and chaotic dynamics using machine learning approach. In: 2022 Fourth International Conference Neurotechnologies and Neurointerfaces (CNN), pp. 46–50. IEEE (2022)
43. Gerasimova, S., et al.: Memristive neural networks for predicting seizure activity. Sovremennye Tehnol. Med. **15**(4), 30 (2023)
44. Gerasimova, S.A., Beltyukova, A., Fedulina, A., Matveeva, M., Lebedeva, A.V., Pisarchik, A.N.: Living-neuron-based autogenerator. Sensors **23**(16), 7016 (2023)

# Numerical Modeling of Water Ice Impact Destruction

Nikita O. Shigaev[1] and Katerina A. Beklemysheva[2](✉)

[1] Ishlinsky Institute for Problems in Mechanics of the Russian Academy of Sciences, Prospekt Vernadskogo, 101-1, Moscow 119526, Russian Federation
[2] Moscow Institute of Physics and Technology, 9 Institutskiy per., Dolgoprudny, Moscow 141701, Russian Federation
amisto@yandex.ru

**Abstract.** Water ice behavior under dynamic loading depends on many conditions, including temperature, freezing process and chemical composition of impurities. Ice specimens produced in laboratory conditions can only imitate natural ice to a certain extent, and strength testing of natural ice is limited by working in the field under severe Arctic conditions. Development of icebreakers and construction of oil production facilities in the Arctic require deep understanding of ice destruction processes. This research applies a hybrid grid-characteristic numerical method on unstructured tetrahedral grids to problems of impact loading of water ice. Viscoelastic material model of ice is supplemented by a discrete destruction model that implements Tsai-Wu failure criterion with linear strengthening. Problem statements with ball-shaped impactor and shock unloading are considered. Three-dimensional dynamic patterns of velocity vector and stress tensor are obtained. Damaged areas are obtained, analyzed and compared to similar experimental statements.

**Keywords:** Numerical modeling · Grid-characteristic method · Material destruction · Ice mechanics

## 1 Introduction

Development of icebreakers and construction of oil production facilities in the Arctic require precise and effective methods of measuring ice properties. Water ice behavior under dynamic loading depends on temperature, freezing process and chemical composition of impurities, which can differ significantly in different geographic regions. A number of conditions can be reproduced in a laboratory, including temperature conditions and salinity, but it sometimes requires complex and expensive equipment. At the same time, testing of natural ice is limited by working in the field under severe Arctic conditions. New technologies and methods of measuring ice properties can be developed based on machine learning and numerical modeling. This approach requires fast and effective numerical

Supported by Russian Science Foundation, grant 23-21-00384.


D. Balandin et al. (Eds.): MMST 2023, CCIS 1914, pp. 253–265, 2024.
https://doi.org/10.1007/978-3-031-52470-7_20

methods as well as an extensive set of experimental data for verification and calibration.

In this article, a hybrid grid-characteristic numerical method on unstructured tetrahedral grids is applied to problems of impact loading of water ice.

## 2 Material Model and Numerical Method

In this research, we use the grid-characteristic numerical method [1]. Generally, this method uses characteristic properties of a hyperbolic system of equations. When applied to a viscoelastic material model, this method allows to model elastic waves with high accuracy end efficiency. Its comparison with discontinuous Galerkin method which is also used for problems of dynamic loading of elastic bodies is presented in [2]. This method was also applied to dynamic problems in seismology [3], railway durability [4] and aircraft composites [5].

A detailed description of the hybrid grid-characteristic method on unstructured tetrahedral grids for destructible anisotropic viscoelastic media is described in [6]. This implementation is based on the method of splitting along spatial directions and physical processes. Viscosity was modeled using the Maxwell model [7]; its detailed description for the anisotropic case is given in [8]. This model allows to track the amount of energy absorbed in each node due to the viscosity of the material and demonstrates a qualitative agreement with the experiments given in [9] and [10]. It should be noted that the choice of a simple viscosity model was due to the lack of ice parameters in open access. The existing software implementation does not impose restrictions on the complexity of the viscosity model, including the calculation of the absorbed energy. Similar to the Maxwell model, this approach allows, for a relatively small number of calculations, to select the empirical values of the internal parameters of the model in the presence of experimental data.

The material destruction model consists of a failure criterion and a model of a destructed node. We used Tsai-Wu, Tsai-Hill and Drucker-Prager failure criteria for a homogeneous orthotropic medium [11]. These criteria use the same set of material parameters, but differ in mathematical formulation and show different results when applied to the same problem statement. The most interesting results were obtained with the Tsai-Wu criterion.

The destructed node uses the discrete failure model – after the failure criterion is met, the node is considered destroyed and is calculated as a perfectly plastic material. This model of a destructed node allows to transmit compression elastic waves but consumes shear and tension waves.

Additionally, a linear strengthening model was implemented – the material strength thresholds at each node are independently linearly dependent on the energy absorbed in this node. This model suggests that a constant part of the mechanical energy absorbed in a node is spent on material strengthening – reduction of dislocations, pores and microscopic cracks in ice. Using notations from [11] and [8], we can write the following formulation for the Tsai-Wu criterion. We consider axes 1, 2 and 3 as X, Y and Z, and the plane of isotropy is X-Y.

$$\begin{aligned}
X^T_{Z\,(n+1)} &= X^T_{Z\,(n)} + X^T_{Z\,(0)} * C_S * \delta E_{(n)} \\
X^C_{Z\,(n+1)} &= X^C_{Z\,(n)} + X^C_{Z\,(0)} * C_S * \delta E_{(n)} \\
X^T_{XY\,(n+1)} &= X^T_{XY\,(n)} + X^T_{XY\,(0)} * C_S * \delta E_{(n)} \\
X^C_{XY\,(n+1)} &= X^C_{XY\,(n)} + X^C_{XY\,(0)} * C_S * \delta E_{(n)} \\
S_{Z\,(n+1)} &= S_{Z\,(n)} + S_{Z\,(0)} * C_S * \delta E_{(n)} \\
S_{XY\,(n+1)} &= S_{XY\,(n)} + S_{XY\,(0)} * C_S * \delta E_{(n)},
\end{aligned} \tag{1}$$

where $X^T_{Z\,(i)}$ – tension strengths along the axis Z on the (i)-th time step,

$X^C_{Z\,(i)}$ – compression strengths along the axis Z on the (i)-th time step,
$X^T_{XY\,(i)}$ – tension strengths along axes X and Y on (i)-th time step,
$X^C_{XY\,(i)}$ – compression strengths along axes X and Y on (i)-th time step,
$S_{Z\,(i)}$ – shear strengths in planes along the Z axis on (i)-th time step,
$S_{XY\,(i)}$ – shear strengths in the XY plane on (i)-th time steps,
$C_S$ – strengthening coefficient,
$\delta E_{(n)}$ – amount of energy, absorbed on the (n)-th time step.

## 3 Ball-Shaped Impactor

### 3.1 Problem Statement

Computational domain comprises ice disk and ball-shaped steel indentor which impacts ice disc from above along its axis. General view is presented on Fig. 1.

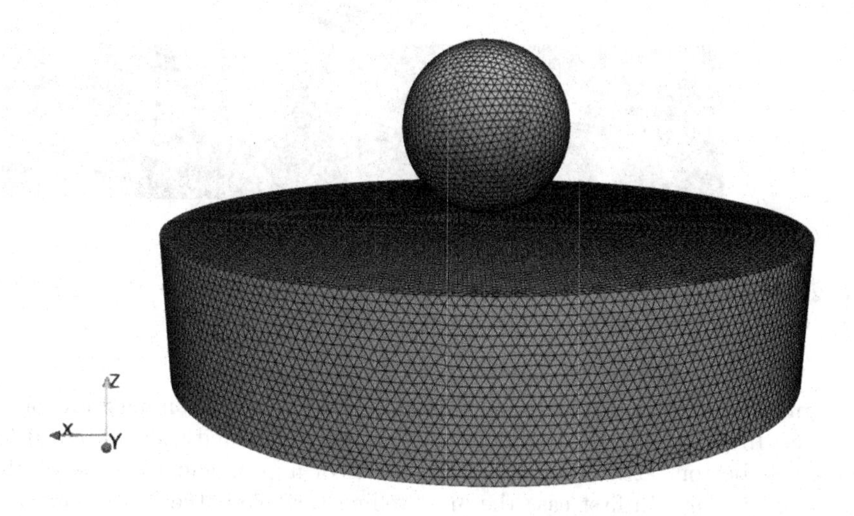


**Fig. 1.** General view of calculation area.

Ice disc has 60 mm of height and 300 mm in diameter. As far as observed destruction is located in small volume under impact area, disc size definitely contains all destruction. Side borders are free. At the bottom of disc there is a fixed border which models metal base of experimental setup. Indentor is a steel ball with a radius of 38 mm. Its initial speed is 2.5 m/s, what leads to impact with 5.6 J of energy. All materials considered viscoelastic, ice is destructible according to the Tsai-Wu criterion. A set of calculations was also performed to compare Tsai-Wu, Tsai-Hill and Drucker-Prager criteria. Steel is indestructible because its strength threshold is much higher than thresholds in ice. Ice properties: density $\rho$ is 910 kg/m$^3$, Lame parameters are $\lambda$ = 3.96 GPa, $\mu$ = 1.87 GPa. Steel properties: density $\rho$ is 7850 kg/m$^3$, Lame parameters are $\lambda$ = 122 GPa, $\mu$ = 81.2 GPa.

### 3.2 Numerical Results

According to the Tsai Wu criterion, destruction depends on compression strength and shear strength. Destruction area depending on different strength thresholds is presented on Figs. 2, 3, 4.

**Fig. 2.** Destruction area for strength thresholds: $X_c = Y_c = X_t = Y_t = 2.7$ MPa; $S = 1.0$ MPa

Signals from surface of ice allow to analyze destructed area impact on course of waves. In particular, three-dimensional patterns of velocity was obtained for two areas: bottom area of ice disk on the line of impact and top area of disc near impact point. In first case the most telling variable is the Z-component of velocity. Graph of this component in comparison with results for indestructible

**Fig. 3.** Destruction area for strength thresholds: $X_c = Y_c = X_t = Y_t = 1.8$ MPa; $S = 0.7$ MPa

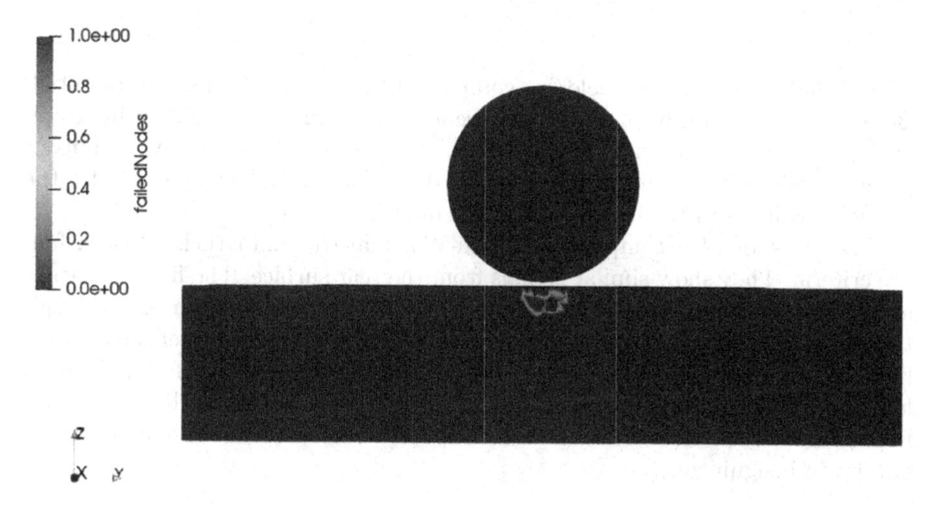


**Fig. 4.** Destruction area for strength thresholds: $X_c = Y_c = X_t = Y_t = 1.2$ MPa; $S = 0.5$ MPa

ice model is presented on Fig. 5. All the graphs obtained for Tsai-Wu criterion occupy approximately the same area. For that reason graph shows only bounding area but not separate lines.

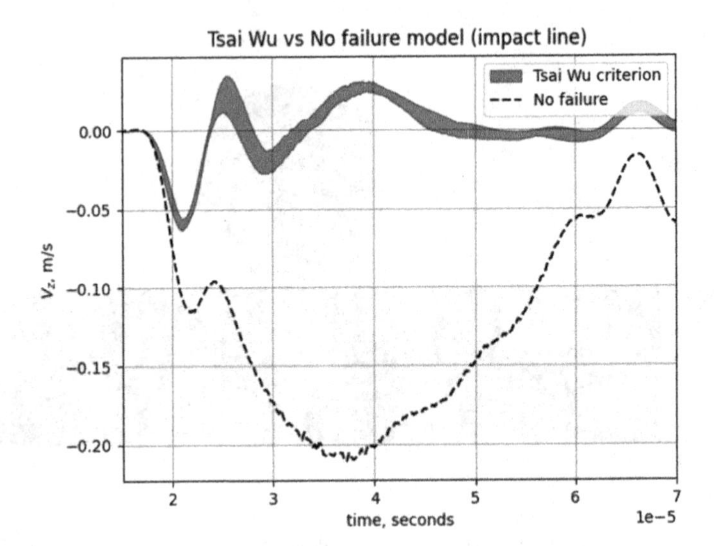


**Fig. 5.** Velocity on impact line.

On the other hand, all velocity components in top of the disc are essential. Because of that graph on Fig. 6 represents magnitude of velocity. The graph on Fig. 7 represents vertical displacement of surface in 30 mm from the impact point. It allows to estimate the residual depth of the indentation left by the impactor which can be directly measured in an experiment.

Figure 8 shows the comparison of Tsai-Wu, Tsai-Hill and Drucker-Prager failure criteria. They show similar signals from the rear surface. The first two peaks are the same, because they are caused by the direct propagation of the longitudinal wave from the impact point. Other peaks are caused by reflected waves that are influenced by the destructed area, and the difference in failure criteria leads to slightly different destruction areas. In the statement with low velocity impact and in the absence of global cracks the difference in considered failure criteria in insignificant.

## 4 Shock Unloading

### 4.1 Problem Statement

This problem statement is based on experimental setup from [13]. An ice cube is subjected to slow compression loading and shock unloading. This loading sequence leads to the appearance of a single crack in a shape of a flat ring with an undestructed area in the center. Previous attempts to model this effects [14] reproduced a flat disc crack, but failed to explain the undestructed area.

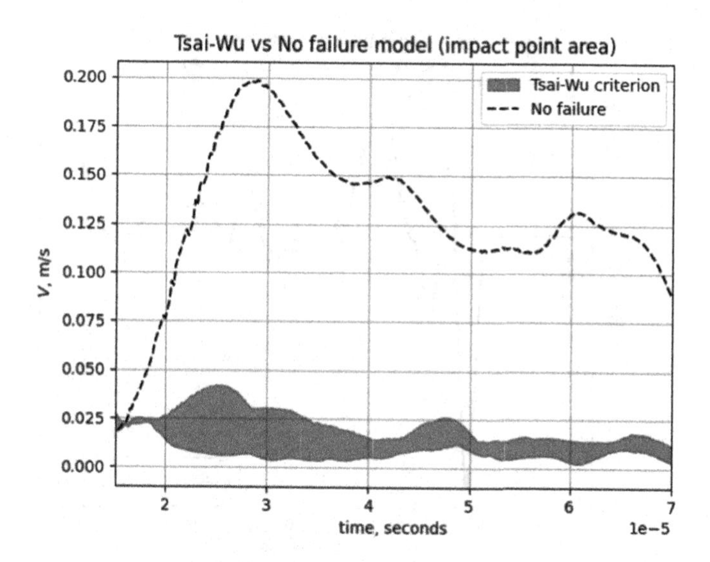


**Fig. 6.** Velocity near impact point.

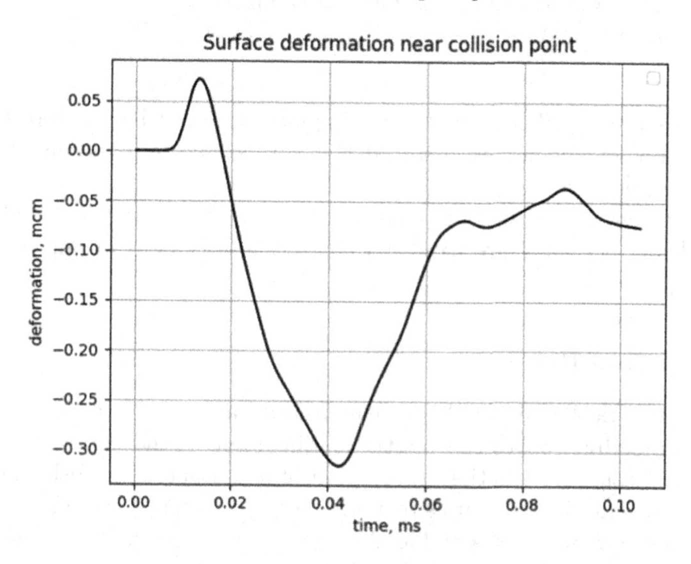


**Fig. 7.** Vertical surface displacement in 30 mm from the impact point.

In this research, we used the linear strenghening model and performed the calculation of a cubic ice sample. The following material parameters were used: Lame parameters $\lambda = 2.9$ MPa and $\mu = 5.7$ MPa, density $\rho = 0.917$ g/cm$^3$. Compressive strength threshold 100 MPa, tensile strength 1.4 MPa, shear strength 1 MPa, coefficient $\tau = 0.001\ s^{-1}$, applied pressure 10 MPa [8].

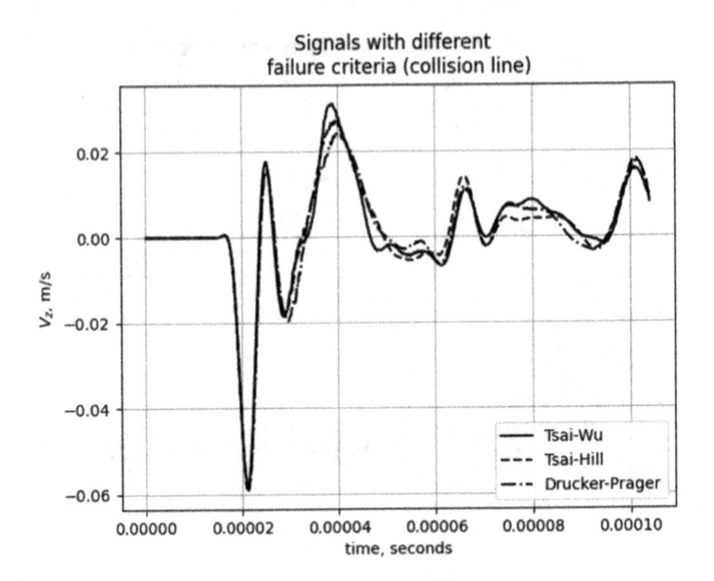


**Fig. 8.** Velocity on impact line for different failure criteria and the same strength thresholds: $X_c = Y_c = X_t = Y_t = 1.8$ MPa; $S = 0.7$ MPa

Strengthening coefficient values in this model do not have a direct physical interpretation and values were sorted through to aquire a reasonable damage area in the specimen.

General view of calculation area is given on Fig. 9. Border conditions on lateral sides of the cube are fixed borders, circle areas on top and bottom are used to apply loading.

### 4.2 Numerical Results

The characteristic shape of failure area without hardening is given in [14]. After the shock unloading, waves go from the loading areas to the center of the sample, interfere and cause destruction in a region of a characteristic disk shape, separated from all sample boundaries by undamaged material. Variations of material strength, destruction criteria and loading area shape lead to different shapes and sizes of the destructed area, but the center of the specimen is destroyed in any case that destroys the surrounding material. In some statements, a ring-shaped form that was obtained in the experiment, is visible, but is accompanied by the damaged area in the center.

The characteristic shape of the failure area with material hardening is given on Fig. 10. Elastic waves behave in the same way as in the previous statement, but the central area has higher strength thresholds due to previous slow compressive loading.

To estimate the shape of the hardened area, we can use the pattern of consumed energy distribution after the initial stage of loading – slow compression

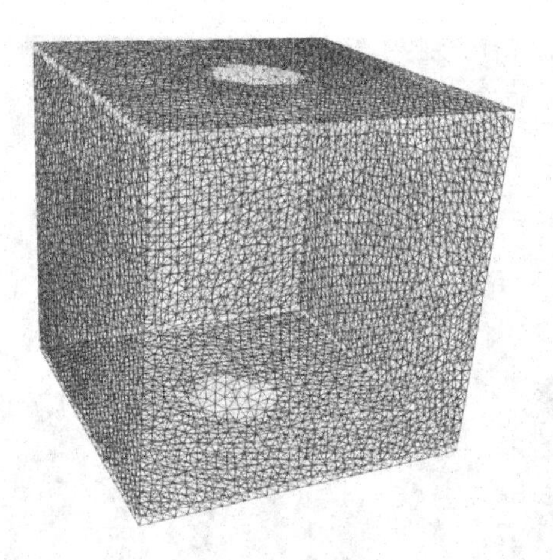

**Fig. 9.** General view of calculation area.

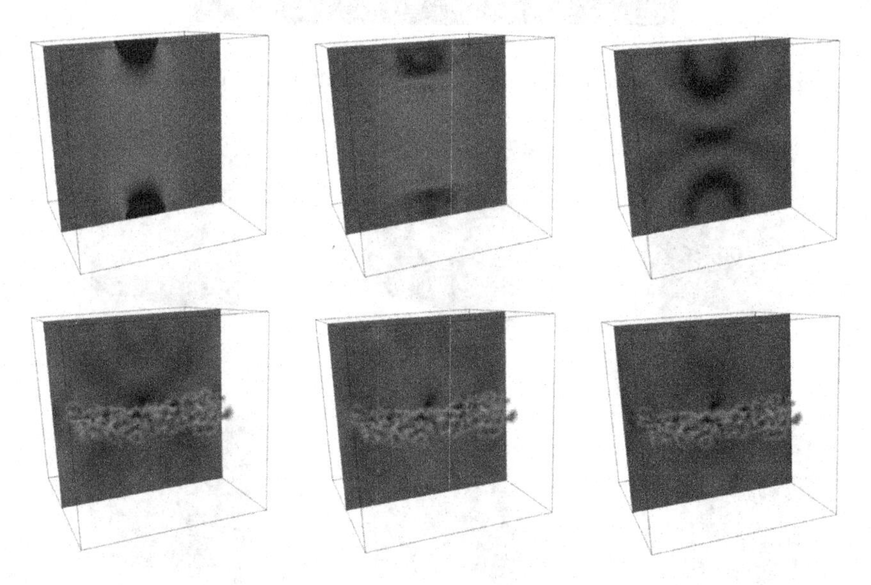

**Fig. 10.** Elastic wave patterns and damaged areas during the shock unloading, consecutive time steps. The vertical section shows the vertical component of the stress tensor, the volume shows nodes that failed according to the Tsai-Wu criterion. Strenghening coefficient value is 71.

(Fig. 11). The hardened area is generally located in the pillar-shaped area in the center, between loading points. Generally, areas closer to a loading point have a higher amount of consumed energy, but the area in the center becomes hardened enough to provide an undestructed area in the center after the shock unloading.

The sensitivity of size and shape of the damaged area to the strengthening coefficient are given on Fig. 12. Generally, the damaged area in these statements has four distinctive parts: two symmetrical areas near the loading points, a small

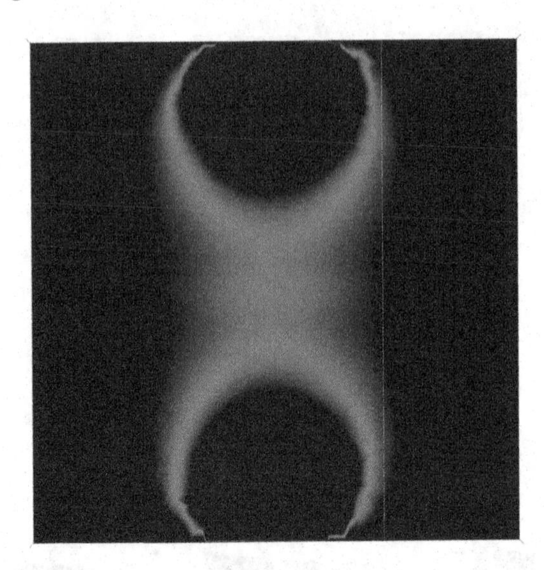

**Fig. 11.** Typical pattern of energy distribution after the initial slow loading.

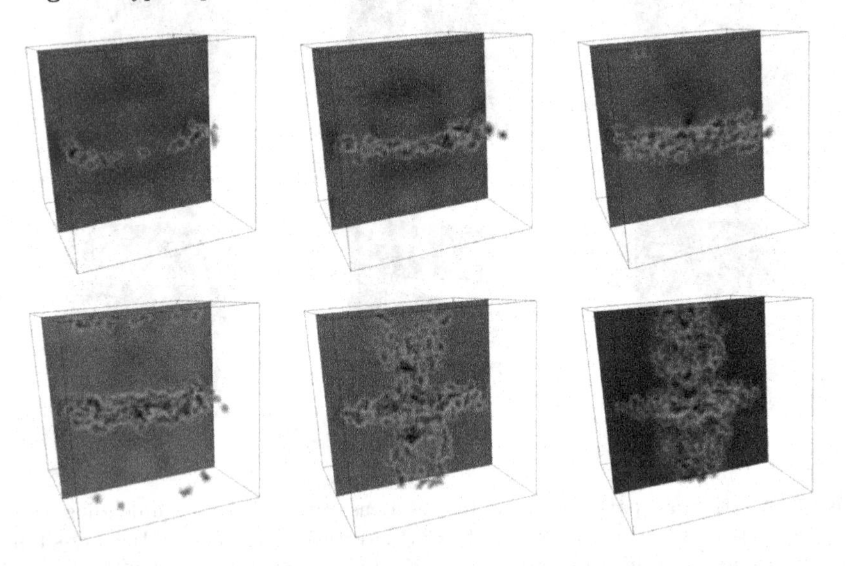

**Fig. 12.** Elastic wave patterns and damaged areas after the shock unloading. The vertical section shows the vertical component of the stress tensor, the volume shows nodes that failed according to the Tsai-Wu criterion. From left to right, different strengthening coefficient value: 200, 125, 90, 58, 5, 0 (no strenghening).

disk-shaped area in the center and a ring-shaped area in the middle of the specimen. The ring-shaped destruction area appears in unstrengthened ice around the strengthened pillar in the center. The increasing of the strengthening coefficient leads to the disappearance of the first three zones, leaving only the ring-shaped damaged area.

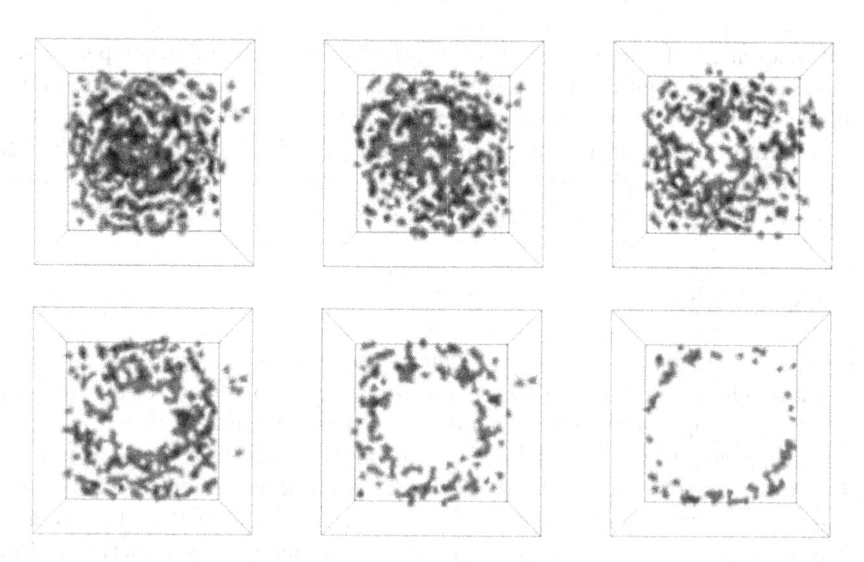

**Fig. 13.** Damaged areas after the shock unloading, view from top. From left to right, different strengthening coefficient value: 50, 58, 71, 90, 125, 200.

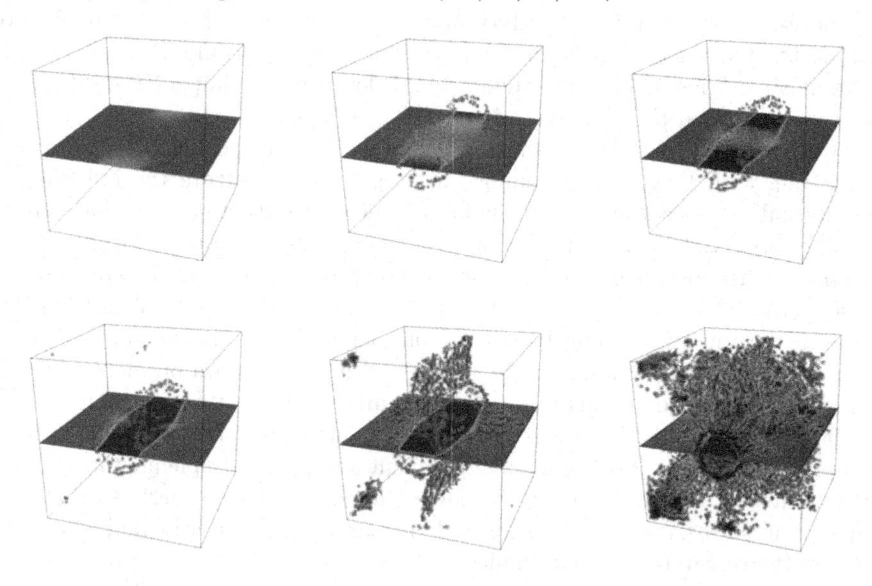

**Fig. 14.** Damaged areas for a statement with slow compression loading and slow unloading. Strengthening coefficient value is 200, maximum pressure is 100 MPa. From left to right, consequent time steps.

Figure 13 shows the view on damaged area from the top. The size of the intact area in the center directly depends on the strengthening coefficient and might be estimated visually.

Another interesting result with a strengthening model was obtained for a similar statement with higher compression and slow unloading. Figure 14 shows the development of damaged area during the loading. The pillar-shaped area in the center remains intact and is surrounded by destructed nodes, while large-scale cracks appear in the remaining volume. It shows certain similarity with columnar fractures from [13] but requires further clarifications in terms of loading regime. Calculations allow us to estimate the shape of the undestructed area in the center and compression regime that can be used to obtain this effect.

## 5 Conclusion

Grid-characteristic numerical method allows to model elastic wave patterns during a low-velocity impact with high precision and implement various material models, including viscosity, failure and linear strengthening.

Two statements were considered in this article. The first one is a fall of a ball-shaped impactor on an ice specimen. Tsai-Wu failure model with different strength values for shear and compression was used to demonstrate the influence of strengh parameters on a signal that can be obtained in the experiment from the surface of the specimen. The significant dependency of signal from strength parameters was not observed. Following steps in this research include direct comparison with experimental data and implementation of various models of the destructed area. Two other failure criteria, Tsai-Hill and Drucker-Prager, were also considered. In the statement with low velocity impact the difference between them and the Tsai-Wu criterion in insignificant.

The second statement is a shock unloading of an ice specimen. Tsai-Wu failure model with linear strenghening was used to demonstrate the influence of an internal model parameter on the large-scale destruction pattern that can be possibly observed in an experiment. Numerical results show qualitative coincidence with experimental data, but require further research. Results in this paper have better coincidence with the experiment than calculations from [14] because the material strengthening is considered. However, the strengthening model under consideration is quite simple and its verification on more complex three-dimensional loading statements is required to assess its applicability.

It should be noted that the setting of the second experiment [13] itself is quite complex and requires both original equipment and precise implementation. This statement can be too difficult to reproduce for measuring properties of natural ice, but it shows complex non-linear behavior of the material and allows to expand the research of ice mechanics.

## References


1. Chelnokov F.B.: Explicit representation of grid-characteristic schemes for the elasticity equations in two-dimensional and three-dimensional spaces. Matematicheskoe Modelirovanie **18**(6), 96–108 (2006). [In Russian]
2. Biryukov, V.A., Miryaha, V.A., Petrov, I.B., Khokhlov, N.I.: Simulation of elastic wave propagation in geological media: intercomparison of three numerical methods. Comput. Math. Math. Phys. **56**(6), 1086–1095 (2016)

3. Muratov, M.V., Petrov, I.B.: Application of fractures mathematical models in exploration seismology problems modeling. In: Smart Modeling for Engineering Systems. GCM50 2018. Smart Innovation, Systems and Technologies, vol. 133, pp. 120–131 (2019)
4. Favorskaya, A., Khokhlov, N.: Modeling the impact of wheelsets with flat spots on a railway track. Procedia Comput. Sci. **126**, 1100–1109 (2018)
5. Beklemysheva, K.A., Golubev, V.I., Petrov, I.B., Vasyukov, A.V.: Determining effects of impact loading on residual strength of fiber-metal laminates with grid-characteristic numerical method. Chin. J. Aeronaut. **34**, 1–12 (2021)
6. Beklemysheva, K.A., Danilov, A.A., Petrov, I.B., Salamatova, V.Y., Vassilevski, Y.V., Vasyukov, A.V.: Virtual blunt injury of human thorax: age-dependent response of vascular system. Russ. J. Numer. Anal. Math. Modell. **30**(5), 259–268 (2015)
7. Christensen, R.M.: Theory of Viscoelasticity. Academic Press, London (1971)
8. Beklemysheva, K.A., Petrov, I.B.: Numerical modeling of high-intensity focused ultrasound with grid-characteristic method. Lobachevskii J. Math. **41**, 2638–2647 (2020)
9. Chan, A.H., Fujimoto, V.Y., Moore, D.E., Martin, R.W., Vaezy, S.: An image-guided high intensity focused ultrasound device for uterine fibroids treatment. Med. Phys. **29**(11), 2611–20 (2002)
10. Maxwell, A., Sapozhnikov, O., Bailey, M.: Disintegration of tissue using high intensity focused ultrasound: two approaches that utilize shock waves. Acoust. Today **8**, 24–36 (2012)
11. Beklemysheva, K.A., Ermakov, A.S., Petrov, I.B., Vasyukov, A.V.: Numerical simulation of the failure of composite materials by using the grid-characteristic method. Math. Models Comput. Simul. **5**(8), 557–567 (2016)
12. Beklemysheva, K.A., Vasyukov, A.V., Kazakov, A.O., Petrov, I.B.: Grid-characteristic numerical method for low-velocity impact testing of fiber-metal laminates. LJM **39**(7), 874–883 (2018)
13. Epifanov, V.P.: Physical mechanisms of ice contact fracture. Dokl. Phys. **52**, 19–23 (2007)
14. Stognii P., Epifanov V., Golubev V., Beklemysheva K., Miryaha V.: The numerical modelling of dynamic processes in the ice samples using the grid-characteristic method. In: Proceedings of the 26th International Conference on Port and Ocean Engineering under Arctic Conditions 14–18 June 2021, Moscow, Russia (2021)

# Dilated W-Net for Geological Inversion Problems

Maksim Nikishin and Alexey Vasyukov(✉)

Moscow Institute of Physics and Technology, Dolgoprudny, Russia
{nikishin.ma,a.vasyukov}@phystech.edu

**Abstract.** The paper is devoted to the application of deep convolutional neural networks for geological inversion problems. The present work studies possible methods for improving the quality of neural network predictions compared to the baseline result obtained earlier using the U-net architecture with default parameters. The following techniques are considered: (a) different tensor padding methods; (b) dilated convolution to extract global features; (c) W-net architecture that follows the original ideas of U-net and extends them. A significant increase in the quality of predictions was obtained when using W-net with dilated convolution for the first sections of the network.

**Keywords:** Geological inversion · Seismic inversion · Convolutional neural networks · U-net · W-net

## 1 Introduction

This paper discusses the usage of convolutional neural networks for geological inversion problems. This class of problems targets the identification of a heterogeneous geological media structure based on an observed wave response from it. This problem belongs to a broad class of coefficient inverse problems. Geological inversion processes the data of seismic surveys during the exploratory search for oil, gas and minerals.

During the seismic survey the scanning pulses are generated on the Earth surface and the response from the underlying structures is recorded by the receivers. This recorded response is the input data for the geological inversion problem. The solver of this problem should provide as the output the structure of the media that caused this response.

Mathematical problem statement can be expressed as follows:

$$\begin{cases} Au(\boldsymbol{r},t) = f(\boldsymbol{r},t), \quad \boldsymbol{r} \in G, t \in [0;T] \\ A = A\left(t, \boldsymbol{r}, \{c_j(\boldsymbol{r})\}\right), j \in \overline{0,N} \\ Pu(t_m, \boldsymbol{r}) = f_m(\boldsymbol{r}), \quad t_m \in [0,T], m \in \overline{0,M} \\ \{c_j(\boldsymbol{r})\} - ? \end{cases}$$

Supported by RSF project 22-11-00142.


D. Balandin et al. (Eds.): MMST 2023, CCIS 1914, pp. 266–279, 2024.
https://doi.org/10.1007/978-3-031-52470-7_21

In this notation $G$ is a domain in $n$-dimensional real space, $u$ is the set of state variables of the system, $A$ is the operator specifying the dynamics of the medium. A is assumed to depend on a finite number of unknown function coefficients $\{c_j(\boldsymbol{r})\}$ $j \in \overline{0, N}$ that should be determined as a solution. Observations $f_m(\boldsymbol{r}), m \in \overline{0, M}$ are known that are related to the vector of variables $u(t_m)$ at various times $t_m \in [0, T]$ with the help of the projection operator $P$. This problem is ill posed from mathematical point of view.

Modern methods for geological inversion problems typically use full wave inversion (FWI) [1]. FWI implies that the distribution of parameters in a medium is reconstructed using an iterative gradient minimization of some loss function measuring the difference between the actual data and the wave response produced by the numerical simulation of the current approximation of media structure. This procedure is considered to be a golden standard, but it is extremely computationally expensive. Classical FWI uses an acoustical approximation of the media, modern variations of the method switch to an elastic approximation, that makes the method even more computationally expensive [2].

This motivates the development of alternative methods, that should be significantly faster compared with classical FWI. These fast methods typically complement classical golden standard approaches rather than replacing them. Recent research in this area is typically based on different machine learning techniques. One can name Bayesian methods [3] and different types of neural networks like convolutional [4] and generative adversarial [5] ones.

The present paper contributes to the field of using convolutional neural networks.

### 1.1 Data

Models of geological media used in this work are taken from [4]. This dataset contains synthetic models in which a large acoustically hard object (salt structure) is located in a layered medium with significantly lower sound velocities. The Fig. 1 below shows examples of media from this set, where the harder media are shown in yellow, and the softer ones, which are background ones, in blue.

Each sample has a length along the OX axis of 3000 m and a depth along the OY axis of 2000 m. The images are 200 by 300 pixels. The speed of sound varies from 2500 m/s in softer media to 4500 m/s in hard media.

Seismograms for these media were computed in [6] and published in machine-readable form [7]. In the present work these samples are used to test various approaches to improve the quality of media structure predictions obtained by convolutional neural networks.

The reason for using synthetic data is the absence of publicly available databases of experimental seismograms and their corresponding geological formations. Machine readable data public availability is extremely rare in this area of research. Only certain individual samples are available, but they cannot be combined into one dataset, since they have fundamentally different parameters,

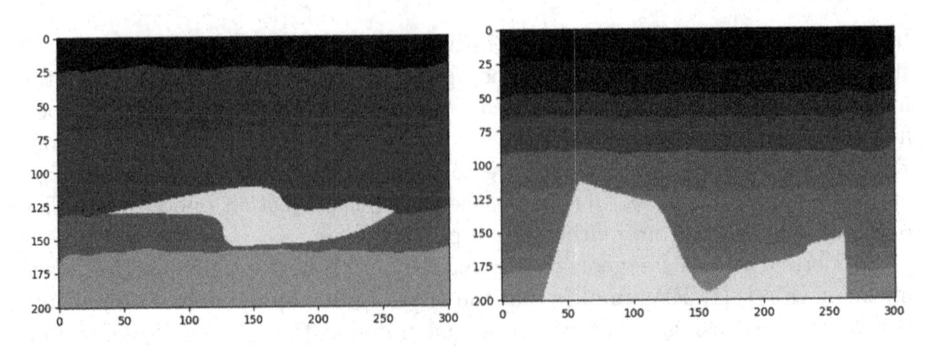


**Fig. 1.** Media samples. (Color figure online)

such as different numbers of pulse source position or different sizes of images of the original geological formations. Working with synthetic data is a commonly used approach when developing tools and methods for geological inversion problems.

The data presented in [7] provides 9 seismograms for each geological medium sample. These seismograms differ in a pulse source position at the upper boundary. Examples of seismograms are shown in Fig. 2:

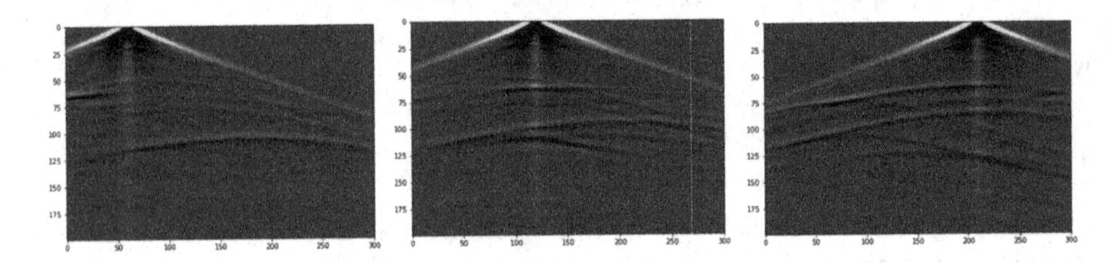

**Fig. 2.** Seismic responses samples.

The neural network receives for training as inputs (a) known distributions of sound velocities in the medium and (b) seismograms obtained by scanning this medium with different locations of the impulse source at the upper boundary of the area. The source positions are the same for all samples. Seismograms for each sample are presented for the network as independent input channels.

The input data of the network during the inference stage is a set of seismograms for an unknown media recorded with different positions of the impulse source (the same positions that were used during the training). The output of the network is the prediction of the distribution of sound speeds in the media under investigation.

# 2 Methods

## 2.1 Network Architecture

The architectures for this work are U-net and its direct improvement W-net [14]. One of main advantages of U-net architecture is low possibility of overfitting [8]. An advantage of W-net architecture is an ability to get a result similar to the basic architecture with a 1-3 times smaller amount of parameters.

A diagram in Fig. 3 shows W-net architecture. W-net architecture is a sequential application of U-net neural networks with output-input communication through convolution.

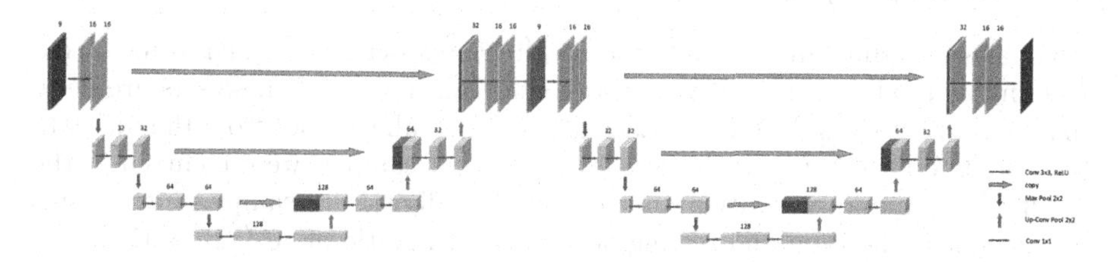


**Fig. 3.** W-net architecture.

Each parallelepiped on the diagram is a tensor of the third rank, the initial tensor is obtained by combining the tensors of the second rank (individual seismograms). The figure shows the number of channels (the number of tensors of the second rank) in each tensor. Operations applied on tensors are indicated by arrows.

The network was implemented using PyTorch deep learning framework [9].

## 2.2 Network Tuning Methods

One of the main methods for improving the quality of convolutional neural network predictions is padding. It is necessary in order to avoid the loss of boundary values when convolving using a kernel larger than a unit one. The main types of padding are: zero padding, reflection padding and replication padding [10].

The previous works demonstrated that reflection padding in all tests gave worse results than replication padding. It was also found that zero padding is more sensitive to changes in values at the borders of the image, while other types of padding can reduce the influence of artifacts at the borders, making the convergence of the convolutional neural network more stable [11].

Another method for improving the quality of the prediction is dilation. It allows to identify larger features at the image without a significant increase in the number of neural network parameters [13].

Recent works show also the positive effect of dilation on noise reduction in convolutional neural networks [12].

### 2.3 Training Method

We used AdamW optimization algorithm [15] to train all convolutional neural networks in this work. The algorithm is a $l_2$ regularization of the adaptive moment estimation method - the Adam algorithm [16].

It should be noted that refutations of the original proof of the convergence of the Adam algorithm have been published [17], but later an alternative proof of the correctness has been proposed [18]. This allows the original algorithm to be used without modification. However, we used Amsgrad fix [17], although it should not significantly contribute to the final accuracy or convergence rate.

### 2.4 Training Pipeline

Comparison of different methods and architectures were performed using 4 fixed training and validation sets created from the full data set. These sets are hereinafter referred to as Set1–Set4. Each set included all samples from the full data set, the difference between the sets is about what samples were included in the groups for training, validation and testing. The division of samples into groups for training, validation and testing occurred in the ratio of 70%/15%/15%.

It should be noted that it was found that the data set contains individual samples with anomalously high velocity values in seismograms. The nature of these anomalies was not considered in this work. Anomalous samples were excluded from consideration. The exclusion was based on filtering by the maximum velocity value modulus in the seismogram. If a certain threshold was exceeded, the sample was excluded. Based on this procedure subsets were obtained for each set. These subsets differ in the level of filtration. The subsets are numbered alphabetically. The subset with the highest threshold received the letter $A$, the one with the middle value received $B$, and the one with the lowest value received $C$. The subset with the letter $B$ is used further in the work if the letter is not mentioned. This subset is considered to have a sufficient degree of filtering for further processing.

AdamW optimization algorithm with the Amsgrad option was used for all the experiments using the gradient descent step $\alpha = 0.001$ and the remaining constants $\epsilon, \beta_1, \beta_2, \lambda$ equal to $10^{-8}, 0.9, 0.99, 0.01$ respectively.

The depth of the neural network was chosen based on the fact that a tensor size of 128 on the bottom layer is desirable. This value was determined empirically - the networks generally showed the best results with this tensor size. Switching to a tensor size of 256 caused the network with W-net architecture to be oversaturated with free parameters. This did not allow the network to converge stably.

### 2.5 Metrics for Evaluation

Before we consider the results, a little should be said about the methods of comparing the obtained predictions and the ground truth images. Analytical similarity may be shown well by using the mean squared error (MSE) metric. This metric was used as a main metric for the present work.

However, MSE doesn't represent a visual similarity of images. So, an additional structural similarity index measure (SSIM) [19] was considered. SSIM as a widely used method for comparing the similarity of images in the problem of geological inversion. Using SSIM allows us to compare our results with the original work [4] that uses the same dataset of geological structures.

## 3 Numerical Experiments

### 3.1 Results for Padding

Padding has made a significant contribution to image quality. The predictions without padding and with different types of padding are presented in Figs. 4 and 5

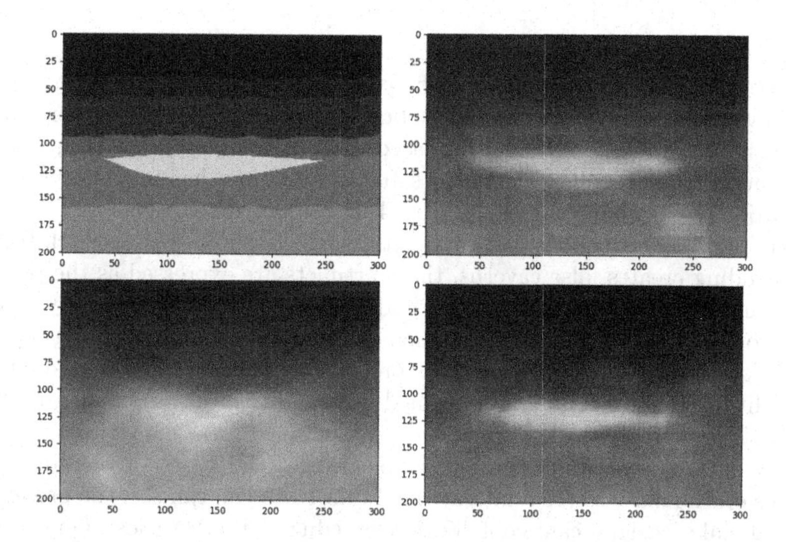


**Fig. 4.** Ground truth image (top left); U-net architecture prediction without padding (bottom left); U-net architecture prediction with zero padding (top right); U-net architecture prediction with padding replicate (bottom right)

One can see that the network without padding (Fig. 4 bottom left) gives a prediction that is weakly consistent with the original image. This happens due to effects at the lower levels of the encoder. The implementation of MaxPool function causes that in some cases some boundary values are not used. This affects critically the ability to restore the image structure in the decoder, if the compression is high.

The present work compares the results for zero-padding and padding replicate.

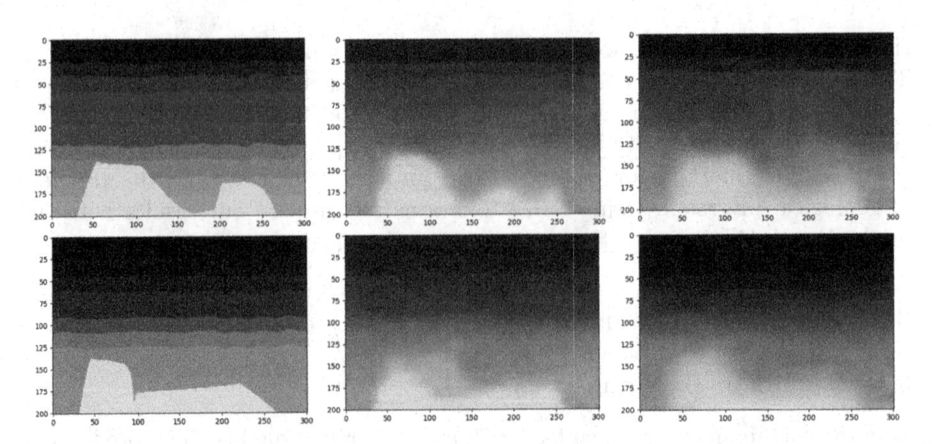


**Fig. 5.** Test images (left); predictions of a neural network based on the W-net architecture with zero padding (middle); neural network predictions based on W-net architecture with padding replicate (right).

U-net architecture with zero-padding features the presence of noisy echos of central values propagating towards the boundaries. However, the harder acoustic area is specified more precisely compared with padding replicate (Fig. 4 on the top right). Padding replicate with the same architectures has small noise velocity fluctuations throughout the image (Fig. 4 on the bottom right).

The Fig. 5 shows that W-net retains the main features of U-net architecture. Zero-padding creates false caverns, these artifacts are expressed as the presence of two small peaks on Fig. 5 at the top middle image and as an imaginary increase in the sound speed near the peak at the bottom middle image. Padding replicate provides less contrast images but gives predictions that are more consistent with the reality. These experiments confirmed the validity of using padding replicate as a basis for further work.

In addition, we will present a comparison of the types of padding or its absence on the most filtered data at the Table 1. As you can see, the presence of padding makes a significant contribution to reducing the MSE loss. On the other hand, padding replicate gives better results than zero padding on well-filtered data, but the difference is not significant.

As the result, we suppose that padding replicate provides less contrast images, but gives predictions that are more consistent with the reality. These experiments confirmed the validity of using padding replicate as a basis for further work.

**Table 1.** The results of training neural networks based on the U-net architecture with different types of padding in training datasets with maximized filtering. NP - No Padding; PZ - Padding Zero; PR - Padding Replicate.

| Model | | UNet NP | UNet PZ | UNet PR | UNet NP | UNet PZ | UNet PR |
|---|---|---|---|---|---|---|---|
| Dataset | Count | Data MSE | | | Data SSIM | | |
| Set1C | mean | 0.004805 | 0.003302 | 0.003250 | 0.886887 | 0.878468 | 0.881388 |
| | std | 0.002106 | 0.001526 | 0.001495 | 0.015326 | 0.012319 | 0.013767 |
| | min | 0.001451 | 0.001261 | 0.001320 | 0.828008 | 0.837255 | 0.834158 |
| | max | 0.011806 | 0.009316 | 0.009594 | 0.917019 | 0.908387 | 0.913253 |
| Set2C | mean | 0.004967 | 0.003642 | 0.003584 | 0.880364 | 0.861616 | 0.869227 |
| | std | 0.002190 | 0.001817 | 0.001730 | 0.013940 | 0.013101 | 0.012165 |
| | min | 0.001860 | 0.001336 | 0.001505 | 0.824468 | 0.826696 | 0.831549 |
| | max | 0.012603 | 0.011275 | 0.014616 | 0.913829 | 0.888999 | 0.898794 |
| Set3C | mean | 0.004474 | 0.003402 | 0.003808 | 0.878570 | 0.870495 | 0.870042 |
| | std | 0.002211 | 0.001312 | 0.001756 | 0.014897 | 0.013957 | 0.014955 |
| | min | 0.001789 | 0.001291 | 0.001578 | 0.819320 | 0.810008 | 0.828690 |
| | max | 0.016657 | 0.009345 | 0.012189 | 0.910683 | 0.899862 | 0.904928 |
| Set4C | mean | 0.004464 | 0.003330 | 0.003165 | 0.883290 | 0.875162 | 0.881773 |
| | std | 0.001960 | 0.001744 | 0.001611 | 0.016219 | 0.013847 | 0.012783 |
| | min | 0.001408 | 0.001081 | 0.001229 | 0.825702 | 0.829256 | 0.833317 |
| | max | 0.013263 | 0.016024 | 0.015055 | 0.918885 | 0.909248 | 0.910429 |

### 3.2 Results for Dilated W-Net

The main method of improving the quality of prediction in the present work is a switch to W-net. This architecture allows to capture global structure of the image and its local features using different sections of the network. Dilation was used to handle global features, while local features were captured using standard convolution.

The first stage of the experiments in this area was a study which features should be extracted first. The results can be seen in Fig. 6.

The images with the primary extraction of local features are shown in the center column of the figure. This neural network has dilation rate 1 on the first two sections and dilation rate 2 on the later two. The right side of the figure shows images for a network with a reverse feature extraction sequence, first global and then local. Accordingly, the structure has dilation rate 2 on the first two sections and rate 1 on the later two.

The Fig. 6 shows clearly that extracting local features first provides blurred resulting image. This indicates that information about the local structure of the image is lost worse is this case compared with determining local features on the last sections of W-net networks.

This draws a conclusion that it is beneficial to extract global features first.

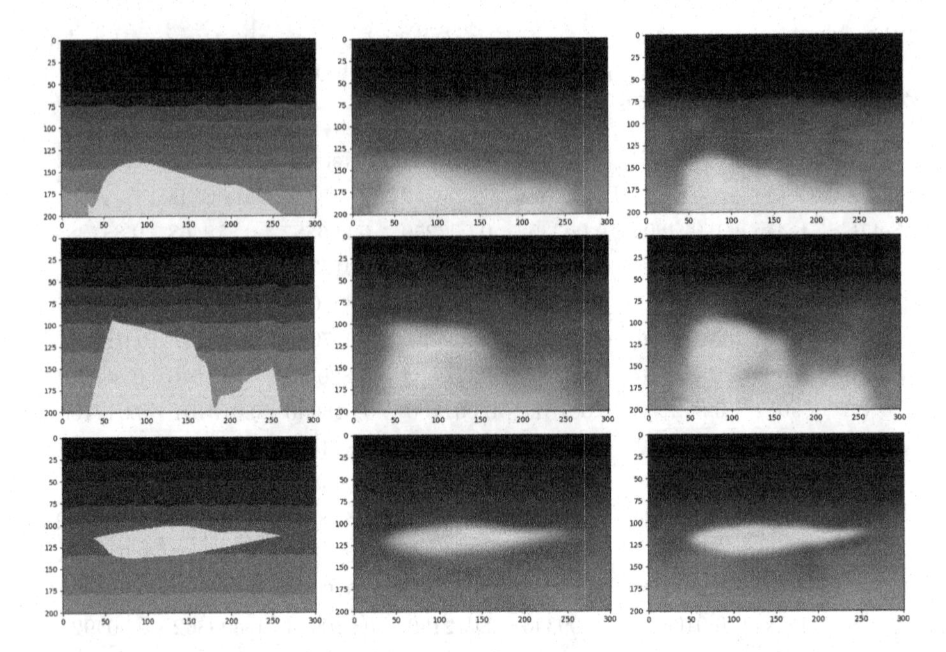

**Fig. 6.** Ground truth images (left); W-net predictions with a dilation rate 1-2 (centered); W-net predictions with a dilation rate 2-1 (right).

Another part of the experiments was the check how the dilatation rate of the convolution core influences the quality of predictions when extracting global features of the image. The networks with dilation rates 2, 3 and 5 were tested. Typical predictions can be seen in the Fig. 7, where dilation rate 2 is in the upper right, rate 3 is in the lower left, and rate 5 is in the lower right corner. It can be seen from the Fig. 7 that the dilation rate 2 is not enough to determine the global features of the image, while the dilation rate 5 causes the loss of local information leading to the smoothing of the boundaries of acoustically dense media.

This draws a conclusion that dilation rate 3-1 should be chosen as the most promising one.

### 3.3 Architectures Comparison

The Table 2 shows the final comparison of different neural networks architectures. The baseline network in the present work is U-net with padding replicate. The presented results indicate that W-net architecture retains the features of the original neural network. The dilation applied on the first sections of the network allows to capture the global features of the data. This provides a significant improvement in predictions quality, determined by the similarity with the ground truth by measuring both SSIM index and MSE loss.

**Table 2.** The results of testing different architectures. Padding is replicative. A neural network with dilatation 3-1 did not converge on the 4th data set (highlighted in red), so it is not taken into account.

| Dataset | | Set 1b | Set 2b | Set 3b | Set 4b |
|---|---|---|---|---|---|
| Model | Value | Data | | | |
| UNet MSE | count | 228.000000 | 228.000000 | 228.000000 | 228.000000 |
| | mean | 0.004130 | 0.004544 | 0.003795 | 0.003881 |
| | std | 0.002013 | 0.002550 | 0.002715 | 0.002140 |
| | min | 0.001749 | 0.001553 | 0.001290 | 0.001224 |
| | max | 0.015220 | 0.018304 | 0.023707 | 0.012676 |
| UNet SSIM | count | 228.000000 | 228.000000 | 228.000000 | 228.000000 |
| | mean | 0.869493 | 0.867275 | 0.872588 | 0.876020 |
| | std | 0.013004 | 0.016399 | 0.013402 | 0.012511 |
| | min | 0.831013 | 0.770303 | 0.826365 | 0.824120 |
| | max | 0.897508 | 0.896708 | 0.901034 | 0.903738 |
| WNet MSE | count | 228.000000 | 228.000000 | 228.000000 | 228.000000 |
| | mean | 0.002651 | 0.002908 | 0.003175 | 0.002807 |
| | std | 0.001450 | 0.001873 | 0.002071 | 0.001523 |
| | min | 0.000946 | 0.000854 | 0.001006 | 0.000861 |
| | max | 0.011143 | 0.013613 | 0.024249 | 0.011993 |
| WNet SSIM | count | 228.000000 | 228.000000 | 228.000000 | 228.000000 |
| | mean | 0.898853 | 0.898582 | 0.898969 | 0.899764 |
| | std | 0.013549 | 0.016072 | 0.013143 | 0.013235 |
| | min | 0.846000 | 0.855861 | 0.861085 | 0.843490 |
| | max | 0.929145 | 0.930040 | 0.938533 | 0.926261 |
| WNet dilation rate 2-1 MSE | count | 228.000000 | 228.000000 | 228.000000 | 228.000000 |
| | mean | 0.002647 | 0.002310 | 0.002649 | 0.002473 |
| | std | 0.001563 | 0.002902 | 0.001856 | 0.001692 |
| | min | 0.000942 | 0.000839 | 0.000886 | 0.000815 |
| | max | 0.010629 | 0.042404 | 0.020321 | 0.013632 |
| WNet dilation rate 2-1 SSIM | count | 228.000000 | 228.000000 | 228.000000 | 228.000000 |
| | mean | 0.901642 | 0.908859 | 0.903897 | 0.906870 |
| | std | 0.014892 | 0.012722 | 0.011002 | 0.013198 |
| | min | 0.848805 | 0.864716 | 0.869478 | 0.857376 |
| | max | 0.930052 | 0.933238 | 0.935458 | 0.938613 |
| WNet dilation rate 3-1 MSE | count | 228.000000 | 228.000000 | 228.000000 | 228.000000 |
| | mean | 0.001832 | 0.002190 | 0.001967 | 0.024634 |
| | std | 0.000985 | 0.002252 | 0.002057 | 0.007062 |
| | min | 0.000706 | 0.000720 | 0.000555 | 0.010504 |
| | max | 0.008034 | 0.024546 | 0.026968 | 0.048186 |
| WNet dilation rate 3-1 SSIM | count | 228.000000 | 228.000000 | 228.000000 | 228.000000 |
| | mean | 0.910902 | 0.907880 | 0.911993 | 0.838094 |
| | std | 0.012978 | 0.014779 | 0.012177 | 0.016390 |
| | min | 0.850412 | 0.859612 | 0.879887 | 0.799805 |
| | max | 0.938352 | 0.934677 | 0.936811 | 0.888724 |

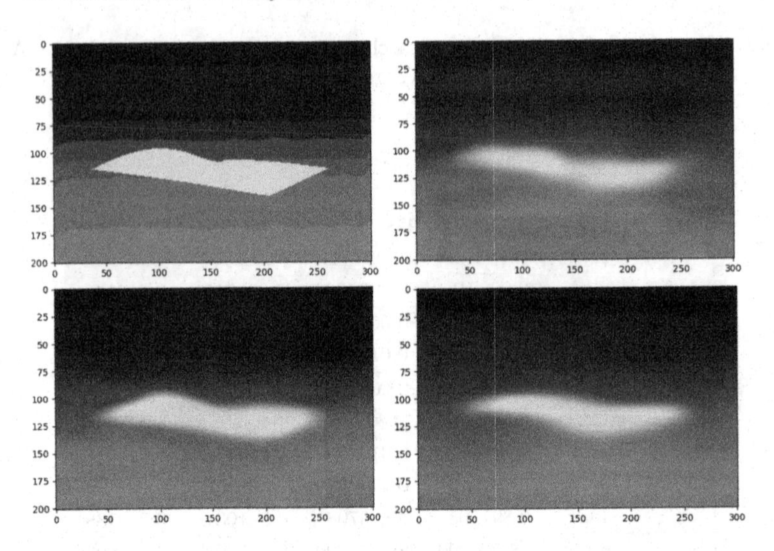


**Fig. 7.** Ground truth image (upper left corner); W-net prediction for dilation rate 2-1 (upper right corner); W-net prediction for dilation rate 3-1 (lower left corner); W-net prediction for dilation rate 5-1 (lower right corner).

### 3.4 Training Nuances

It is worth mentioning the features observed during the training of the networks. The Figs. 8, 9, 10 show how the metrics change during the training. It can be seen that by the 50th epoch the validation loss exceeds the training error. This indicates the beginning of overfitting.

The Fig. 8 shows a typical graph for training a network based on U-net architecture with different padding. The training without padding features causes a plateau at the beginning of training, which indicates the presence of a barrier in training such models.

A similar feature (but to a lesser extent) is demonstrated by W-net and dilated W-net with a 5-1 dilation rate (Fig. 9). In the first case it is happened due to the presence of two similar combinations of encoder and decoder sections. In the second case it is happened due to the waste of some local information in the first combination of encoder and decoder sections caused by the usage of a greater degree of dilation than was necessary.

W-net models with dilatation rate 2-1 and 3-1 (Fig. 10) reach the overfitting plateau quite quickly (in about 15 epochs). This indicates the rapid convergence of these architectures and ensures stable learning.

It is worth mentioning separately a non-converged neural network based on the W-net architecture with dilation rate of 3-1 (Fig. 11). After the start of successful training this neural network fell out of the convergence area and couldn't return. This indicates that using this architecture requires several models to be trained to confirm the convergence.

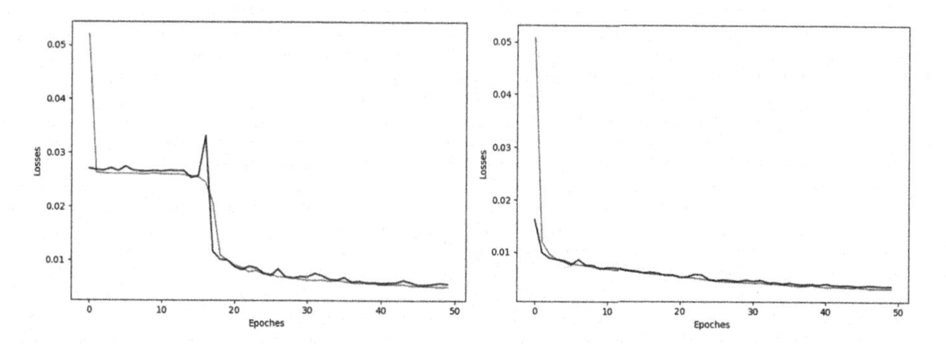


**Fig. 8.** Loss evolution during the training for U-net architecture with different padding. Red is the training loss, green is the validation loss. Left image is for the model without padding, right image is for the model with padding. (Color figure online)

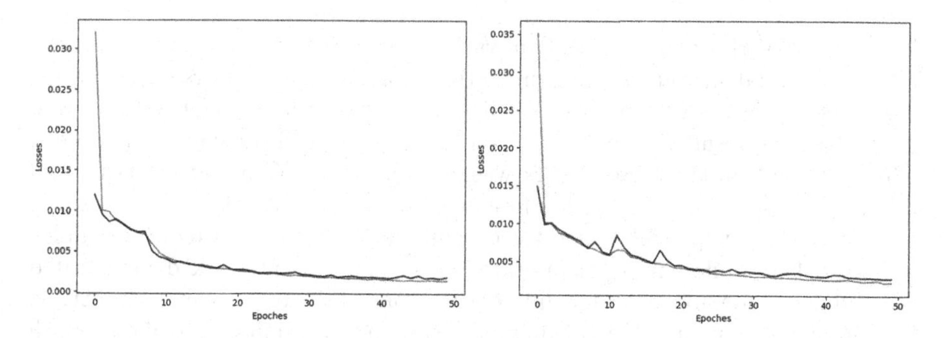


**Fig. 9.** Loss evolution during the training for W-net networks. Red is the training loss, green is the validation loss. Left image is for the model with dilatation rate 5-1, right image is for the model without dilatation. (Color figure online)

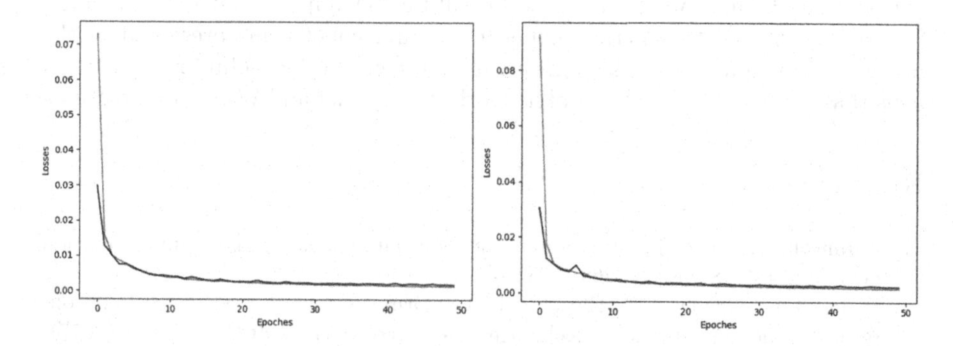


**Fig. 10.** Loss evolution during the training for W-net networks. Red is the training loss, green is the validation loss. Left image is for the model with dilation rate 2-1, right image is for the model with dilation rate 3-1. (Color figure online)

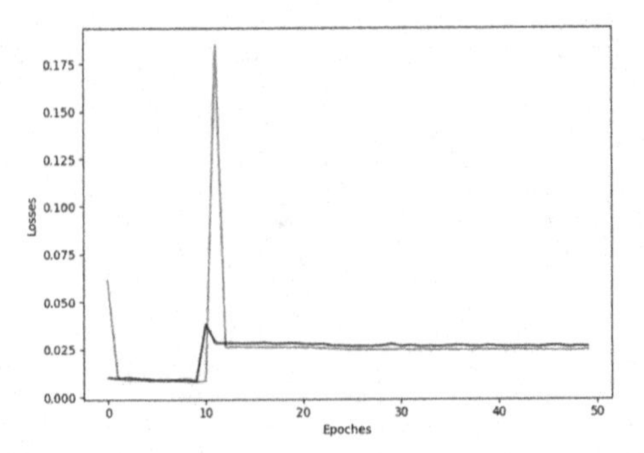


**Fig. 11.** Loss evolution during the training for non-converged W-net with dilation rate 3-1. Red is the training loss, green is the validation loss. (Color figure online)

## 4 Conclusions

The present work provides a study of various methods for improving the quality of convolutional neural network predictions for geological inversion problems. The paper describes the options to achieve a significant improvements compared to the baseline result that used the U-net architecture with default parameters. The best result in the present work was obtained using W-net architecture and dilated convolution applied to the first sections of the network.

The original paper [4] that set a baseline in this area reported SSIM index around 0.4 for different problem statements. The present work demonstrated SSIM index to be above 0.9 for different versions of dilated W-net architecture. This improvement caused an adequate increase in the training time of one epoch from 90 s on the U-net architecture to 190 s on the W-net architecture.

These results show that commonly used convolutional networks can be significantly improved if finetuned for a specific problem. Open datasets and common metrics are only starting to be adopted for seismic inversion problems, so we expect that the next works in this field will further improve our recent results.

We suppose that numerical experiments and conclusions presented in this paper are not limited to seismic inversion and can also be useful for other problems that use convolutional neural networks with U-net and W-net architectures.

## References

1. Tarantola, A.: Inversion of seismic reflection data in the acoustic approximation. Geophysics **49**(8), 1259–1266 (1984)
2. Wang, T.F., Cheng, J.B.: Elastic full waveform inversion based on mode decomposition: the approach and mechanism. Geophys. J. Int. **209**(2), 606–622 (2017)
3. Grana, D., Fjeldstad, T., Omre, H.: Bayesian Gaussian mixture linear inversion for geophysical inverse problems. Math. Geosci. **49**, 493–515 (2017)
4. Fangshu, Y., Jianwei, M.: Deep-learning inversion: a next-generation seismic velocity model building method. Geophysics **84**(4), R583–R599 (2019)

5. Wu, B., Meng, D., Zhao, H.: Semi-supervised learning for seismic impedance inversion using generative adversarial networks. Remote Sens. **13**(5), Article no. 909 (2021)
6. Stankevich, A., Nechepurenko, I., Shevchenko, A., Gremyachikh, L., Ustyuzhanin, A., Vasyukov, A.: Learning velocity model for complex media with deep convolutional neural networks. arXiv:2110.08626 (2021)
7. Stankevich, A., Nechepurenko, I., Shevchenko, A., Gremyachikh, L., Ustyuzhanin, A., Vasyukov, A.: Numerical nine-shot seismo records for 1600 acoustic impedance distributions [Data set]. https://doi.org/10.5281/zenodo.5515485. Accessed 30 Aug 2023
8. Ronneberger, O., Fischer, P., Brox, T.: U-Net: convolutional networks for biomedical image segmentation. In: Navab, N., Hornegger, J., Wells, W.M., Frangi, A.F. (eds.) MICCAI 2015. LNCS, vol. 9351, pp. 234–241. Springer, Cham (2015). https://doi.org/10.1007/978-3-319-24574-4_28
9. Paszke, A., et al.: PyTorch: an imperative style, high-performance deep learning library. In: Advances in Neural Information Processing Systems, vol. 32, pp. 8024–8035 (2019)
10. PyTorch documentation on torch.nn.modules.padding. https://pytorch.org/docs/stable/_modules/torch/nn/modules/padding.html. Accessed 30 Aug 2023
11. Liu, G., et al.: Partial convolution based padding. arXiv:1811.11718 (2018)
12. Spuhler, K., Serrano-Sosa, M., Cattell, R., DeLorenzo, C., Huang, C.: Full-count PET recovery from low-count image using a dilated convolutional neural network. Med. Phys. **47**(10), 4928–4938 (2020)
13. PyTorch documentation on torch.nn.Conv2d. https://pytorch.org/docs/stable/generated/torch.nn.Conv2d.html. Accessed 30 Aug 2023
14. Xia, X., Kulis, B.: W-Net: a deep model for fully unsupervised image segmentation. arXiv:1711.08506 (2017)
15. Loshchilov, I., Hutter, F.: Decoupled weight decay regularization. arXiv:1711.05101 (2017)
16. Kingma, D.P., Adam, B.: A method for stochastic optimization. arXiv:1412.6980 (2014)
17. Reddi, S.J., Kale, S., Kumar, S.: On the Convergence of Adam and Beyond. arXiv:1904.09237 (2018)
18. Bock, S., Goppold, J., Weiß, M.: An improvement of the convergence proof of the ADAM-Optimizer. arXiv:1804.10587 (2018)
19. Wang, Z., Bovik, A.C., Sheikh, H.R., Simoncelli, E.P.: Image quality assessment: from error visibility to structural similarity. IEEE Trans. Image Process. **13**(4), 600–612 (2004)

# The Choice of Evaluation Metrics in the Prediction of Epileptiform Activity

Nikolay Gromov[1], Albina Lebedeva[1], Ivan Kipelkin[1], Oksana Elshina[2], Konstantin Yashin[2], Lev Smirnov[1], Tatiana Levanova[1](✉), and Svetlana Gerasimova[1]

[1] Lobachevsky University, Nizhny Novgorod 603950, Russia
tatiana.levanova@itmm.unn.ru
[2] Privolzhsky Research Medical University, Nizhny Novgorod 603022, Russia

**Abstract.** In this study, we investigate the problem of prediction of epileptiform activity from EEG data using a deep learning approach. We implement LSTM deep neural network and study how the quality of the prediction depends on the choice of measures of observational error, such as MAPE and RMSE. We show that comparison of results obtained using different metrics is important to obtain a comprehensive assessment of the problem.

**Keywords:** epileptiform activity · EEG · seizure prediction · long short-term memory (LSTM) · MAPE · RMSE

## 1 Introduction

Epilepsy is a neurological disorder that affects over 50 million people worldwide [1]. A seizure is an abnormal and transient electrical activity in the brain which can cause lapses in attention and memory, sensory hallucinations, and full-body convulsions. These persistent seizures present a risk of injury, reduce the mobility, and can lead to both economic hardship and social isolation [2]. Approximately every third patient with epilepsy cannot be fully cured with antiepileptic drugs [3].

Electroencephalogram (EEG) can detected the patients' brain activities and therefore be used for epileptiform activity prediction and analysis [4]. Numerous methods have already been proposed to address this problem, including linear methods [5], nonlinear dynamical theory [6], statistical approach [7]. Among best nonlinear measures compared to other nonlinear features one can point out the mean phase coherence [8], Shannon entropy index [9], and conditional probability index [10].

Machine learning approaches that proved their efficacy in many prediction tasks are also widely used for prediction of epileptic seizures [11–14]. The problem of predicting epileptic seizures can be formalized in two main settings. The first setting is a classification problem; the second one is time series prediction.


D. Balandin et al. (Eds.): MMST 2023, CCIS 1914, pp. 280–293, 2024.
https://doi.org/10.1007/978-3-031-52470-7_22

Depending on the type of formalization, different machine learning architectures can be used. In case of classification task, the standard pipeline in this case consist of several independent steps, such as pre-processing, feature extraction, classifier training, and post-processing. In case of time series prediction, various deep learning approaches can be used. Note that in this case one can predict not only the future seizure (ictal stage), but all future activity (including pre-ictal and inter-ictal activity), which can help to gain more insights in the nature of epileptiform activity.

Deep neural networks (DNN) have a great potential to learn temporal and spatial dependencies. Also, DNN can be effectively used for EEG signal pre-processing and automatic feature extraction tasks [15–19]. Thus, it came as no surprise that most widely used deep learning approaches here are convolutional neural networks [20], long short-term memory networks [21,22] and transformers [23–25].

Nevertheless, the problem of seizure prediction based on EEG signal is yet to be fully solved [26]. Namely, the results of prediction can depend on various factors, including the selected features, type of formalization (particular machine learning problem setting), choice of statistical technique for formalization, system design parameters (e.g. the choice of lead seizures, prediction window etc.), splitting the data set into train and test parts, and, of course, performance metrics. This uncertainty combines with the fact that seizure prediction models are patient-specific, which leads to ill-posedness of the problem. In a number of studies, see e.g. [27–31], several seizure prediction systems were proposed.

In our study we demonstrate the importance of properly chosen performance metrics and system design parameters for meaningful seizure prediction comparisons. The paper is organized as follows. In Sect. 2 we describe a data set and methods of EEG data preprocessing applied to it. In Sect. 3 we describe an architecture of an LSTM network we used to predict epileptic seizures. We present the comparison of results for different measures of observational error in Sect. 4. In Sect. 5 we discuss our findings, before we draw our conclusions in Sect. 6.

## 2 EEG Data

### 2.1 Data Source

As a source of EEG data we used CHB-MIT Scalp EEG Database freely available at Physionet [32]. The data is collected from Children's Hospital Boston (CHB), composed of EEG recordings observed for several days. This is the only open database that contains continuous long-term scalp EEG recordings suitable for the purpose of our study.

The recording is grouped into 23 cases collected from 22 subjects (17 females of age 1.5–19 years old and five males aged 3–22 years old [33]. The signals are sampled at 256 samples/sec of 16-bit resolution, which includes 23 EEG signals.

EEG electrode position and nomenclature (International 10–20 systems) are also recorded. In some cases, other signals are also recorded (ECG signal). The EEG signals are in easy to read format where the dummy signals are ignored. The files are composed of 664 .edf files, and the records (seizures) list the files with more seizures [33]. The records include 198 seizures (182 original sets with 23 cases). The long sequence of EEG samples contains seizure intervals, the beginning and the end of which are annotated in separate .seizure annotation files. These seizure intervals are termed as ictal stages. The example of EEG data that contains recorded seizures is shown in Fig. 1.

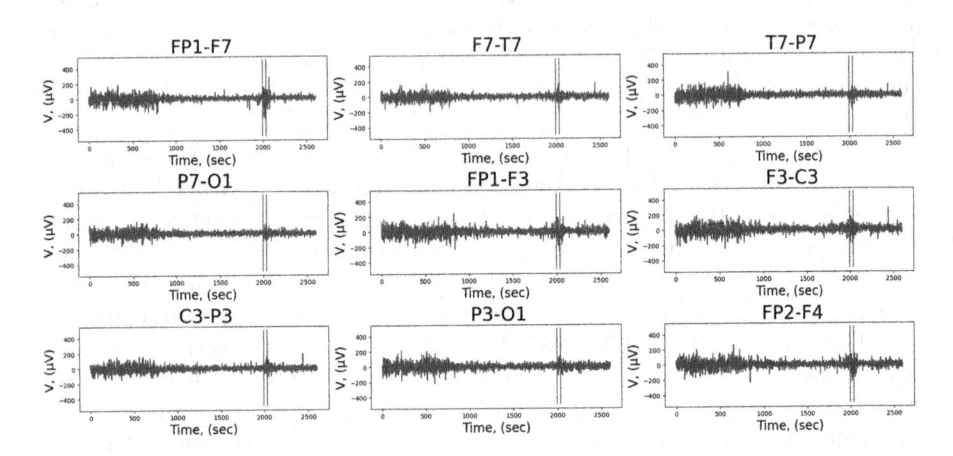


**Fig. 1.** Example of epileptic EEG signals from CHB-MIT Scalp EEG Database: frontal area (left column), occipital area (middle column), and other areas (right column).

In total, 18 channels are consistent across all 24 cases: FP1-F7, F7-T7, T7-P7, P7-O1, FP1-F3, F3-C3, C3-P3, P3-O1, FP2-F4, F4-C4, C4-P4, P4- O2, FP2-F8, F8-T8, T8-P8, P8-O2, FZ-CZ and CZ-PZ. We conducted two types of numerical experiments: (i) single-channel prediction, for which data only from FP1-F7 channel were used, and (ii) multi-channel prediction of epileptiform activity using all channels. The first approach (single-channel prediction) allows one to reduce the workload to the total number of electrode points. The second one (multi-channel prediction) allows to take into account all obtained recordings from 18 channels and dependencies between them.

### 2.2 Data Preprocessing

For numerical experiments, we normalized the data by converting them to zero mean and unit variance. After that, the noise was filtered with a Gaussian filter with $\sigma = 3$ and with $radius = 5$. The data was then separated into inputs and outputs for the model ("time sequence-response" format). The input data consisted of 20 time samples, and the output was the 21-st sample. The optimal number of time samples in the input data was determined experimentally during the optimization process. After normalizing and splitting the data, we split it into train and test sets in a 4:1 ratio.

## 3 The Model

### 3.1 LSTM Architecture

Long short-term memory networks (LSTM) are capable of learning long term dependencies in given data. This is the reason why LSTM networks are first choice networks for time series prediction, including epilepsy prediction from EEG data [21,22]. Also, in our previous study LSTM network allowed to obtain qualitative prediction of extreme events from artificially generated time series [34] and local field potential (LFP) data collected in experiments with mice [35].

In our study, we used 2-layer LSTM architecture as in [35]. The model includes a first linear input layer of size $1 \times 100$ with the Relu activation function. After it 2 LSTM cells are placed, each with 100 neurons in hidden state, and the output from these cells are the last hidden state from both cells. The result of these operations is projected by the last linear output layer with a weight matrix of size $200 \times 1$. This final output layer allows us to obtain the predicted value.

To train the weight matrices and shifts using error back propagation method we employ a mean squared error (MSE) loss function:

$$MSE = \frac{1}{n}\sum_{i=1}^{n}(y_i - \hat{y}_i)^2, \quad (1)$$

where n is a number of training sequences, $y_i$ is true sample and $\hat{y_i}$ is predicted sample.

### 3.2 Measures for the Observational Error

In previous studies researchers used different observational error metrics to examine the quality of a deep learning models by quantitatively measuring the accuracy in terms of RMSE, co-efficient of determination ($R^2$), MSE and MAE [36]. As the errors are not generalized and specific to the problem, thus overall prediction performance can be better judged using a combination of prediction methods.

We compared the applicability and the efficiency of 2 well known metrics for the observational error for the problem of epileptic seizure prediction. Our choice of evaluation metrics is conditioned by the fact that both metrics have very intuitive interpretation in terms of relative error.

The first one is root-mean-square error (RMSE):

$$RMSE = \sqrt{\frac{1}{n}\sum_{i=1}^{n}(y_i - \hat{y}_i)^2}. \tag{2}$$

Here the meaning of parameters is the same as in case of MSE, see Eq. (1). RMSE metric shows how far the predicted value is from the true value in absolute measure.

The second metric is mean absolute percentage error (MAPE):

$$MAPE = \frac{1}{n}\sum_{i=1}^{n}|\frac{y_i - \hat{y}_i}{y_i}| * 100 \tag{3}$$

This metric is quite similar to RMSE, but the measure of difference between true value and predicted value is relative and is computed in percents.

## 4 Results

Proposed LSTM network was trained on sequences "time sequence – response". Here the response of the model to the previous step was iteratively added to the sequence to the network input. We compared the results of usage of two performance metrics for single channel prediction and multi-channel prediction.

In Fig. 2 examples of true and predicted EEG signal are shown for single channel and multi-channel prediction. It can be seen from Fig. 2 that high-amplitude events are predicted with better accuracy in case of multi-channel prediction than for single channel prediction.

In Fig. 3 the dependencies of the value of evaluation metrics on the prediction step length (prediction horizon) for RMSE (Fig. 3(a)) and MAPE (Fig. 3(b)) are shown. As it can be seen from Fig. 3(b), MAPE metric is sufficiently increasing with the increase in the prediction step length, especially for multi-channel prediction. When predicting up to 5 steps forward with accuracy 95%, one can see that high-amplitude deviations are still well detected. This fact lead us to the conclusion that it is reasonable to take into account MAPE evaluation metric for prediction horizon of the specified length. Note that in this case MAPE metric does not exceed 100%, which means that high-amplitude events will be detected fairly accurately. Single high deviations (or high amplitude events) make a stronger contribution to the MAPE metric, then to the RMSE metric. In EEG time series with epileptiform activity there are a lot of high amplitude deviations, and the model does not need to predict them all perfectly to make an effective prediction of ictal stage. This means that RMSE metric also could be relevant in this task. Using both evaluation metrics to configure deep learning model can help to gain useful insights in the model's behaviour and reliability.

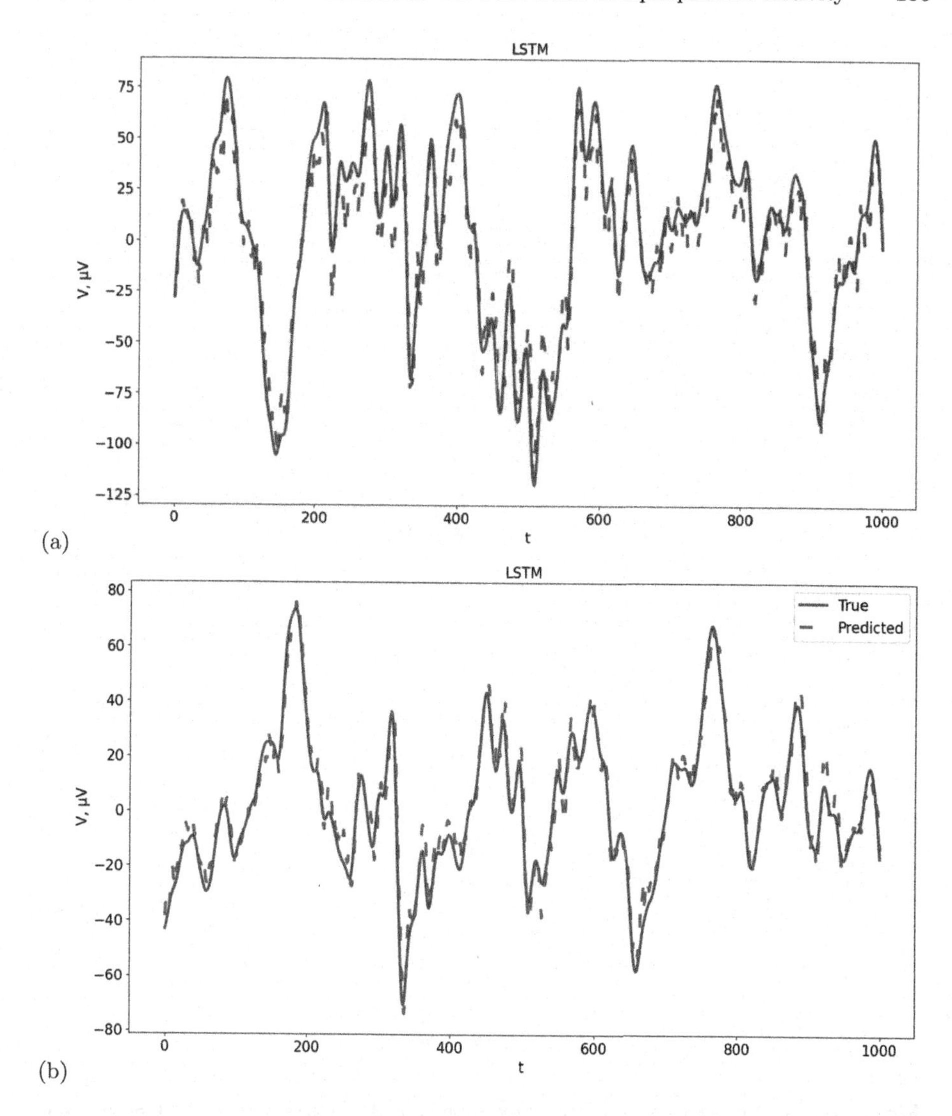


**Fig. 2.** True (blue solid line) and predicted (red dashed line) value for 8 steps of prediction. (a) Single channel prediction. (b) Multi-channel prediction. (Color figure online)

The high variability of brain dynamics for different persons as well as for a single person over time, imply the need in the obtaining of additional information that can be useful for epileptic seizure prediction. As it was shown in

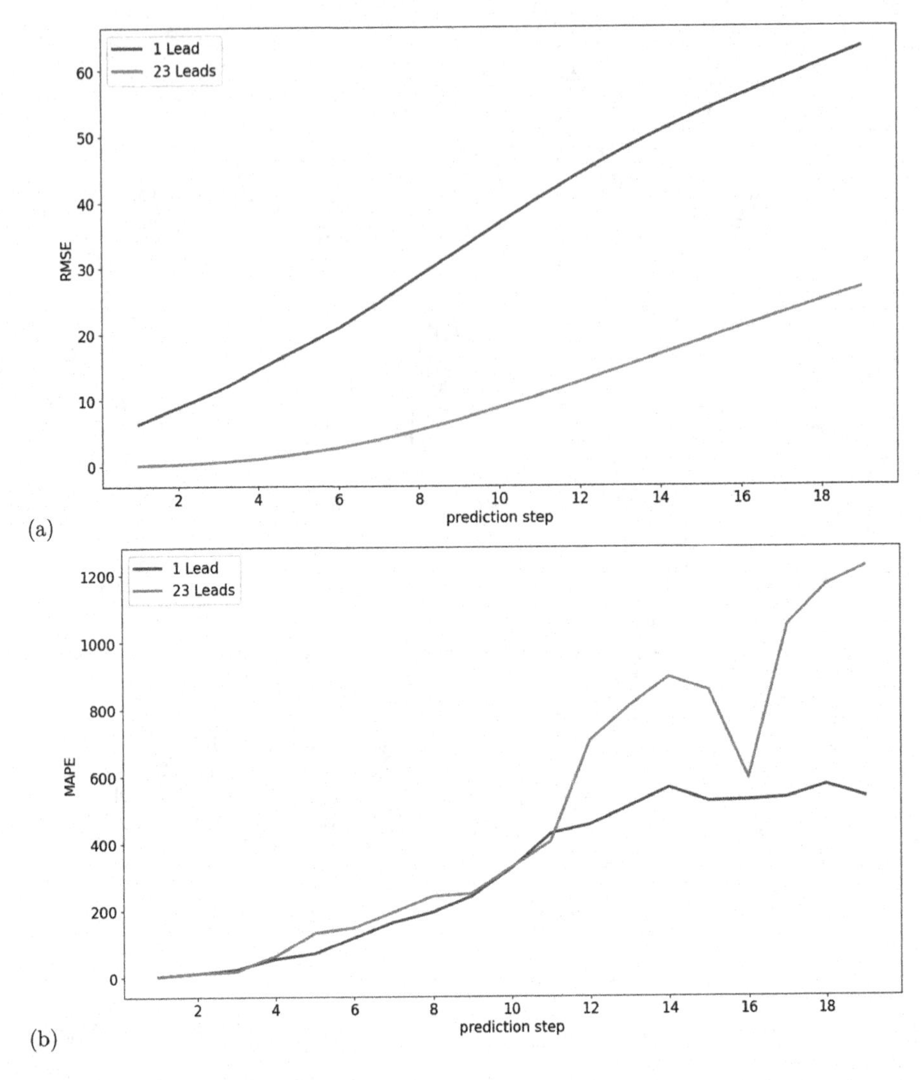


**Fig. 3.** The dependence of the value of evaluation metrics on the prediction step length (prediction horizon). (a) RMSE. (b) MAPE. Blue marker corresponds to single channel prediction, yellow marker corresponds to multi channel prediction. (Color figure online)

previous studies [37], in this context it is utterly important to study spatiotemporal correlation structure of the EEG signals across a range of temporal scales. Nevertheless, epileptic seizure prediction approaches described in literature [38]

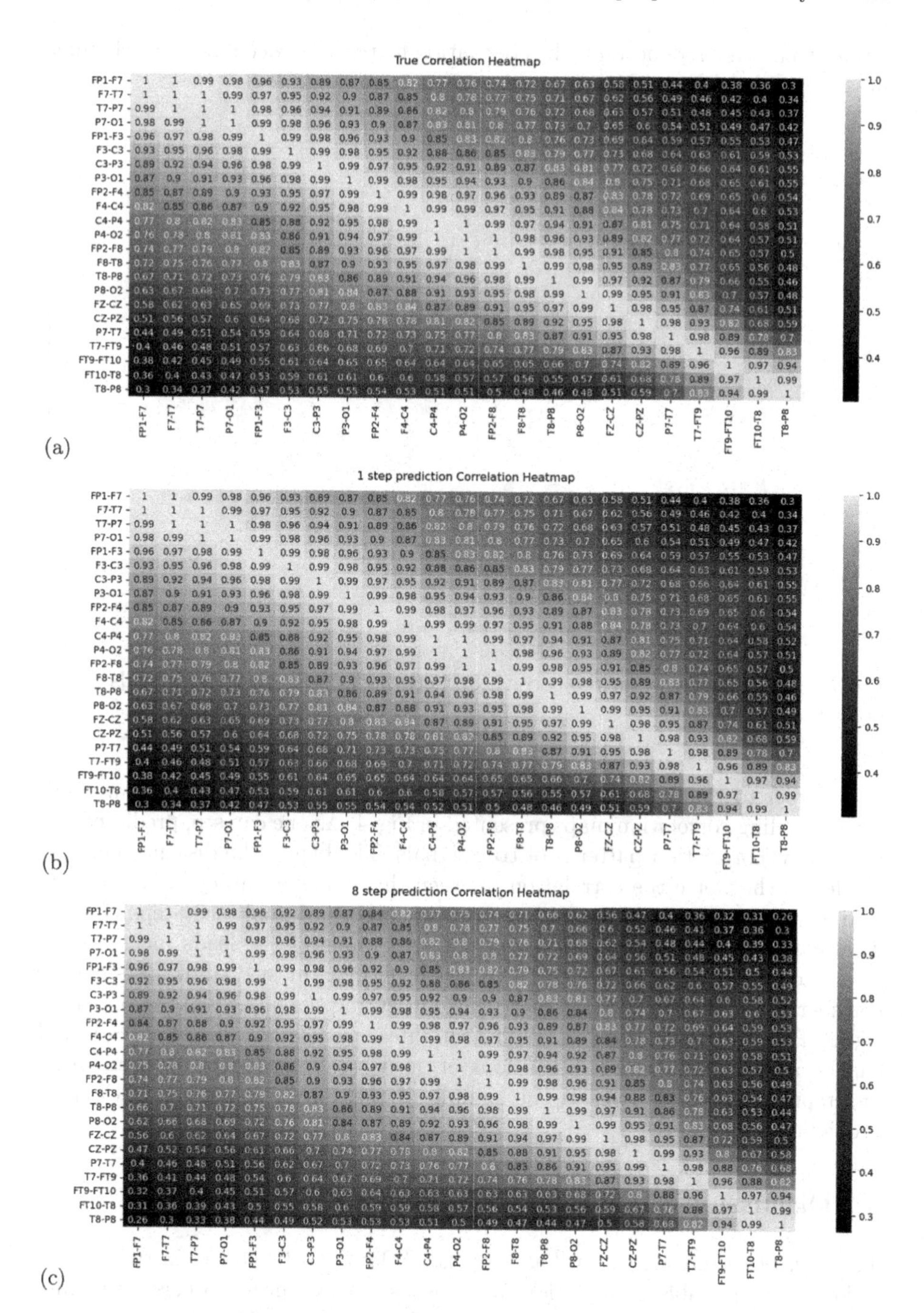


**Fig. 4.** The correlation map between all the leads in record. (a) True signal. (b) Predicted signal in 1 step prediction mode. (c) Predicted signal in 8 step prediction mode.

often do not take into account the correlation features between multiple channels in the recorded EEG signals.

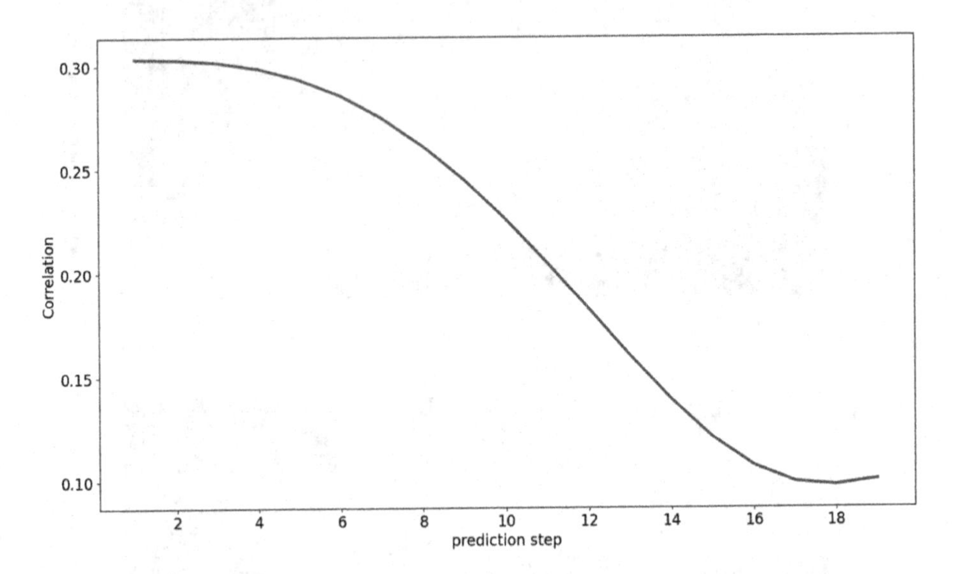


**Fig. 5.** The dependence of the correlation between FP1-F7 and T8-P8 leads on the prediction step length.

In order to study correlation structure of true and predicted signals, we built corresponding correlation maps presented in Fig. 4. As one can see, predicted signals follow correlation patterns of true signals. The bigger the distance between the leads, the worse the correlation between them. The accuracy of the correlation also decreases with the increase in the number of predicted steps (Fig. 5). This pattern is most obviously seen in the example of the leads most distant from each other. In this case the correlation between most distant leads can be considered as a new evaluation metric for measuring the quality of the predicted signal. Proposed evaluation metric incorporates spatial dependence of leads and can be used to reduce the dimensions of deep learning models by eliminating less representative leads from the sample, if the correlation varies little with prediction step length.

## 5 Discussion

The task of seizure detection based on EEG data can be solved in two formulations. One possible formulation is the classification into two classes: seizure (ictal) and the non-seizure (pre-ictal, inter-ictal) activity [39,40]. Another possible formulation is the prediction of full time series that contain epileptiform activity, where the main task is to increase the prediction horizon without compromising accuracy of the prediction.

The first formulation, a supervised classification, is based on manual or automatic feature selection. A number of studies recently proposed various approaches to tackle the problem, see e.g. the review [41] and the references therein. Supervised classification can suffer from the overfitting and class imbalance, which comes from the rare occurrence of seizures [42]. To deal with imbalanced data, one can artificially balance the number of examples from both classes in the training data set. Another promising way to deal with the class imbalance is to construct feature space where two classes will be far from each other. In many cases, the class imbalance leads to overfitting. The overfitting is an undesirable behavior of the model that is observed when the model gives accurate predictions for training data but fails to deal with new data, which makes prediction unreliable. To address this issue one can choose interpretable features of seizures [43] common for the most patients, e.g. using one-class SVM for outlier detection [44]. Unsupervised anomaly detection techniques are also can be used to treat epileptic seizures as rare events with their own distinctive patterns [45]. This approach harks back to the study of epileptic seizures as extreme events [46–48]. The formation of epileptic seizures possess the features of extreme events [3,49,50], as it was shown for different animal [46,51] and human [48] models of epilepsy. If the problem of seizure prediction is formulated in terms of standard binary classification, then evaluation of proposed model can be done using standard sensitivity and specificity [52–55].

The second formulation implies the prediction of the whole EEG signal with epileptiform activity. This approach allows not only to predict the ictal stage itself, but to see earlier precursors of epileptic seizure. This approach could also be useful for detecting another relevant conditions using EEG [56]. In this case features that indicate the beginning and development of the pre-ictal stage and the imminent onset of the seizure are learned automatically by the DNN model and are a part of its internal representation. In the proposed paper we used this approach. In order to guarantee the high-quality prediction of the EEG signals, we propose to use several evaluation metrics, such as RMSE, MAPE and correlation between leads, together. Each of these evaluation metrics helps to take into account certain aspects of the signals.

## 6 Conclusion

This study compares various evaluation metrics that can be used to assess the accuracy of building deep learning models in prediction of time series with epileptiform activity. Our methodology proposes the combination of evaluation metrics for better analysis of the deep learning model in task of seizure prediction from EEG data. We hope that obtained results will contribute to a deeper understanding of the theoretical and practical aspects of epileptiform activity prediction, as well as help to obtain a number of practically important results, on the basis of which new effective methods for epilepsy treatment will be developed.

Note that the deep neural networks used have a large number of parameters (weights) adjusted in the process of training the networks. This, in turn, leads to large computational costs. To solve this problem, hardware implementations based on memristors have been widely used in various neural networks [57], including LSTM networks [58]. A memristor can perform calculations in memory, and a memristor array can speed up vector-matrix multiplication, so memristor-based neural network implementations are a promising way to speed up calculations and reduce power consumption. Memristive deep neural networks can be a promising new hardware platform for deep neural networks that will allow to create in near future portable device for epilepsy prediction and prevention.

**Acknowledgements.** This work was supported by Russian Science Foundation (Project No. 22-71-00112).

## References

1. Ioannou, P., et al.: The burden of epilepsy and unmet need in people with focal seizures. Brain Behav. **12**(9), e2589 (2022)
2. Jette, N., Engel, J.: Refractory epilepsy is a life-threatening disease: lest we forget (2016)
3. Lehnertz, K.: Epilepsy: extreme events in the human brain. In: Albeverio, S., Jentsch, V., Kantz, H. (eds.) Extreme Events in Nature and Society. FRONTCOLL, pp. 123–143. Springer, Heidelberg (2006). https://doi.org/10.1007/3-540-28611-X_6
4. Wang, C., Zou, J., Zhang, J., Wang, M., Wang, R.: Feature extraction and recognition of epileptiform activity in EEG by combining PCA with ApEn. Cogn. Neurodyn. **4**, 233–240 (2010)
5. Salant, Y., Gath, I., Henriksen, O.: Prediction of epileptic seizures from two-channel EEG. Med. Biol. Eng. Comput. **36**, 549–556 (1998)
6. Iasemidis, L.D., Chris Sackellares, J., Zaveri, H.P., Williams, W.J.: Phase space topography and the Lyapunov exponent of electrocorticograms in partial seizures. Brain Topogr. **2**, 187–201 (1990)
7. Aarabi, A., Fazel-Rezai, R., Aghakhani, Y.: EEG seizure prediction: measures and challenges. In: 2009 Annual International Conference of the IEEE Engineering in Medicine and Biology Society, pp. 1864–1867. IEEE (2009)
8. Mormann, F., Kreuz, T., Andrzejak, R.G., David, P., Lehnertz, K., Elger, C.E.: Epileptic seizures are preceded by a decrease in synchronization. Epilepsy Res. **53**(3), 173–185 (2003)
9. Rosenblum, M., Tass, P., Kurths, J., Volkmann, J., Schnitzler, A., Freund, H.: Chaos in brain? (2000)
10. Mormann, F., et al.: On the predictability of epileptic seizures. Clin. Neurophysiol. **116**(3), 569–587 (2005)
11. Alvarado-Rojas, C., et al.: Slow modulations of high-frequency activity (40–140 Hz) discriminate preictal changes in human focal epilepsy. Sci. Rep. **4**(1), 4545 (2014)
12. Bandarabadi, M., Rasekhi, J., Teixeira, C.A., Karami, M.R., Dourado, A.: On the proper selection of preictal period for seizure prediction. Epilepsy Behav. **46**, 158–166 (2015)

13. Direito, B., Teixeira, C., Ribeiro, B., Castelo-Branco, M., Sales, F., Dourado, A.: Modeling epileptic brain states using EEG spectral analysis and topographic mapping. J. Neurosci. Methods **210**(2), 220–229 (2012)
14. Teixeira, C.A., et al.: Epileptic seizure predictors based on computational intelligence techniques: a comparative study with 278 patients. Comput. Methods Programs Biomed. **114**(3), 324–336 (2014)
15. Freestone, D.R., Karoly, P.J., Cook, M.J.: A forward-looking review of seizure prediction. Curr. Opin. Neurol. **30**(2), 167–173 (2017)
16. Khan, H., Marcuse, L., Fields, M., Swann, K., Yener, B.: Focal onset seizure prediction using convolutional networks. IEEE Trans. Biomed. Eng. **65**(9), 2109–2118 (2017)
17. Kiral-Kornek, I., et al.: Epileptic seizure prediction using big data and deep learning: toward a mobile system. EBioMedicine **27**, 103–111 (2018)
18. Abdelhameed, A.M., Bayoumi, M.: Semi-supervised deep learning system for epileptic seizures onset prediction. In: 2018 17th IEEE International Conference on Machine Learning and Applications (ICMLA), pp. 1186–1191. IEEE (2018)
19. Daoud, H., Bayoumi, M.A.: Efficient epileptic seizure prediction based on deep learning. IEEE Trans. Biomed. Circuits Syst. **13**(5), 804–813 (2019)
20. Zhang, Y., Guo, Y., Yang, P., Chen, W., Lo, B.: Epilepsy seizure prediction on EEG using common spatial pattern and convolutional neural network. IEEE J. Biomed. Health Inform. **24**(2), 465–474 (2019)
21. Tsiouris, K.M., Pezoulas, V.C., Zervakis, M., Konitsiotis, S., Koutsouris, D.D., Fotiadis, D.I.: A long short-term memory deep learning network for the prediction of epileptic seizures using EEG signals. Comput. Biol. Med. **99**, 24–37 (2018)
22. Singh, K., Malhotra, J.: Two-layer LSTM network-based prediction of epileptic seizures using EEG spectral features. Complex Intell. Syst. **8**(3), 2405–2418 (2022)
23. Yan, J., Li, J., Xu, H., Yu, Y., Xu, T.: Seizure prediction based on transformer using scalp electroencephalogram. Appl. Sci. **12**(9), 4158 (2022)
24. Hussein, R., Lee, S., Ward, R.: Multi-channel vision transformer for epileptic seizure prediction. Biomedicines **10**(7), 1551 (2022)
25. Maddineni, S., Janapati, S., Kosana, V., Teeparthi, K.: A hybrid deep transformer model for epileptic seizure prediction. In: 2022 International Conference on Advancements in Smart, Secure and Intelligent Computing (ASSIC), pp. 1–6. IEEE (2022)
26. Chen, H.H., Cherkassky, V.: Performance metrics for online seizure prediction. Neural Netw. **128**, 22–32 (2020)
27. Gadhoumi, K., Lina, J.M., Mormann, F., Gotman, J.: Seizure prediction for therapeutic devices: a review. J. Neurosci. Methods **260**, 270–282 (2016)
28. Mormann, F., Andrzejak, R.G., Elger, C.E., Lehnertz, K.: Seizure prediction: the long and winding road. Brain **130**(2), 314–333 (2007)
29. Schelter, B., et al.: Testing statistical significance of multivariate time series analysis techniques for epileptic seizure prediction. Chaos: Interdisc. J. Nonlinear Sci. **16**(1) (2006)
30. Snyder, D.E., Echauz, J., Grimes, D.B., Litt, B.: The statistics of a practical seizure warning system. J. Neural Eng. **5**(4), 392 (2008)
31. Winterhalder, M., Maiwald, T., Voss, H., Aschenbrenner-Scheibe, R., Timmer, J., Schulze-Bonhage, A.: The seizure prediction characteristic: a general framework to assess and compare seizure prediction methods. Epilepsy Behav. **4**(3), 318–325 (2003)

32. Goldberger, A.L., et al.: Physiobank, physiotoolkit, and physionet: components of a new research resource for complex physiologic signals. Circulation **101**(23), e215–e220 (2000)
33. Shoeb, A.H.: Application of machine learning to epileptic seizure onset detection and treatment. Ph.D. thesis, Massachusetts Institute of Technology (2009)
34. Gromov, N., Gubina, E., Levanova, T.: Loss functions in the prediction of extreme events and chaotic dynamics using machine learning approach. In: 2022 Fourth International Conference Neurotechnologies and Neurointerfaces (CNN), pp. 46–50. IEEE (2022)
35. Gerasimova, S., et al.: Memristive neural networks for predicting seizure activity. Sovremennye Tehnologii v Medicine **15**(4), 30 (2023)
36. Hussain, L., et al.: Regression analysis for detecting epileptic seizure with different feature extracting strategies. Biomed. Eng./Biomedizinische Technik **64**(6), 619–642 (2019)
37. Williamson, J.R., Bliss, D.W., Browne, D.W., Narayanan, J.T.: Seizure prediction using EEG spatiotemporal correlation structure. Epilepsy Behav. **25**(2), 230–238 (2012)
38. Qi, N., Piao, Y., Yu, P., Tan, B.: Predicting epileptic seizures based on EEG signals using spatial depth features of a 3D–2D hybrid CNN. Med. Biol. Eng. Comput. 1–12 (2023)
39. Birjandtalab, J., Pouyan, M.B., Nourani, M.: Unsupervised EEG analysis for automated epileptic seizure detection. In: First International Workshop on Pattern Recognition, vol. 10011, pp. 124–128. SPIE (2016)
40. Tzallas, A.T., Tsipouras, M.G., Fotiadis, D.I., et al.: Automatic seizure detection based on time-frequency analysis and artificial neural networks. Comput. Intell. Neurosci. **2007** (2007)
41. Walther, D., Viehweg, J., Haueisen, J., Mäder, P.: A systematic comparison of deep learning methods for EEG time series analysis. Front. Neuroinform. **17**, 1067095 (2023)
42. Mazurek, S., Blanco, R., Falcó-Roget, J., Argasiński, J.K., Crimi, A.: Impact of the pre-processing and balancing of EEG data on the performance of graph neural network for epileptic seizure classification. In: Rutkowski, L., Scherer, R., Korytkowski, M., Pedrycz, W., Tadeusiewicz, R., Zurada, J.M. (eds.) ICAISC 2023. LNCS, vol. 14126, pp. 258–268. Springer, Cham (2023). https://doi.org/10.1007/978-3-031-42508-0_24
43. Karpov, O.E., et al.: Detecting epileptic seizures using machine learning and interpretable features of human EEG. Eur. Phys. J. Spec. Top. **232**(5), 673–682 (2023)
44. Karpov, O.E., et al.: Extreme value theory inspires explainable machine learning approach for seizure detection. Sci. Rep. **12**(1), 11474 (2022)
45. Karpov, O.E., et al.: Evaluation of unsupervised anomaly detection techniques in labelling epileptic seizures on human EEG. Appl. Sci. **13**(9), 5655 (2023)
46. Pisarchik, A., et al.: Extreme events in epileptic EEG of rodents after ischemic stroke. Eur. Phys. J. Spec. Top. **227**, 921–932 (2018)
47. Rings, T., Mazarei, M., Akhshi, A., Geier, C., Tabar, M.R.R., Lehnertz, K.: Traceability and dynamical resistance of precursor of extreme events. Sci. Rep. **9**(1), 1744 (2019)
48. Karpov, O.E., et al.: Noise amplification precedes extreme epileptic events on human EEG. Phys. Rev. E **103**(2), 022310 (2021)
49. Mishra, A., Saha, S., Vigneshwaran, M., Pal, P., Kapitaniak, T., Dana, S.K.: Dragon-king-like extreme events in coupled bursting neurons. Phys. Rev. E **97**(6), 062311 (2018)

50. Olenin, S.M., Levanova, T.A.: Extreme events in small ensemble of bursting neurons with chemical and electrical couplings. In: 2023 International Joint Conference on Neural Networks (IJCNN), pp. 1–6. IEEE (2023)
51. Frolov, N.S., et al.: Statistical properties and predictability of extreme epileptic events. Sci. Rep. **9**(1), 7243 (2019)
52. Brinkmann, B.H., et al.: Crowdsourcing reproducible seizure forecasting in human and canine epilepsy. Brain **139**(6), 1713–1722 (2016)
53. Mirowski, P.W., LeCun, Y., Madhavan, D., Kuzniecky, R.: Comparing SVM and convolutional networks for epileptic seizure prediction from intracranial EEG. In: 2008 IEEE Workshop on Machine Learning for Signal Processing, pp. 244–249. IEEE (2008)
54. Mirowski, P., Madhavan, D., LeCun, Y., Kuzniecky, R.: Classification of patterns of EEG synchronization for seizure prediction. Clin. Neurophysiol. **120**(11), 1927–1940 (2009)
55. Shiao, H.T., et al.: SVM-based system for prediction of epileptic seizures from iEEG signal. IEEE Trans. Biomed. Eng. **64**(5), 1011–1022 (2016)
56. Rasheed, K., et al.: Machine learning for predicting epileptic seizures using EEG signals: a review. IEEE Rev. Biomed. Eng. **14**, 139–155 (2020)
57. Yakopcic, C., Alom, M.Z., Taha, T.M.: Memristor crossbar deep network implementation based on a convolutional neural network. In: 2016 International Joint Conference on Neural Networks (IJCNN), pp. 963–970. IEEE (2016)
58. Li, C., et al.: Long short-term memory networks in memristor crossbar arrays. Nat. Mach. Intell. **1**(1), 49–57 (2019)

# Astrocyte Controlled SNN Dynamic Induced by Sensor Input

Sergey V. Stasenko[1,2(✉)] and Victor B. Kazantsev[1,2]

[1] Lobachevsky State University of Nizhny Novgorod, Nizhny Novgorod 603950, Russia
stasenko@neuro.nnov.ru
[2] Moscow Institute of Physics and Technology, Moscow 117303, Russia

**Abstract.** We examined a mathematical model with SNN and astrocyte networks to represent a 2D video image as a spatio-temporal spiking pattern. The SNN maintains excitation-inhibition balance with excitatory and inhibitory neurons. Astrocytes regulate synaptic transmission strength. Introducing excitatory pulses representing the image helped prevent hyperexcitation, and when properly adjusted, enhanced image representation in the SNN.

**Keywords:** spiking neural network · synchronization · neuron-glial network

## 1 Introduction

SNNs, known as spiking neuron networks, have attracted attention for their potential in information processing [1–3]. Unlike classical neurons and artificial neuron networks (ANNs), SNNs incorporate a more complex model of local neurons, synapses, and network architectures. However, training and learning in SNNs pose challenges due to the intricate dynamics of neurons and synapses, resulting in diverse activities in time and space. To tackle these challenges, researchers have explored the integration of additional cell layers, particularly astrocytes, into SNNs [4–6]. Astrocytes play a role in synaptic transmission by influencing both pre- and postsynaptic compartments through the release of gliotransmitters [7–10]. Mathematical models have been developed to study the functional impact of astrocytes on neural dynamics, including effects on neural excitability, frequency selectivity, and presynaptic functions [11–14]. The tripartite synapse model has demonstrated how astrocytes contribute to spike-timing-dependent plasticity and learning [15–17]. Astrocytic networks are characterized by calcium wave propagation and synchronization [18,19]. Incorporating astrocytes into SNNs has shown advantages in enhancing short-term memory performance, recognizing overlapping patterns, and representing colored patterns.


D. Balandin et al. (Eds.): MMST 2023, CCIS 1914, pp. 294–301, 2024.
https://doi.org/10.1007/978-3-031-52470-7_23

However, the understanding of information representation in spiking patterns and the management of disturbances to the excitation-inhibition balance caused by external stimulation, which can impact encoded information, still present ongoing challenges [20].

In this study, we examined the dynamics of an SNN augmented with astrocytes, focusing on their role in regulating the excitation-inhibition balance when a stimulation pattern is introduced to the network. Our findings reveal that the astrocytes not only prevent excessive network excitation but also enhance the representation of images in the spatiotemporal firing of the SNN.

## 2 The Model

In this study, we utilized the Izhikevich model [21] to capture the dynamics of a single neuron. The model is described by the following equations:

$$\begin{cases} C_m \frac{dV_m}{dt} = k(V_m - V_r)(V_m - V_t) - U_m + I_{ext} + I_{syn} + I_{stim} \\ \frac{dU_m}{dt} = a(b(V_m - V_r) - U_m) \end{cases} \quad (1)$$

If $V_m \geq V_{peak}$, then

$$\begin{cases} V_m = c \, U_m = U_m + d \end{cases} \quad (2)$$

The neuron model involves parameters $a$, $b$, $c$, $d$, $k$, and $C_m$, representing the dynamics of the membrane potential $V_m$ and the recovery variable $U_m$. The resting potential is determined by $b$ ($-70$ to $-60$ mV), $a$ controls the recovery variable timescale, and $V_{peak}$ limits spike amplitude. Parameters $c$ and $d$ define values after a burst. External current is denoted by $I_{ext}$, and the model exhibits excitability, generating spikes. The specific parameter values are: $a = 0.02$, $b = 0.5$, $c = -40$ mV, $d = 100$, $k = 0.5$, and $C_m = 50$.

The synaptic current $I_{syn}$ is the sum of synaptic currents from connected neurons, given by $I_{syn} = \sum y_{ij} w_{ij}$. Excitatory synapses have positive weights ($w_{ij}$), while inhibitory synapses have negative weights. $y_{ij}$ represents the output signal from the $i$-th neuron to the $j$-th neuron generating $I_{syn}$.

Synaptic weights range randomly from 20 to 60. When a presynaptic neuron spikes, the postsynaptic neuron experiences a decaying jump in synaptic current. The neurotransmitter concentration $y_{ij}$ changes with time, governed by specific conditions. Glutamatergic synapses promote synchronized firing, while GABAergic neurons maintain an excitation-inhibition balance. Glutamate dynamics are simplified using a mean-field approximation [22,23]. Parameter values include relaxation times $\tau_X = 100$ ms and $\tau_y = 4$ ms.

Extrasynaptic glutamate binds to astrocyte processes, activating metabotropic glutamate receptors. This leads to gliotransmitter release by the astrocyte.

In our simplified mathematical model [24–31], the relationship between neurotransmitter and gliotransmitter concentrations is described by:

$$\frac{dY}{dt} = -\alpha_Y Y + \frac{\beta_Y}{1 + exp(-X + X_{thr})} \tag{3}$$

Here, $Y$ represents gliotransmitter concentration, $\alpha_Y$ is the clearance rate, and $\beta_Y$ and $X_{thr}$ are parameters. Parameter values used are: $\alpha_Y = 120$ ms, $\beta_Y = 0.5$, and $X_{thr} = 3.5$.

Astrocytes affect neurotransmitter release, modulating synaptic transmission. In our model, for glutamatergic synapses, we introduce synaptic depression as:

$$I_{syn} = \sum y_{ij} w_{ij} (1 - \frac{\gamma_Y}{1 + exp(-Y + Y_{thr})}) \tag{4}$$

Here, $I_{syn}$ is the sum of synaptic currents, $w_{ij}$ is the weight of glutamatergic synapses, and $\gamma_Y$ represents astrocyte influence. We use $Y_{thr} = 2$ for illustration.

The input image is taken from the MNIST database [32]. It is a 189 by 189 pixel image encoded from 0 to 1, with 0 representing colorless pixels and 1 representing black. To simplify stimulation, the image is zero-padded and centered to match the 300-neuron network. The stimulation current equation is:

$$I_{stim} = S \times A_S \tag{5}$$

Each column of the stimulation current matrix is applied to each neuron for 1 ms, totaling the image width of 189 ms.

## 3 Results

To study the astrocytic modulation of the dynamics of the spiking neural network during stimulation with an input image, several cases were considered: 1) the dynamics of the spiking neural network under the influence of uncorrelated noise, 2) the dynamics of the spiking neural network when stimulated by a sensory signal, and 3) the dynamics of the spiking neural network in the presence of sensory input and astrocytic modulation of synaptic transmission. In our model, we consider astrocytic depression of synaptic transmission. Additionally, we investigated the influence of astrocytic modulation on the quality of sensory input representation.

We conducted simulations of an SNN without astrocyte action or external stimulation. The presence of inhibitory neurons balanced the network firing, preventing hyperexcitation. Initially, there was a high-frequency burst followed by rare spiking dynamics (Fig. 1a). The untrained SNN exhibited irregular spiking due to uncorrelated noise.

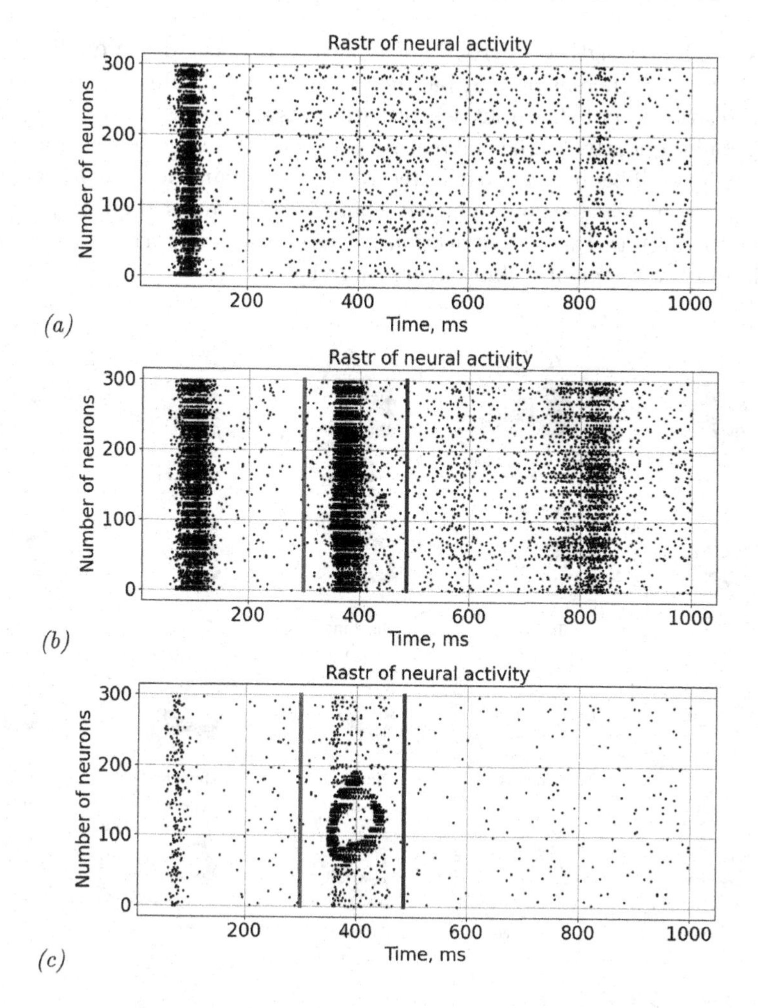


**Fig. 1.** The figure shows rasters of neural activity: (a) SNN without sensor input and astrocyte regulation, (b) SNN with only sensor input, (c) SNN with sensor input and astrocyte regulation. The red line represents the stimulus start, and the blue line represents the stimulus end. (Color figure online)

Next, we trained the SNN by providing the input image using Eq. (5). The dynamics of the stimulated SNN are shown in Fig. 1b. However, without astrocyte feedback, the SNN responded with hyperexcitation, diffusing the image features. It can also be noticed that burst activity begins to be observed in the dynamics of the spike neural network.

When astrocyte feedback was activated (depression of synaptic plasticity), quasi-synchronous bursts were suppressed, and the SNN displayed normal dynamics (Fig. 1c). The presence of the image was evident in the spike pattern. By tuning astrocytic parameters, we controlled the inhibition and excitation processes. The strength of astrocyte feedback, $\gamma_Y$, affected the quality of SNN training (Fig. 2). Increasing the parameter $\gamma_Y$ to a certain value results in a gradual restoration of the input sensory stimulus in the raster of spiking neural activity. Subsequently, further increases in $\gamma_Y$ lead to image blurring and deformation (Fig. 2). This may be attributed to the significant depression of synaptic transmission, which disrupts synaptic connections, causes their thinning, and disrupts the architecture of connections. These effects distort the image representation in the raster of spiking neural activity.

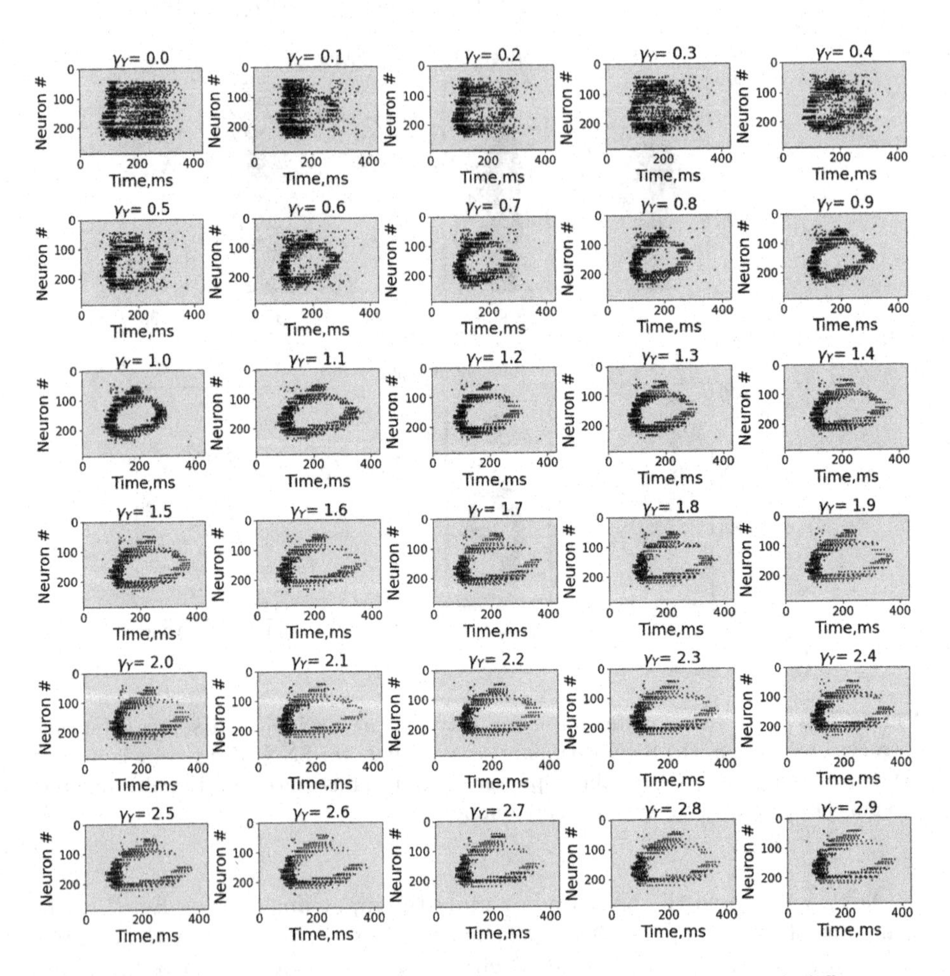


**Fig. 2.** Neural activity (upper left) representing the supplied pattern with different $\gamma_Y$ values.

## 4 Conclusion

We examined an SNN model with astrocytic feedback and trained it to represent video images accurately. The image was captured in the spatio-temporal dynamics of the SNN, with the astrocyte feedback preventing excessive neuron excitation. Adjusting the astrocyte parameters affected image clarity, mainly due to SNN hyperexcitation during stimulation.

**Acknowledgements.** The study of the model on the MNIST data was supported by a grant from the Russian Science Foundation (grant number 23-11-00134), while the numerical calculations were supported by project 0729-2021-013, which is carried out within the framework of the State task for the performance of research work by laboratories that have passed the competitive selection within the framework of the national project "Science and Universities", in respect of which the decision of the Budget Commission of the Ministry of Education and Science of Russia (dated September 14, 2021 No. BC-P/23) on the provision of subsidies from the federal budget for the financial support of the state task for the implementation of research work.

## References

1. Ghosh-Dastidar, S., Adeli, H.: Improved spiking neural networks for EEG classification and epilepsy and seizure detection. Integr. Comput.-Aided Eng. **14**, 187–212 (2007)
2. Dora, S., Kasabov, N.: Spiking neural networks for computational intelligence: an overview. Big Data Cogn. Comput. **5**, 67 (2021)
3. Lobov, S., Chernyshov, A., Krilova, N., Shamshin, M., Kazantsev, V.: Competitive learning in a spiking neural network: towards an intelligent pattern classifier. Sensors **20**, 500 (2020)
4. Gordleeva, S., et al.: Modeling working memory in a spiking neuron network accompanied by astrocytes. Front. Cell. Neurosci. **15**, 631485 (2021)
5. Abrego, L., Gordleeva, S., Kanakov, O., Krivonosov, M., Zaikin, A.: Estimating integrated information in bidirectional neuron-astrocyte communication. Phys. Rev. E **103**, 022410 (2021)
6. Tsybina, Y., et al.: Astrocytes mediate analogous memory in a multi-layer neuron-astrocyte network. Neural Comput. Appl. **34**, 9147–9160 (2022)
7. Araque, A., Parpura, V., Sanzgiri, R., Haydon, P.: Glutamate-dependent astrocyte modulation of synaptic transmission between cultured hippocampal neurons. Eur. J. Neurosci. **10**, 2129–2142 (1998). http://www.ncbi.nlm.nih.gov/pubmed/9753099
8. Araque, A., Parpura, V., Sanzgiri, R., Haydon, P.: Tripartite synapses: glia, the unacknowledged partner. Trends Neurosci. **22**, 208–215 (1999). https://linkinghub.elsevier.com/retrieve/pii/S0166223698013496
9. Wittenberg, G., Sullivan, M., Tsien, J.: Synaptic reentry reinforcement based network model for long-term memory consolidation. Hippocampus **12**, 637–647 (2002). http://www.ncbi.nlm.nih.gov/pubmed/12440578
10. Wang, X.: Synaptic basis of cortical persistent activity: the importance of NMDA receptors to working memory. J. Neurosci.: Off. J. Soc. Neurosci. **19**, 9587–603 (1999). http://www.ncbi.nlm.nih.gov/pubmed/10531461

11. Nadkarni, S., Jung, P.: Dressed neurons: modeling neural-glial interactions. Phys. Biol. **1**, 35–41 (2004). http://www.ncbi.nlm.nih.gov/pubmed/16204820
12. Nadkarni, S., Jung, P.: Modeling synaptic transmission of the tripartite synapse. Phys. Biol. **4**, 1–9 (2007). http://www.ncbi.nlm.nih.gov/pubmed/17406080
13. Volman, V., Ben-Jacob, E., Levine, H.: The astrocyte as a gatekeeper of synaptic information transfer. Neural Comput. **326**, 303–326 (2007). http://www.mitpressjournals.org/doi/abs/10.1162/neco.2007.19.2.303
14. De Pitta, M., Volman, V., Berry, H., Ben-Jacob, E.: A tale of two stories: astrocyte regulation of synaptic depression and facilitation. PLoS Comput. Biol. **7**, e1002293 (2011). http://dx.plos.org/10.1371/journal.pcbi.1002293
15. Postnov, D., Ryazanova, L., Sosnovtseva, O.: Functional modeling of neural-glial interaction. Bio Syst. **89**, 84–91 (2007). http://www.ncbi.nlm.nih.gov/pubmed/17320272
16. Amiri, M., Bahrami, F., Janahmadi, M.: Functional contributions of astrocytes in synchronization of a neuronal network model. J. Theor. Biol. **292**, 60–70 (2011). http://www.ncbi.nlm.nih.gov/pubmed/21978738
17. Wade, J., McDaid, L., Harkin, J., Crunelli, V., Kelso, J.: Bidirectional coupling between astrocytes and neurons mediates learning and dynamic coordination in the brain: a multiple modeling approach. PLoS One **6**, e29445 (2011). http://dx.plos.org/10.1371/journal.pone.0029445
18. Ullah, G., Jung, P., Cornell-Bell, A.: Anti-phase calcium oscillations in astrocytes via inositol (1, 4, 5)-trisphosphate regeneration. Cell Calcium **39**, 197–208 (2006). http://www.ncbi.nlm.nih.gov/pubmed/16330095
19. Kazantsev, V.: Spontaneous calcium signals induced by gap junctions in a network model of astrocytes. Phys. Rev. E Stat. Nonlinear Soft Matt. Phys. **79**, 010901 (2009)
20. Perea, G., Navarrete, M., Araque, A.: Tripartite synapses: astrocytes process and control synaptic information. Trends Neurosci. **32**, 421–431 (2009)
21. Izhikevich, E.: Dynamical systems in neuroscience: the geometry of excitability and bursting. dynamical systems. First 441 (2007). http://www.amazon.com/Dynamical-Systems-Neuroscience-Excitability-Computational/dp/0262090430
22. Stasenko, S., Lazarevich, I., Kazantsev, V.: Quasi-synchronous neuronal activity of the network induced by astrocytes. Procedia Comput. Sci. **169**, 704–709 (2020). https://linkinghub.elsevier.com/retrieve/pii/S1877050920302982
23. Lazarevich, I., Stasenko, S., Kazantsev, V.: Synaptic multistability and network synchronization induced by the neuron-glial interaction in the brain. JETP Lett. **105**, 210–213 (2017)
24. Stasenko, S., Hramov, A., Kazantsev, V.: Loss of neuron network coherence induced by virus-infected astrocytes: a model study. Sci. Rep. **13**, 1–11 (2023)
25. Stasenko, S., Kazantsev, V.: Dynamic image representation in a spiking neural network supplied by astrocytes. Mathematics **11**, 561 (2023)
26. Stasenko, S., Kazantsev, V.: Astrocytes enhance image representation encoded in spiking neural network. In: Kryzhanovsky, B., Dunin-Barkowski, W., Redko, V., Tiumentsev, Y. (eds.) NEUROINFORMATICS 2022. SCI, vol. 1064, pp. 200–206. Springer, Cham (2022). https://doi.org/10.1007/978-3-031-19032-2_20
27. Gordleeva, S., Stasenko, S., Semyanov, A., Dityatev, A., Kazantsev, V.: Bi-directional astrocytic regulation of neuronal activity within a network. Front. Comput. Neurosci. **6**, 92 (2012)
28. Barabash, N., Levanova, T., Stasenko, S.: STSP model with neuron-glial interaction produced bursting activity. In: 2021 Third International Conference Neurotechnologies And Neurointerfaces (CNN), pp. 12–15 (2021)

29. Stasenko, S., Kazantsev, V.: 3D model of bursting activity generation. In: 2022 Fourth International Conference Neurotechnologies and Neurointerfaces (CNN), pp. 176–179 (2022)
30. Barabash, N., Levanova, T., Stasenko, S.: Rhythmogenesis in the mean field model of the neuron-glial network. Eur. Phys. J. Spec. Top. (2023)
31. Olenin, S., Levanova, T., Stasenko, S.: Dynamics in the reduced mean-field model of neuron-glial interaction. Mathematics. **11**, 2143 (2023)
32. LeCun, Y., Cortes, C., Burges, C.: The MNIST database of handwritten digits. The Courant Institute of Mathematical Sciences (2010)

# Experimental Data Compression for GPU-Based Solution of Inverse Coefficient Problem for Vibrational Testing Data

Stepan Lavrenkov(✉) and Alexey Vasyukov

Moscow Institute of Physics and Technology, Dolgoprudny, Russia
{lavrenkov.sa,a.vasyukov}@phystech.edu

**Abstract.** The paper is devoted to an inverse problem of acquiring elastic properties of a thin plate from vibrational testing data. A direct problem is solved as a boundary value problem for transverse movement of a plate with known geometry. The direct problem computes an amplitude-frequency characteristic as a result. The inverse problem of acquiring the elastic parameters from an experimentally obtained frequency response is solved as a nonlinear optimization problem. The computations are performed on GPU. The present work investigates how the loss function depends on the parameters of the material. Based on this study the paper proposes the approach to optimize memory requirements and computation time for the target inverse problem. The convergence study is performed to ensure that the proposed data compression approach provides a stable solution.

**Keywords:** Inverse problem · Non-destructive testing · Vibrational testing

## 1 Introduction

The present paper concentrates on a method to reconstruct elastic parameters of thin structure using the data obtained from its vibrational testing. This work aims to contribute to the field of non-destructive testing for structural engineering problems. Advanced materials are extensively developed for modern-day engineering projects, including materials tailor-made for a certain task. However, the experimental status of such materials may lead to lack of reliable data on their elastic properties, crucial for structural design problems [1,2]. During the production processes the material may change its properties. So, it's desirable to have the technique to obtain the elastic parameters of the material of the final sample having complex geometry.

The present work contributes to the field by providing an approach to solve an inverse coefficient problem for a thin plate using a limited amount of memory. The algorithm proposed in this paper provides a stable solution using a limited

Supported by RSF project 22-11-00142.


D. Balandin et al. (Eds.): MMST 2023, CCIS 1914, pp. 302–309, 2024.
https://doi.org/10.1007/978-3-031-52470-7_24

number of frequencies. This allows to compress experimental data effectively and perform all computations using widely available GPU hardware.

## 2 Mathematical Model

Model describes the following process: thin plate is clamped with one end in the vibration stand, i.e. generator of sinusoidal oscillations, and another end moving freely. Two accelerometers are placed on each end allowing to receive an amplitude-frequency characteristic (AFC) of the plate. The problem statement is close to the one considered in [3]. So, the forward problem is to get the AFC of the plate with known properties. Thus, the inverse problem is to compute the elastic moduli of a material, having the AFC for a plate with given geometry and accelerometer position. The approach to the problem is similar to the one described in [4].

### 2.1 Forward Problem

Using the Kirchhoff-Love plate theory the equation for the transverse movement of a thin two-dimensional plate in 3D space was obtained [5,6]:

$$2e\rho\ddot{u} - \frac{2e^3}{3}\rho\Delta\ddot{u} - \text{div div } \mathbf{M} = f,$$

where $u$ is the transverse movement of the plate, $e = \frac{1}{2}h$ half of the plate thickness, $f$ is used to describe external forces, $\mathbf{M}$ is the tensor of moments:

$$\begin{pmatrix} M_{xx} & M_{xy} \\ M_{xy} & M_{yy} \end{pmatrix} = \int_{-e}^{e} \begin{pmatrix} \sigma_{xx} & \sigma_{xy} \\ \sigma_{xy} & \sigma_{yy} \end{pmatrix} z^2 \, dz.$$

For a linear elastic material:

$$\begin{pmatrix} M_{xx} \\ M_{yy} \\ M_{xy} \end{pmatrix} = \begin{pmatrix} D_{11} & D_{12} & D_{16} \\ D_{12} & D_{22} & D_{26} \\ D_{16} & D_{26} & D_{66} \end{pmatrix} \begin{pmatrix} \frac{\partial^2}{\partial x^2} u \\ \frac{\partial^2}{\partial y^2} u \\ \frac{\partial^2}{\partial x \partial y} u \end{pmatrix} = D \begin{pmatrix} 1 & \nu & 0 \\ \nu & 1 & 0 \\ 0 & 0 & \frac{1-\nu}{2} \end{pmatrix} \begin{pmatrix} \frac{\partial^2}{\partial x^2} u \\ \frac{\partial^2}{\partial y^2} u \\ \frac{\partial^2}{\partial x \partial y} u \end{pmatrix},$$

where $D = \frac{Eh^3}{12(1-\nu^2)}$ is flexural rigidity.

Considering the oscillations as harmonic and introducing the damping with loss factor $\beta$: $\hat{\mathcal{D}}_{ij} = D_{ij}(1+i\beta)$, we arrive at the equation of motion in frequency domain with boundary conditions on the clamped end (2), which oscillates with amplitude $g_\omega$, and on the free end (3):

$$-\rho\omega^2 \left( u - \frac{1}{3}e^2 \Delta u \right) - \frac{1}{2e} \text{div div } \hat{\mathcal{D}} \nabla\nabla u = \frac{1}{2e} f \tag{1}$$

$$\begin{cases} u = g_\omega \\ \frac{\partial u}{\partial n} = 0 \end{cases} \quad on \ \Gamma_c \tag{2}$$

$$\begin{cases} -\frac{2}{3}e^3\rho\omega^2 \frac{\partial}{\partial n} u + \text{div } \mathbf{M} \cdot \mathbf{n} + \frac{\partial}{\partial \tau}(\mathbf{Mn} \cdot \tau) = 0 \\ \mathbf{Mn} \cdot \tau = 0 \end{cases} \quad on \ \Gamma_f. \tag{3}$$

The equation is being solved using finite element method with Morley basis on a triangular mesh [7].

### 2.2 Inverse Problem

Let $\theta$ be the vector of unknown model parameters (for an isotropic material $\theta = (D,\ \nu,\ \beta)^T$), $\{(\omega_k,\ u_k^{exp})\}_{k=1}^N$ - AFC points acquired in the experiment. Then, the inverse problem is equal to finding the minimum of a loss functional (4):

$$\min_\theta L(\theta) = \min_\theta \frac{1}{N} \sum_{k=1}^{N} \left(u(\theta, \omega_k) - u_{exp}(\omega_k)\right)^2,\ \theta \in \Theta, \tag{4}$$

where $u(\theta, \omega_k)$ is a solution of a forward problem with parameters $\theta$ and frequency $\omega_k$, $\Theta$ is the admissible set of parameters.

The solver is implemented using Python programming language with FreeFem library for solving Eq. (1), jax package for JIT-compilation of Python code, parallel GPU computations and automatic differentiation technique [8], used in optimization algorithm based on trust-region method [9].

## 3 Results and Discussion

### 3.1 Forward Problem Solution

We obtained experimental AFCs for solid isotropic rectangular plates made from steel X10CrNiTi18-10 and titanium alloy ST-A90, the parameters are presented in Table 1.

**Table 1.** Properties of steel and titanium plates.

| | Steel | Titanium |
|---|---|---|
| Dimensions, [mm] | 99.9 × 20.0 × 2.0 | 99.6 × 20.0 × 3.0 |
| Density $\rho$, [kg · m$^{-3}$] | 7920 | 4550 |
| Young's modulus $E$, [GPa] | 198.0 | 106.0 |
| Shear modulus $G$, [GPa] | 68.5 | 40.0 |
| Flexural rigidity $D$, [Pa · m$^3$] | 164.6 | 266.7 |
| Poisson's ratio $\nu$ | 0.44 | 0.33 |
| Loss factor $\beta$ | 0.006 | 0.007 |
| Accelerometer mass, [g] | 1.7 | 1.7 |
| Accelerometer radius, [mm] | 3.8 | 3.8 |

The forward problem was solved using these parameters (Fig. 1).

One can see two additional peaks on experimental AFCs, which are not present in computed ones. Positions and shapes of the peaks do not differ for both materials, so the peaks are caused by the behaviour of the experimental setup and are not connected to the material properties.

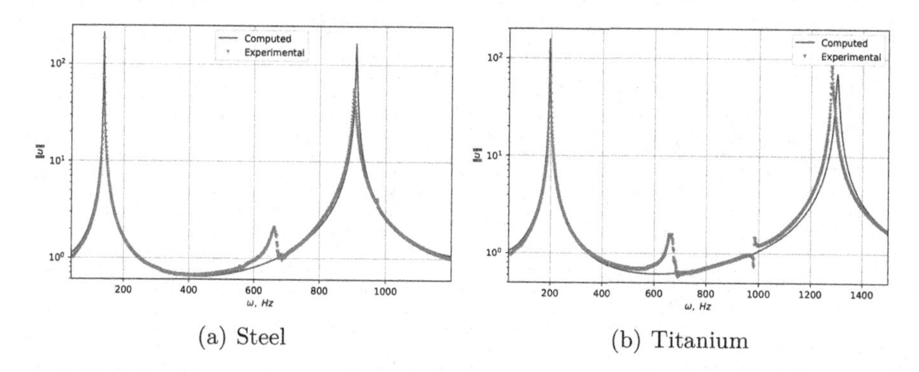


(a) Steel (b) Titanium

**Fig. 1.** Forward problem solution for steel (a) and titanium (b) in comparison to the experimental data.

### 3.2 Inverse Problem Solution

The experimental dataset contains around $10^3$–$10^4$ points, but solving the inverse problem using GPU with 12 GB of VRAM is possible only for datasets containing less than 300 points. Thus, the problem of developing data compression algorithm with minimal accuracy loss arises.

In order to find the optimal algorithm we conducted a series of numerical experiments. The AFC for the specimen made of steel was divided into three sections: interval that includes the first peak 40–200 Hz, interval between peaks 200–800 Hz, second peak interval 800–1100 Hz. Two parameters from the set $\{D, \nu, \beta\}$ were fixed on the true value, while the third one was changed in range from 50% to 150% from the true value. On each iteration the value of the loss functional for each interval was recorded. Intervals where the functional changed the most are significant for finding material properties, so selecting points from experimental dataset should start from these segments.

The results are shown in Fig. 2.

The results demonstrate that intervals with peaks contain more information about elastic properties than the space between them. This observation corresponds with the theory of oscillations, as both resonant frequencies and amplitudes depend on the parameters of an oscillating system. Thus, we propose the following compression algorithm: points are evenly selected from each peak starting from the highest point. An example is shown in Fig. 3.

To study the efficiency of this algorithm we used the following procedure: 50 iterations of optimization algorithm were made from the initial condition $(1.04 \cdot D, 0.96 \cdot \nu, 1.00 \cdot \beta)$ for different amounts of selected points, after that the value of loss functional was recorded (Fig. 4). We compare the results of the numerical experiment with the result of the same procedure on uniformly distributed points (Fig. 3).

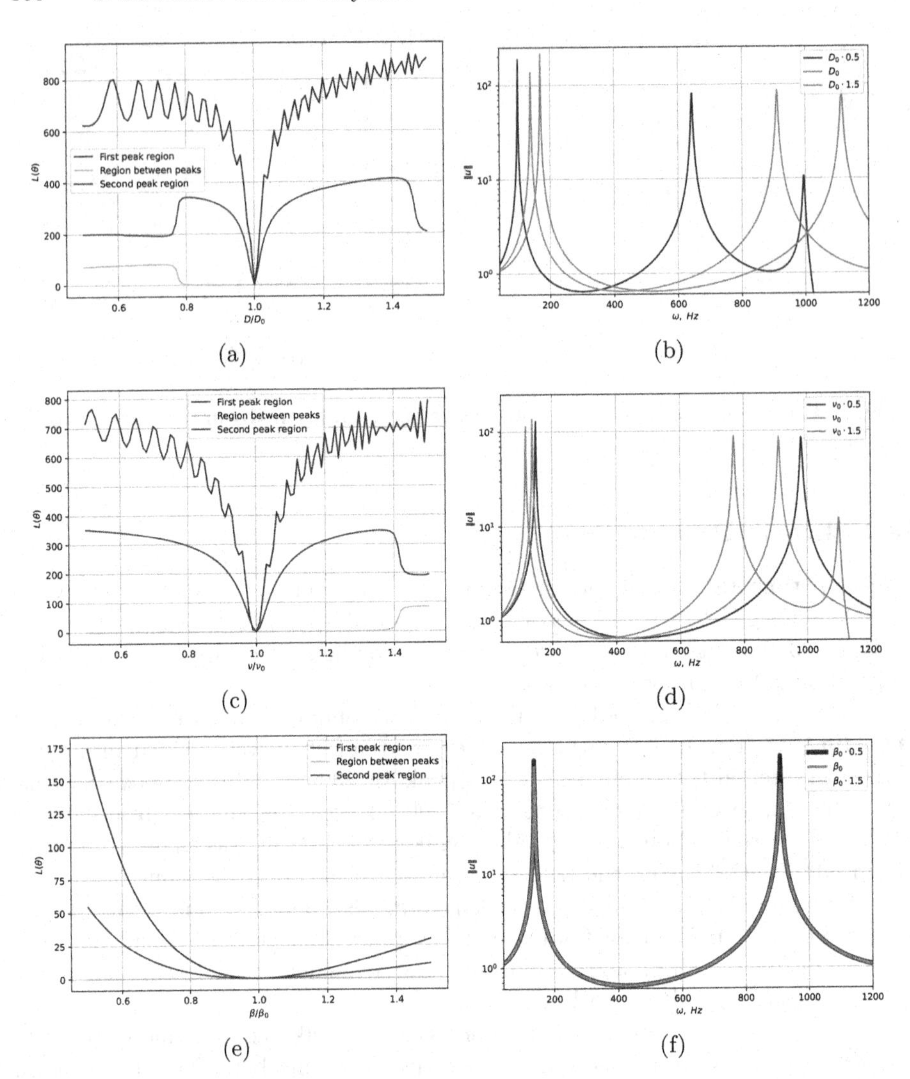


**Fig. 2.** Dependency between loss functional value and particular parameter change (left column), AFCs corresponding to two extreme parameter values and the real value (right column).

One can see that algorithm gives more accurate and stable solution when the size of the dataset is sufficient.

Solving the inverse problem for described plates from steel and titanium produces the results presented in Table 2.

Corresponding to each set of parameters AFCs are shown in Fig. 5.

Results demonstrate that for these experimental AFCs only Young's modulus $E$ can be received with good accuracy. To study the possible ways to correct this

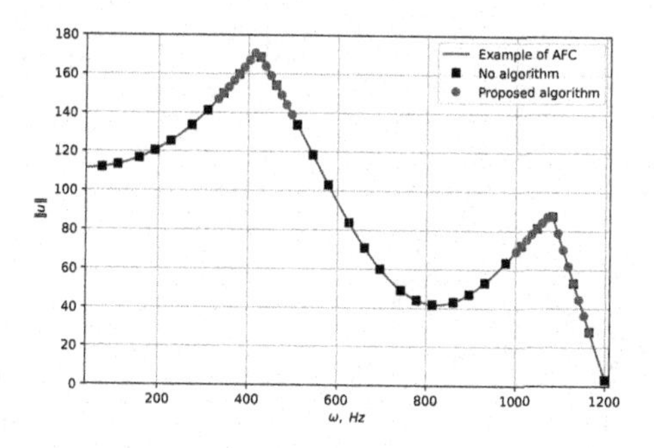


**Fig. 3.** Example of evenly distributed and selected by the algorithm points.

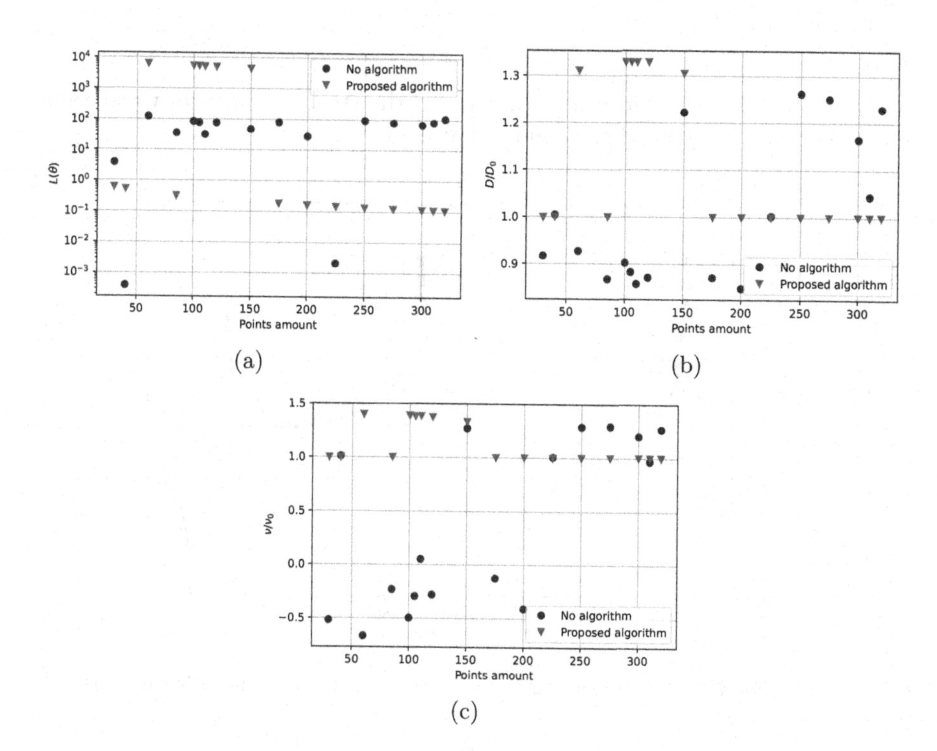


**Fig. 4.** Dependency between optimization result and the amount of points.

**Table 2.** Ratio of parameters to their corresponding true values.

| | Steel | | | | | Titanium | | | | |
|---|---|---|---|---|---|---|---|---|---|---|
| | $D$ | $\nu$ | $\beta$ | $E$ | $G$ | $D$ | $\nu$ | $\beta$ | $E$ | $G$ |
| Initial condition | 1.04 | 0.96 | 1.5 | 1.06 | 1.07 | 1.04 | 0.96 | 0.5 | 1.05 | 1.06 |
| After optimization | 1.02 | 1.03 | 1.57 | 1.007 | 0.999 | 1.05 | 1.17 | 0.66 | 1.006 | 0.97 |

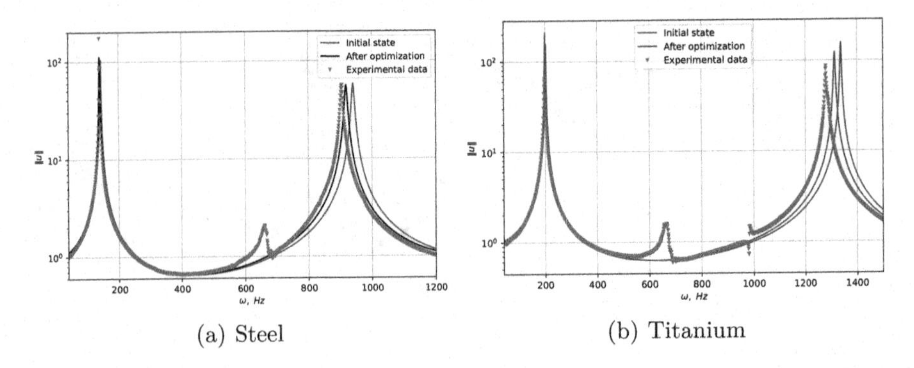


**Fig. 5.** Solution of the inverse problem in comparison to the initial condition and the experimental dataset.

behaviour we solved the forward problem for the steel specimen in wider range of frequencies while changing $E$ and $G$ parameters.

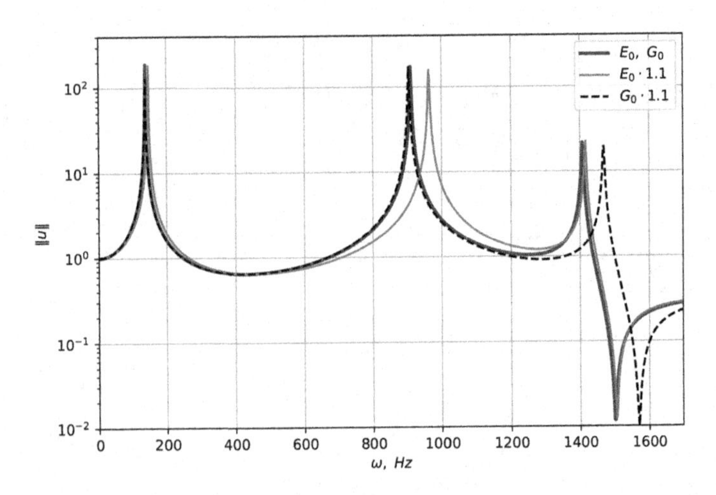


**Fig. 6.** AFCs for the steel specimen with lowered by 10% $E$ and with increased by 10% $G$.

In Fig. 6 one can see that changing shear modulus slightly affects two resonant peaks. At the same time, the antiresonant peak's position changed dramatically. For Young's modulus the result is the exact opposite. So, we suggest extending experimental frequency interval to reach more accurate solution.

## 4 Conclusions

In this work the comparison between experimentally obtained AFCs and the results of the numerical modelling of oscillating thin metal plate was provided.

Forward problem solution is near to the experimental data, but noises from the experimental setup should be eliminated as they can lead to accuracy loss.

The algorithm of data compression for the inverse problem solution was suggested and validated, allowing to use GPU for accelerated computations. It was ascertained that peaks contain the information about material parameters. Also, peaks differentiation was discovered: resonant ones depend mostly on Young's modulus, antiresonant ones are depend on shear modulus. This fact grounds the suggestion of extending experimental frequency range for the production of more accurate results.

## References

1. Nishiyama, K., Shimizu, Y., Kuninaka, H., Miyamoto, T., Fukuda, M., Nakai, T.: Research and development status of microwave discharge ion thruster $\mu$20. In: 29th International Electric Propulsion Conference, pp. 3–5 (2005)
2. Madeev, S.V., Lovtsov, A.C., Laptev, I.N., Sitnikov, N.N.: Determination of the operational characteristics of structural materials from a carbon-carbon composite for electrodes of ion engines (in Russian). Compos. Nanostruct. **8**(2), 140–150 (2016)
3. Tuan, P.H., et al.: Exploring the resonant vibration of thin plates: reconstruction of Chladni patterns and determination of resonant wave numbers. J. Acoust. Soc. Am. **137**(4), 2113–2123 (2015)
4. Aksenov, V., Vasyukov, A., Beklemysheva, K.: Acquiring elastic properties of thin composite structure from vibrational testing data. ArXiv:2203.15857 (2022)
5. Reffy, J.N.: Mechanics of Laminated Composite Plates and Shells: Theory and Analysis. CRC Press, Boca Raton (2003)
6. Bonaldi, F., Geymonat, G., Krasucki, F., Vidrascu, M.: Mathematical and numerical modeling of plate dynamics with rotational inertia. J. Numer. Math. **26**(1), 21–33 (2018)
7. Morley, L.S.D.: The triangular equilibrium element in the solution of plate bending problems. Aeronaut. Q. **19**(2), 149–169 (1968)
8. Evtushenko, Y.: Computation of exact gradients in distributed dynamic systems. Optim. Methods Softw. **9**(1–3), 45–75 (1998)
9. Gonglin, Y., Shide, M., Zengxin, W.: A trust-region-based BFGS method with line search technique for symmetric nonlinear equations. Adv. Oper. Res. **2009**, 909753 (2009)

# Spiking Reservoir Neural Network for Time Series Classification

Maxim Kostyukov(✉), Dmitry Rostov, and Grigory Osipov

Lobachevsky State University of Nizhny Novgorod, 23 Gagarin Avenue, Nizhny Novgorod 603022, Russia
maximkostyukov@yandex.ru

**Abstract.** Brain-inspired reservoir computing methods attracted a great attention due to their reduced computation complexity by using fixed internal synaptic strengths. We consider a reservoir neural network of spiking neurons supervised trained to use it as a feature extraction layer for solving time-series classification tasks. The original time series input is first encoded into the multiple spike streams to feed it into the spiking reservoir layer. We conducted experiments on ambulatory ECG recordings. The considered approach demonstrates competitive accuracy and robustness over other existing methods.

**Keywords:** Reservoir computing · Spiking neural network · Time series classification

## 1 Introduction

Reservoir computing appeared as an extension of the idea of recurrent neural networks. The approach is based on signal processing using a nonlinear system - a reservoir. The idea of using a nonlinear system originally appeared on the basis of biological neural networks [3].

The relevance of this approach is due to its lower energy consumption compared to modern classical neural network architectures, which consist of many layers, each of which consists of many neurons with complex dynamics (LSTM, Transformer). At the same time, current researches suggests that a single reservoir neural network can be applied to multiple tasks [8].

## 2 Preliminaries

### 2.1 Echo State Network

Reservoir computing is a new approach to creating and training recurrent neural networks. In 2001, independently, Wolfgang Maas and Herbert Jaeger [4] proposed similar approaches, which consisted of the following principles:

- A reservoir network consists of a single inner layer that is set randomly and does not change during training;


D. Balandin et al. (Eds.): MMST 2023, CCIS 1914, pp. 310–316, 2024.
https://doi.org/10.1007/978-3-031-52470-7_25

- The output signal of is a combination of the neural signals of the reservoir, obtained by logistic or ridge regression;
- The reservoir state is calculated from the previous one, previous input and output signals.

Maas has called his approach an Echo State Network. The Echo State Networks are universal for time series, real-time systems. Also reservoir networks have good extensibility, so to apply the network to a new task it is enough to add new output layer. Since echo networks have only one hidden layer with predefined weights, they have rather low computational complexity.

### 2.2 Liquid State Machine

In this paper, the main model considered will be the spiking reservoir network model or Liquid State Machine [6].

Liquid state machine (LSM) is a type of reservoir neural network that uses a spiking neural network as a reservoir. Spiking neurons process information via discrete spike streams, as opposed to traditional analog neuron models that compute and transmit continuous-valued signals. But this approach imposes the necessity to encode input signals into spike streams for their further transmission to the reservoir.

## 3 Proposed Model

### 3.1 Spiking Reservoir Neural Network

In this paper we use spike reservoir computing as a feature extractor to solve classification problems.

The original time series input is first encoded into the multiple spike streams to feed it into the spiking reservoir layer. These spike streams are then continuously input to the reservoir. The time series features are obtained by reservoir neurons spikes sums. Once the feature image of input time series is extracted, the final classification computed by logistic regression layer.

The input weight matrix and the matrix of coupling strengths inside the reservoir will follow the rules:

$$\begin{aligned} W_{in} &\sim N(0,1) \\ W_{in} &\leftarrow \alpha W_{in} \end{aligned} \tag{1}$$

$$\begin{aligned} W_{res} &\sim N(0,1) \\ W_{res} &\leftarrow \frac{W_{res}}{\lambda_{max}(W_{res})} \end{aligned} \tag{2}$$

where $\alpha$ denotes input scaling, and $\lambda_{max}(W_{in})$ is the spectral radius of $W_{in}$. The input scaling parameter is responsible for contribution of the input values to the reservoir dynamics. And the spectral radius matrix varies previous reservoir state contribution to the dynamics. These parameters determine the achievement of the echo state property of the reservoir (Fig. 1).

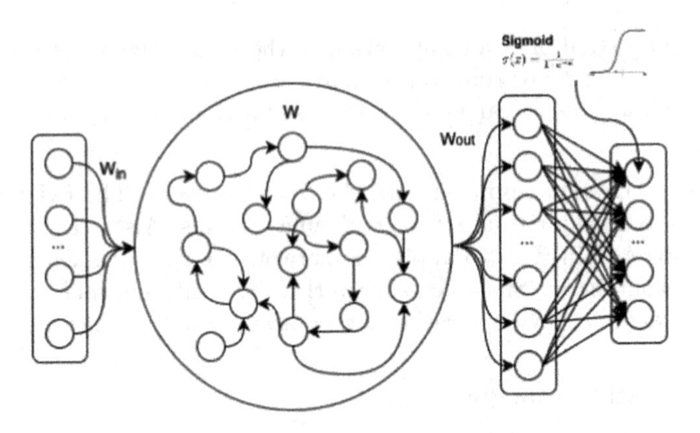


**Fig. 1.** The proposed model architecture

### 3.2 Spiking Neuron Models

Neuron in spiking neural networks are inspired by biological spiking neurons, whose membrane potential varies in response to an input stimulus. When a potential threshold is reached, neuron emits "spike", and its membrane potential is reset to the recovery potential.

In this work, we consider and compare the LIF neuron model and the model based on the Adler equation.

**Leaky integrate-and-fire neuron** is the most common neuron model for spiking neural networks.

It follows the evolution format as

$$C_{mem}\frac{dV_{mem}}{dt} = -\alpha V_{mem} + I(t) \tag{3}$$

$$if\ V_{mem} > V_{thresh} : V_{mem} = V_{reset} \tag{4}$$

where $V_{mem}$ denotes the neuron membrane potential, $V_{thresh}$ denotes the spiking threshold, $V_{res}$ denotes the recovery potential, $C_{mem}$ denotes the membrane capacitance, I is the input stimulus, and $\alpha$ is the leaky rate.

**Model based on Adler equation** has been chosen due its feature of modeling the most important characteristic of a biological neuron - phase behavior.

It follows the evolution format as

$$\frac{dV_{mem}}{dt} = \omega - \sin\left(V_{mem}\right) + I(t) \tag{5}$$

where $\omega$ denotes the natural frequency. In the case of $\omega > 1$ neuron becomes auto-oscillating.

### 3.3 Spike Encoding

The method proposed by Bochte et al. [2] is used in this study to encode the input data into spike streams. This encoding method is based on the overlapping Gaussian receptive fields (Fig. 2).

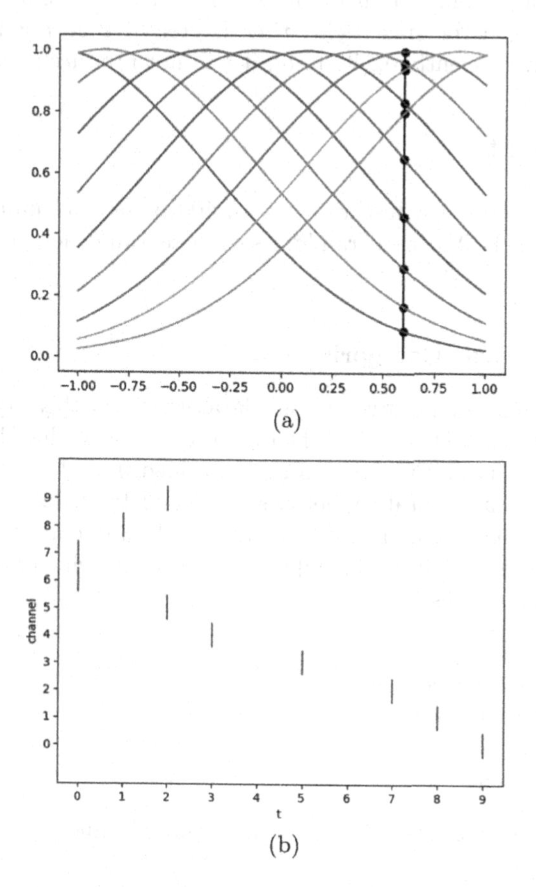


**Fig. 2.** Example of encoding an analog value $x = 0.6$ with $H = 10$, $I_{min} = -1$, $I_{max} = 1$, $\gamma = \frac{8}{25}$ into 10 spike signals.

We use a set of Gaussian functions with a mean of $\mu_i$ and a variance of $\sigma_i$

$$\mu_i = I_{min} + \frac{2i-3}{2} \cdot \frac{I_{max} - I_{min}}{H-2} \tag{6}$$

$$\sigma_i = \frac{1}{\gamma} \cdot \frac{I_{max} - I_{min}}{H-2} \tag{7}$$

where $i = 1, ..., H$ with $H$ being the number of Gaussian functions, $\gamma$ being a designed parameter, and $I_{min}$ and $I_{max}$ are the lower and upper bound of input data, respectively.

So each value is converted into $H$ channels and these variables are then converted into a delay time from $t = 0$ to $t = 10$. A delay time is the response latency of a neuron spike to an input signal, and it is rounded to the nearest internal time-step. If the spike delay time is greater than $t = 9$, the neuron is assumed to be insufficiently excited and it will not produce spike.

# 4 Experiment

In experiments, we used a reservoir size of 500 nodes and input layer with 10 spike input channels. We used random search to tune the spectral radius and input scaling.

## 4.1 ECG Hearbeat Categorization

This dataset is composed of two famous datasets in heartbeat classification, the MIT-BIH Arrhythmia Dataset [7]. Database contains 48 half-hour excerpts of two-channel ambulatory ECG recordings, obtained from 47 subjects studied by the BIH Arrhythmia Laboratory between 1975 and 1979. These recordings were preprocessed, segmented and cropped, with each segment corresponding to a heartbeat. It consists of 109446 samples of heartbeat, which are required to be categorized into 5 classes:

- N - Normal beat
- S - Supraventricular ectopic beat
- V - Venticular ectopic beat
- F - Fusion beat
- Q - Unknown beat

In the experiments, the following results were obtained (Tables 1, 2 and 3):

**Table 1.** Neuron models average accuracy comparison.

| Spiking neuron model | Input scaling | Spectral radius | Average accuracy |
|---|---|---|---|
| LIF neuron | 0.1356 | 0.9398 | 93% |
| Adler equation | 0.0472 | 0.8651 | 92% |

**Table 2.** Neuron models precision and recall comparison.

| Spiking neuron model | Class | Precision | Recall |
|---|---|---|---|
| LIF neuron | N | 0.93 | 0.99 |
| | S | 0.82 | 0.12 |
| | V | 0.94 | 0.68 |
| | F | 0.1 | 0.1 |
| | Q | 0.93 | 0.82 |
| Adler equation | N | 0.89 | 0.96 |
| | S | 0.86 | 0.21 |
| | V | 0.92 | 0.75 |
| | F | 0.07 | 0.11 |
| | Q | 0.98 | 0.86 |

**Table 3.** Average accuracy comparison with other models.

| Work | Approach | Average accuracy |
|---|---|---|
| This paper | LSM with LIF neuron | 92.9% |
| Acharya et al. [1] | Augmentation + CNN | 93.5% |
| Li et al. [5] | DWT + random forest | 94.6% |

## 5 Conclusion

In this paper we proposed a method for time series classification based on Liquid State Machine approach. Unlike most existing model-based methods, which feed time series data directly into machine learning models, our approach first converts the time series to multi-channel spike stream signals. According to the results the considered method demonstrates competitive accuracy and robustness over other existing methods. Although both spiking neuron models performed well, LIF neuron showed better accuracy in the heartbeat classification task.

## References

1. Acharya, U.R., et al.: A deep convolutional neural network model to classify heartbeats. Comput. Biol. Med. **89**, 389–396 (2017). https://doi.org/10.1016/j.compbiomed.2017.08.022, https://www.sciencedirect.com/science/article/pii/S0010482517302810
2. Bohte, S., La Poutre, H., Kok, J.: Unsupervised clustering with spiking neurons by sparse temporal coding and multilayer RBF networks. IEEE Trans. Neural Netw. **13**(2), 426–435 (2002). https://doi.org/10.1109/72.991428

3. Buonomano, D.V., Merzenich, M.M.: Temporal information transformed into a spatial code by a neural network with realistic properties. Science **267**(5200), 1028–1030 (1995). https://doi.org/10.1126/science.7863330, https://www.science.org/doi/abs/10.1126/science.7863330
4. Jaeger, H.: The "echo state" approach to analysing and training recurrent neural networks-with an erratum note. German National Research Center for Information Technology GMD Technical Report, vol. 148, Bonn, Germany (2001)
5. Li, T., Zhou, M.: ECG classification using wavelet packet entropy and random forests. Entropy **18**(8) (2016). https://doi.org/10.3390/e18080285, https://www.mdpi.com/1099-4300/18/8/285
6. Maass, W., Natschläger, T., Markram, H.: Real-time computing without stable states: a new framework for neural computation based on perturbations. Neural Comput. **14**(11), 2531–2560 (2002). https://doi.org/10.1162/089976602760407955
7. Moody, G., Mark, R.: The impact of the MIT-BIH arrhythmia database. IEEE Eng. Med. Biol. Mag. **20**(3), 45–50 (2001). https://doi.org/10.1109/51.932724
8. Pugavko, M., Maslennikov, O., Nekorkin, V.: Multitask computation through dynamics in recurrent spiking neural networks. Sci. Rep. **13** (2023). https://doi.org/10.1038/s41598-023-31110-z

# Author Index





D. Balandin et al. (Eds.): MMST 2023, CCIS 1914, pp. 317–318, 2024.
https://doi.org/10.1007/978-3-031-52470-7

Printed in the United States
by Baker & Taylor Publisher Services